FIRE AND ICE

TALES FROM AN ALASKAN VOLUNTEER FIRE CHIEF

"Whetsell recounts his adventures in an especially amusing voice.....bubbles with punchy remininiscence..." *Anchorage Daily News*

"In writing *Fire and Ice*, Chief Whetsell has done an incredible job of combining experience, wisdom and wit. It doesn't matter if you are a firefighter or Fire Chief, ditch digger or Executive VP of a major corporation, the insights in this book will help you to be better at whatever you do, especially if you already know everything..." *David L. Tyler, Alaska State Fire Marshal*

"Chief Whetsell's *Fire and Ice* not only exudes his ever present wit and wisdom but it showcases what takes place in communities all across Alaska. The Alaskan fire service using their ingenuity and adaptability to respond in extraordinary ways to serve their fellow citizens ..." *Carol Reed, president, Alaska State Firefighters Association*

"Just a few weeks after the Exxon Valdez accident I was sent up to Alaska by BP (British Petroleum) to work on a new oil spill response plan for the Prince William Sound. One of the places I visited to discuss these plans was Cordova. Everyone there was complaining about Alyeska and the oil companies and they were not very interested in plans for the future. Except Dewey; he was different. Instead of complaining he lectured me on the merits of the Incident Command System. I had never heard of ICS and thought he was crazy, but the next morning it dawned on me what he was saying and with his help we introduced ICS to the oil industry. It is now the standard organizational concept for most oil spill response plans as well as other disasters. Thanks, Dewey..." *Mike Williams, Master Mariner, Fellow of the Nautical Institute, Retired Vice President of Alyeska Pipeline Service Company, Alaska*

"Whetsell spent half his lifetime responding to shipboard fires, blazing houses, car wrecks, medical emergencies, underwater incidents, and other disasters...he was the heart of the fire department." *Jill Fredston, author of "SnowStruck" and "Jouneys Along the Arctic Edge" and winner of the 2002 National Outdoor Book Award.*

"Whetsell's contribution to the safety of mankind is legendary in Alaska." *The Muncie Star (article excerpt 3/83)*

"Fire Chief Dewey Whetsell is one of the finest small-town chiefs in the country." *American Fire Journal (article excerpt 3/97)*

"I love what he wrote about FEMA!" *Chief (ret.) Bud Rotroff, senior member of Alaska-DNR's Overhead Team*

"After reading it, I bought another 10 copies to give to my officers, because of the tactics and strategies in the book." *Chief Sam Bunge, Petersburg (Alaska) Fire Department*

FIRE AND ICE

TALES FROM AN ALASKAN VOLUNTEER FIRE CHIEF

CHIEF DEWEY G. WHETSELL

NORTHBOOKS
Eagle River, Alaska

Photo Credits: Cordova Volunteer Fire Department (CVFD)
Cordova Times
Ron Niebrugge (p. xiv)
U.S. Coast Guard (p. 23)
National Park Service (p. 29)
Mike Hicks (p. 42)
Joan Jackson (p. 69)
Real TV video clip (p. 158)
Homer Volunteer Fire Department video clip (p. 168)
Leigh Gallagher (p. 328)
Robert Varnam (p. 329)
Personal collection of author

Published by:

ƊORƮƕBOOKS

17050 N. Eagle River Loop Road, # 3
Eagle River, Alaska 99577
www.northbooks.com

Printed in the United States of America

ISBN 978-0-9789766-8-2

Library of Congress Control Number: 2007942683

DEDICATION

To the volunteers of the Cordova (Alaska) Volunteer Fire Department. And, of course, to my wife Louise and our children, who somehow managed to tolerate my 34-year obsession and absences.

Contents

Operations

Katalla Dive 2

Cordova and PWS 12

Salvage Operation at Mile 13 19

Rescue on Mt. St. Elias 24

Costello and the Dekabrist. 30

Fire Aboard the Alaska Swede 33

The Sea Lark 36

Bristol Monarch 38

Marine Firefighting Symposium. 43

Chugach Cannery Fire 47

North Pacific Processors Cannery Fire 56

   Concealed Space Fires #1 61

   Concealed Space Fires #2 62

   Concealed Space Fires #3. 64

   NIOSH Concealed Space Fire Fatality. 66

   Concealed Space Backdrafts 68

Unique Approaches to Fire Attacks. 70

   Straight Streams on Rollovers. 71

   Wheeled Extinguishers on Structural Fires 76

Drowning in the Omnipotent Written Word. 80

The Ice Auger Amputation.. 83

Drowning and the Small Town Curse. 85

The Shootout at the Episcopal Church 87

Working with Cops and Miscellaneous Gun Calls 93

Air Medevacs. 99

Dive/Rescue Operations. 103

   The Alganik River Recovery 106

   Collapse of the Copper River Bridge 108

   R.I.P. to "What's-His-Name" 110

   Trapped Under a Capsized Boat. 113

THE DEATH OF A RESCUER .115

THE FRANK HANSEN OPERATION 121

THE FLORIAN RESCUE—HASTY TEAM/SUPPORT TEAM CONCEPT 130

BLOWN OFF THE BRIDGE. .133

FALL FROM MOUNT EYAK 136

AVALANCHE AT THE TUNNEL PROJECT 142

AVALANCHE IN A RESIDENTIAL NEIGHBORHOOD 150

SAR ANECDOTES 159

THE SECRETS OF "SAR CENTRAL" 162

DISASTERS

DISASTERS AND THE RYAN AIR CRASH 166

THE CRASH OF COAST GUARD HELO #1471 169

CRASH OF SWISSAIR 111 AND CONVERGENCE ON NOVA SCOTIA. 172

THE EXXON-VALDEZ OIL SPILL AND CONVERGENCE 177

A WORD ABOUT FEMA. 208

MANAGEMENT

SCABS GALORE 210

AMMO FOR THE UNION. 218

FIRE SERVICE SAFETY: FACT VERSUS HYPE 223

WELL-MEANING OUTSIDE AGENCIES 228

EMERGENCY MEDICAL SERVICES: FOSTER PARENTING TO ADOPTION . . 233

"FIREMARK" AND MANAGEMENT FUN 240

LEADERSHIP

CULTURAL BASIS FOR GROUP CONFLICTS 246

    TRIBAL AFFILIATIONS 246

    NEGATIVITY AND NIT-PICKERS. 252

    PROCESS FIXATION .253

    MEAN-WORLD SYNDROME 256

    THE ROSENBERG EXPERIMENT. 261

SETTING THE GROUNDWORK TO MINIMIZE PERSONNEL PROBLEMS 264

SOURCES OF GROUP CONFLICT. 268

    OLD GUARD VS. YOUNG TURKS. 268

    MANAGEMENT VS. UNION 268

    CENTRAL OFFICE VS. FIELD .269

    MY TERRITORY VS. YOUR TERRITORY 269

    YOUR RULES VS. MY NEEDS 269

    PLANNING THE CONFRONTATION. 274

    CONDUCTING THE CONFRONTATION.. 277

LEADERSHIP SUMMARY. 284

TRAINING

NOT ENOUGH HOURS AVAILABLE .286

    MAXIMIZING TRAINING TIME 287

    TRAINING IN STATIONS. 269

    TACTICS SCENARIOS 291

    SINGLE LARGE SCENARIO 284

EXPANDING POLITICAL POWER
FIGHTING THE POLITICAL BATTLES AT THE STATE LEVEL

FIRE SERVICE ASSOCIATIONS 300

    MANAGEMENT PRINCIPLES. .315

WHO WE ARE

THE VOLUNTEER FIRE/RESCUE SERVICE. 329

FOREWORD

My fire service career now spans almost a half a century. I have lost track of the tens of thousands of firefighters that I have had a chance to make acquaintance with. I can't remember half of the fire chiefs that I had worked on projects with. Yet, when the name Dewey Whetsell is mentioned in a conversation I have this instant image and total recall of my exposure to him. Dewey Whetsell is not only unforgettable but inimitable.

When he gave me a copy of this book to review I was looking forward to it for the simple reason that I expected it to be unusual. When I sat it down, I was even more fascinated by what it did to reinforce my perception about the man and his life. Dewey is not only the kind of person who tells you the way things are; but he also tells you the way it ought to be.

I will never forget my first opportunity to meet Dewey at a training conference in Anchorage, Alaska, some twenty years ago. Coming up from the Lower 48 I had an impression in my own mind of who I was going to be talking to and what I was going to talk to him about. In a five minute conversation with Dewey, I was disabused from any idea that I was dealing with the back woods—instead I was confronted with a combination of real people who possessed almost renaissance-like skills.

Therefore, it was a pleasure and a unique honor to be able to write a foreword of this book. If you complete this book without having an understanding of the passion, the commitment and the competency of this man and his peers in the great State of Alaska, I would be moderately surprised.

Ronny J. Coleman
Senior Vice President
Emergency Services Consulting, Inc.
Retired California State Fire Marshal

Chief Coleman's career also includes a term as president of the International Association of Fire Chiefs, chief deputy director of the California Department of Forestry and Fire protection, and author of numerous textbooks, manuals, and more than 200 "Chief's Clipboard" columns in Fire Chief magazine.

ACKNOWLEDGMENTS

I would like to acknowledge editor and publisher Ray Holmsen and his wife, Jan, for the monumental task of editing this voluminous pile of my yammerings and painstakingly deleting most of the expletives I had included when I was feeling either a bit cranky or cavalier.

The incidents in this book are all true, although some of the names were changed to keep me from getting shot.

PREFACE

The people in your city have placed a trust in you. You defend them by whatever means you have; whether it's in the streets, the back country, the chambers of your city council, or the legislative halls of your state capital.

This country's volunteer fire/rescue service has been doing its thing for 275 years. Our toys have improved, but nothing can improve who we are.

You know, you can follow mankind's bloody footprints from the caves to the settling dust of the Twin Towers, and not find much that speaks well of human nature. Yet, you can visit a fire station and note the character of those who will not stoically watch the suffering of others, will not turn a deaf ear to their needs.

And even though the media may electronically inject into our brains the sights of Ground Zero, one can look inside any fire station and see 275 years of a steady increasing testimony of what's good in man.

If people cannot depend fully on their fire service, then nothing is sacred. Remember, if a tragic event puts your neighbor up against the wall—go get him. That's who you are.

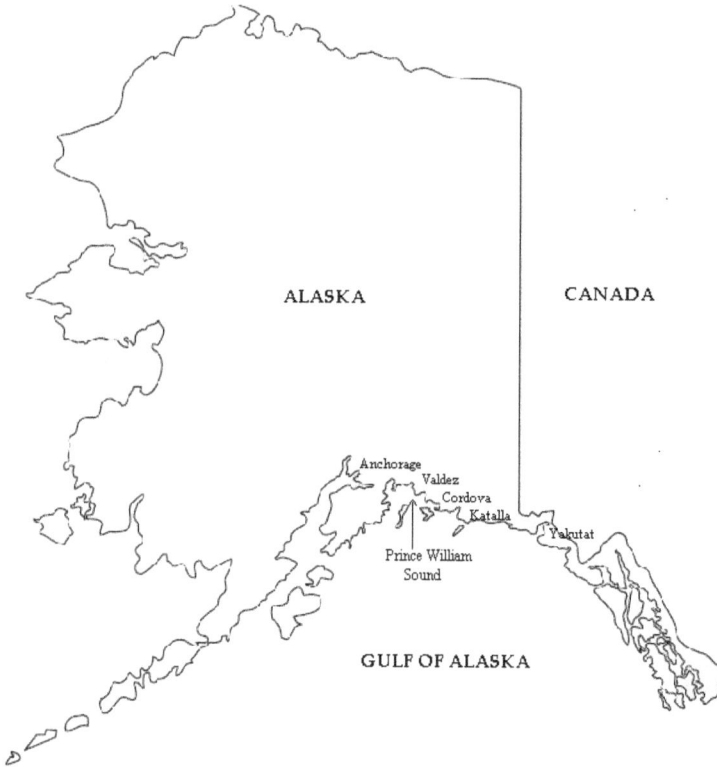

ALASKA

CANADA

Anchorage
Valdez
Cordova
Katalla
Prince William
Sound
Yakutat

GULF OF ALASKA

AERIAL VIEW OF CORDOVA, ALASKA

Photo By Ron Niebrugge

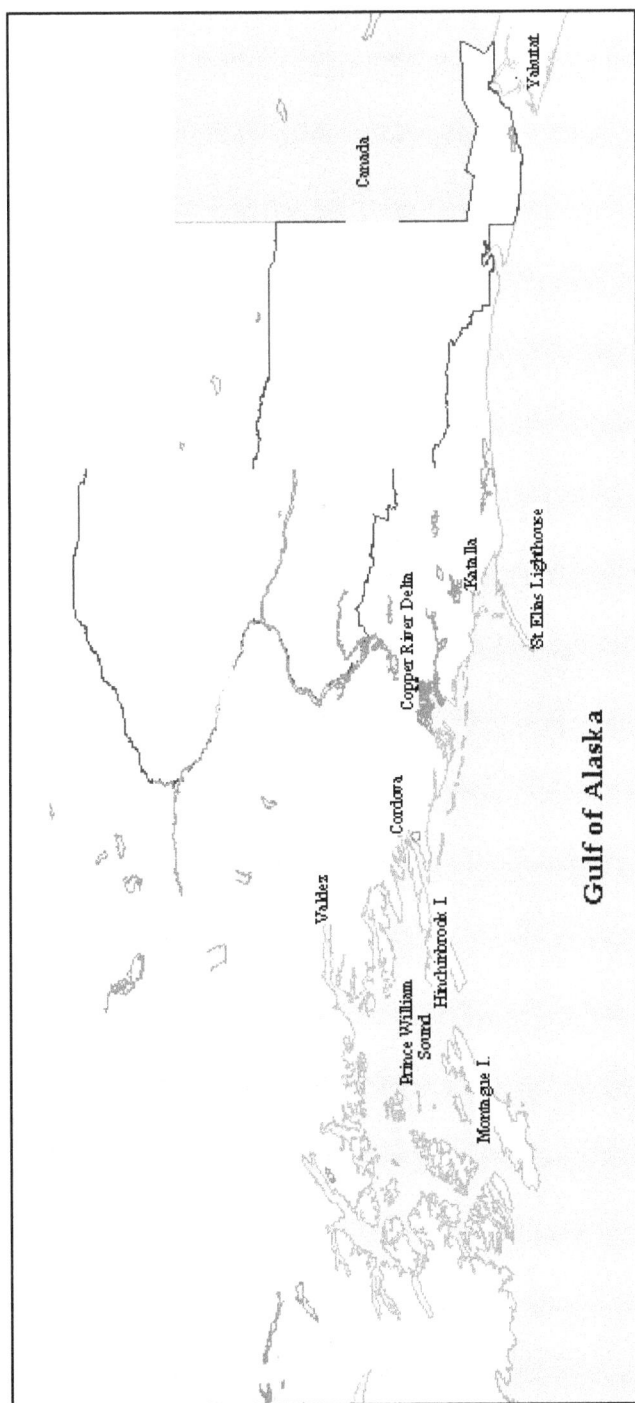

Gulf of Alaska

Canada

Yakutat

St Elias Lighthouse

Katalla

Copper River Delta

Cordova

Valdez

Hinchinbrook I.

Prince William Sound

Montague I.

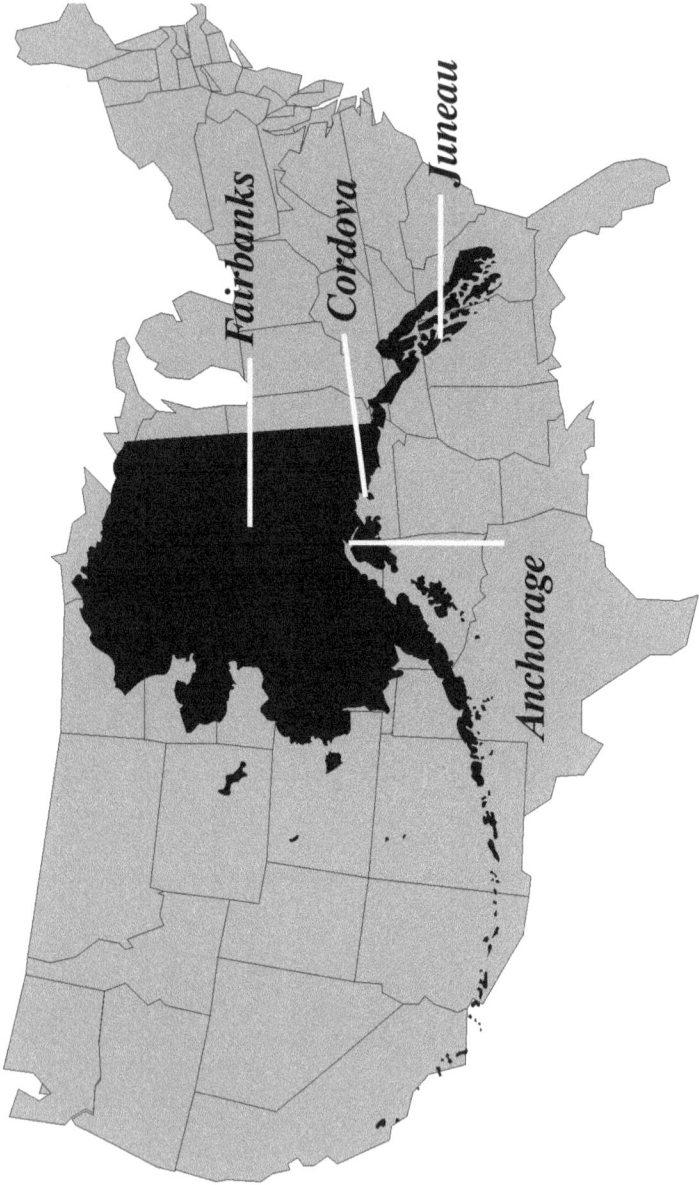

HOW BIG IS ALASKA?

SECTION I

OPERATIONS

KATALLA DIVE

It was a gray Sunday late in May when Alaska State Trooper, Gerry Denison, called me at home and asked if I would "do a dive" for him. It was a body-recovery operation. Four bodies, actually. Of course I agreed, but explained that since commercial fishing season had begun, as well as other summer activities—like construction projects—I would try to avoid asking other divers in the volunteer fire department to help, if possible. Naturally, if it were a rescue operation, that would be different.

"Well, you can't do this alone" he said.

"What happened?" I asked.

"It's near Katalla." Oh, Christ. Katalla was a town site 50 miles east of Cordova, halfway between us and the town of Yakutat. Really, it no longer existed. It was the site of Alaska's first oil field after the turn of the century, but the oil field was destroyed by fire in 1933 after three decades of production. The town remained for some years after, but slowly disappeared. The only things remaining now were a few trappers' cabins buried in the trees, west of the Katalla River near the ocean. The ocean beach, east of the Katalla River, is known for its abundance of razor clams, and clam diggers travel there exclusively in wheeled planes and land on the hard-packed beach at low tides. Like Cordova, the only ways in or out are by air or sea.

That morning, a plane from Cordova was already there, and the party was digging clams when they looked up to see three planes from Valdez come down out of the skies. The first two landed without incident. The third plane, carrying the Joy family, circled and watched the other planes navigate the tricky crosswind as they landed. Mr. Joy, a novice pilot, had arranged to accompany the other planes so he could have this family outing with his wife and their two young sons.

The crosswind that came down out of the mountains blew across the beach at about a 45° angle and on out to sea. Mr. Joy circled the area four or five times, sizing it up. Finally, he approached in, gripping the wheel and pushing pedals in anticipation of gusts. As the wheels touched down and he eased off the power, a sudden gust of wind slammed the plane and lifted the left side up, pushing the wing into the air.

Reacting, Joy gunned the throttle and banked to the right toward the ocean surf. He poured on the power, trying to recover and stabilize. He powered up, but the heavy gusting tailwinds proved too tough for him. The plane yawed left and right, the wings dipped, and the engine and propeller screamed. The onlookers watched helplessly.

Heading straight out to sea, he could not pull up much above the tops of the breakers and in fact, snipped the tops of some of the waves on his way out. Under full power, he tried turning the plane, and water sprayed up from the wingtip or wheel hitting it. Then one wingtip dug into the water and the plane began to cartwheel across the ocean. It stopped in a spray of water and bobbed momentarily upsidedown.

The witnesses could see the belly of the plane on occasion between large waves. They saw what may have been someone clinging to it, or it could have been one of the wheels. They couldn't tell for sure. They ran down the beach searching for something that might float that they could use to get out to the plane, but there was nothing. They tried entering the water, but because of the heavy pounding waves and frigid temperature, they couldn't. Then, the object disappeared from the plane. One of the Cordovans, Bill Bernard (an old co-worker of mine from Cordova's Public Works Department), ran back to the Cordova plane and radioed Chisum Air in Cordova for help.

Chisum called the public safety dispatcher for an Emergency Medical Technician (EMT). Volunteer EMT Stan Shafer (captain of our EMS division) boarded Chisum's DeHavilland Beaver and they were off. After toning out for Shafer, the dispatcher—as per our protocols for incidents outside the city limits—notified the state trooper.

In the 45 minutes it took the Beaver to reach the area, the winds on the far side of the Ragged Mountains had picked up and it was beginning to rain. The ceiling dropped. Reaching the site of the downed plane, the Chisum pilot dipped his wing way down and nearly spun on an axis over the area of the oil slick. In the bucking wind, squinting through the rain, Stan, who had been on many of these trips in snotty weather, suddenly wasn't feeling well. It was a combination of the motion, the steep, unrelenting sideways angle, the roar of the engine, the rattle and vibration of the interior, and the hot oil smell of the engine. He leaned over and told the pilot, "I think we'd better land soon." The pilot sympathetically compromised by making the circles wider and more gradual. Stan was appreciative. He just quietly threw up in his hat and kept looking.

Shortly after, they spotted the body of one of the boys washed up on the beach. They landed, retrieved the boy and flew back to Cordova. There, Stan briefed the trooper who then called me.

Knowing the time of submersion, or even time in the water, the suit-up time and flight time, any thought of a cold-water drowning save was out of the question. Gerry called his post headquarters in Glennallen while I drove to the station to meet him.

Gerry understood that because of the season, I was low on manpower and couldn't leave the town unprotected to do a recovery operation. But city police

3

officer, Fred Brady, said that he would dive with me. Gerry (not a diver himself) contacted Trooper Jim Alexander from Valdez who would come over.

"Well, in that case," I said, "maybe the Valdez Fire Department dive team could come, too. No sense in flying a near-empty plane over." I called the Valdez fire department and they agreed to send over four of their divers. The Valdez Volunteer Fire Department is a combination department with about 12 paid, and 60 volunteers. Valdez isn't much bigger than Cordova, but it has a hell of a tax base—The Trans-Alaska oil pipeline terminus and one of the world's largest petroleum tank farms. Hence—12 paid firefighters.

It was an hour's flight from Valdez, so in the interim we looked at maps, called for weather reports, and talked to the Chisum pilot that would take us to Katalla. Incidentally, we had an agreement with Chisum Air that they would always drop whatever else they were doing and provide air transportation—either fixed-wing, or chopper—for our search and rescue operations in exchange for $50,000 annually. We also purchased a radio for dispatch to communicate directly with their aircraft. For the most part, however, any expenses we incurred on these out-of-town operations were reimbursed by the Alaska State Troopers (AST).

Anyway, Gerry said that the clam diggers had found an old, red crab-pot buoy and used it to mark the spot where the plane was last seen. Sort of. They walked inward from the surf to the edge of the brush and tied the buoy down. It was the best they could do.

About an hour later, Gerry, patrolman Fred Brady, and I drove to the city airstrip on the edge of Lake Eyak, where the Valdez crew would be landing. I was glad to see Trooper Alexander leading the party. Young, tall, and muscular, he knew his business in search and rescue. A real outdoorsman. Even though his hello was pleasant, it was plain he wasn't ecstatic about being here. On the other hand, the crew of Valdez firefighters disembarking from the plane, young (in their early twenties), tough and nervy, all waved and yelled, "Hello", and said that Tom (Chief McAlister) said to say hello and to be careful out there.

I knew the Valdez crew from previous conferences, schools, and state functions. They had also come to Cordova the previous year when we hosted the annual Alaska State Firefighters Association (ASFA) and Alaska Fire Chiefs Association (AFCA) conference here. Alaska is a huge state, but the population is small (about 600,000, at the time of this writing), so it doesn't take long to meet lots of people in your field.

We stuffed the assembled equipment into the Beaver aircraft: air tanks, lead weight belts, inflatable boats, an outboard motor, a gas tank, ropes, and body bags. The pilot took three guys and headed for Katalla.

An hour and a half later he was back. Fred Brady, the Blackburn brothers, (Valdez firefighters), and I seat-belted ourselves in the plane and we were on our way. By the time we passed the Sheridan and Miles Glaciers, the dull marsh lands below and drone of the engine drugged us passengers into silent staring. By the time we reached the Copper River, 27 miles east of Cordova, and moving into an ever decreasing ceiling, the day was dripping. The rivers and streams were brown with run-off from the interior and a dungeon-gray shroud of mist now descended halfway down the mountain range on the other side of the river. The dreariness was thick in the cockpit.

The pilot banked right and swung out over the gulf to make a wide arc around the mountain range. The wind picked up. He flew over the oil slick and pointed, then pointed at the red crab-pot buoy used as a marker. It was high tide, and there would be no landing on the beach. He headed back for the landing strip which was just on the other side of the Katalla River. Approaching it, we could see a couple of cabins in the trees up the river a short distance. The landing strip was little more than a small gravel patch that ran parallel to the beach. One end started in the trees, the other end was a six-foot bank that fell into the river.

The plan, according to Dennison's recommendations, was to wade across this so-called "creek," walk down the beach approximately a quarter of a mile to the marker, and then swim out and recover the bodies.

The Zodiac boats were already inflated and were being loaded. The 5-horse kicker (outboard) was mounted on one. Brady and I grabbed the one without the kicker, walked it into the water and started walking it across the river. Alexander and one of the firefighters climbed in the other one, started the kicker and began cruising downstream. The other three firefighters started forging across.

I don't know if it was because of the rain or if the river always runs like that, but the current was getting very powerful and the river got deeper with each step. About midway across, with the buoyancy of my suit, my feet left the bottom. "'Creek' my ass! I'm not touching bottom," I informed Brady. Brady, being taller, naturally started laughing at me. A moment later he lost the bottom too, and we were drifting downstream, clinging to the Zodiac heaped with equipment. It was tough flutter-kicking—finless—to the opposite shore. We eventually found it shallow enough again to finish walking to the other side.

We watched with envy as Alexander, perched comfortably on his Zodiac, and moving effortlessly, passed us heading for the mouth of the river. As soon as the river met the surf, the boat dunked under a couple of times and the kicker was finished. They dragged the boat to the beach, took off the kicker, pulled the boat back out, and started rowing parallel to the beach toward the area of the oil slick.

Fred and I pulled ours down the beach in about a foot of water, taking turns. Even though we had to lean into the 45-mph rainy wind and wade, it was still easier than rowing. We reached the area marked by the buoy in about an hour, and still about half an hour ahead of Alexander and his diver.

We spread our gear out on the beach, but unable to see the oil slick from this angle, we took just fins, masks with snorkels, and markers, and swam out in search of the slick. We hoped whoever found it could free-dive down and tie a marker on it.

It was impossible to keep oriented since the swells were coming in at a slight angle to the beach and the wind and rain were hitting at a 45° angle to that. It was impossible to see the crab-pot buoy from out there. Swimming through the waves and breakers, compensating for the tide and wind was taxing and none of us could find the slick.

When Alexander made it to the area in the boat, he spotted the slick and the firefighter dove down, but the slick was coming up some distance from the plane. The underwater visibility was zero.

Our searching went on for several hours. Dusk was approaching and with the tide low again, the plane came over from the strip and loaded the equipment into it to take it back across the river. He asked which of us wanted to ride back across the river? Brady said he'd rather ride than walk anytime and headed to the plane. The rest of us looked at one another and started walking back toward the river carrying only our fins (for the swim across the river).

Of course, we had just barely started walking when the plane passed over us and landed on the other side of the river and waited. When we finally made it back, we stashed all the gear and three guys in the plane (which maxed out the weight capacity) and it took off for town.

We learned there were four cabins in the area and we were invited to stay in one of them to await the return of the plane. We were given a pot of coffee and some pumpkin cake. We needed it. Our energy tanks were on "empty" since we'd been running full RPMs for hours. I was so fatigued my body felt like it was a huge water balloon.

Before Brady, the two Valdez divers, and I and finally climbed in the plane for the trip back to Cordova, we had to push the plane as far back into the trees as possible, because the pilot said he needed as much "runway" as possible. He rapped the engine as high as he could while holding tight on the brakes. When he released the brakes, he gunned the engine as fast as he could to get maximum air speed before the airstrip abruptly ended at the river. It was like a take-off from an aircraft carrier. As soon as we ran out of airstrip, the plane dropped down toward the river but kept on flying and started climbing and banking to the right toward the ocean. Turbulence seemed more dangerous while banking hard, and air pockets dropping us down suddenly made most of us gasp loudly. Brady, riding in the front seat next to the pilot turned and laughed, "Don't scream—you're scaring the pilot." We sat in embarrassed silence the rest of the trip.

Arriving in Cordova at 10 p.m., we decided to return early the next morning and be on site at low tide. Since it would be a minus tide, we hoped some of the plane might be visible.

At 6:30 the next morning we headed for Katalla. The weather had improved. That is, the winds had slackened but the ground swells had picked up due to a storm brewing several hundred miles to the southwest. When we landed, this time on the Katalla beach because the tide was out, we were not the first plane to set down there. Alexander walked up to the clam diggers and asked if they'd seen any debris from the plane or anything. The storm that was centered way out in the gulf generated heavy swells that pounded the airplane all night, may have broken it up or dislodged its contents. Literally, without looking up from his clam digging, one of them nodded "yes" he had seen something. Pointing back in the direction of the cliffs by the river, he mumbled "body" and kept digging.

The pilot flew Alexander and Brady to the beach on the other side of the river where they retrieved the body of Mrs. Joy and flew back to Cordova.

Even though the ground swells pushed waves up higher than the day before, we didn't have to begin the day by dragging a loaded Zodiac across a river and quarter-mile down the beach. But about the third time out and back through the waves, and it was yesterday all over again. However, it seemed to me that the other guys (in their early twenties) didn't seem affected by it. I was pushing 40.

About the time the tide began flooding again, someone spotted an object protruding from the bottom of a swell. It was so far out that we swam out in relays using markers tied together to reach the plane. It had taken about half an hour to reach the plane and tie it on, and that was the last we saw of it for that day. As the diver was tying line around it, a receding swell drew him out and an incoming one lifted him high and rolled him over the jagged metal edge of the plane. He managed to quickly attach the end of small nylon twine to it.

The evening low tide would not be low enough to expose the plane again. The subsequent attempts to dive down to the bodies on the plane were pretty hairy, but no one got hurt. We kept trying. We did learn that there was no way to dive straight down without rolling over and reaching bottom nowhere near where you started down. It was hard to navigate the undertow. And you had to do it blind—only outstretching your arms hoping to swim into something while tumbling. I was really afraid of being slammed into the jagged wreckage and getting cut up.

I'm not sure how long we were at this. It seemed like moments; it felt like hours. At times, we dived in a blind rage, determined to swim through it by sheer power and will. By now we were all in pain from exhaustion.

7

Eventually, Brady and Alexander came back. There are official things that must be done by law enforcement people when they have a body. I never envied them. They have to change persona so quickly and so often. I briefed Alexander. If the other two bodies were still in the plane, we could not get them with the tide flooding. And, quite frankly, I was too tired to go on any longer. The others confirmed that they had nothing left inside.

The tide for the next day would be lower still, and the lowest for the month, so on the flight back to town, Brady and I suggested that since the plane had been located and marked, the Valdez crew could be excused to go home if they wanted. We could come out the next morning and retrieve the bodies. Alexander said that was fine, but that he would stay also. Being a state trooper, this operation was in his jurisdiction and felt it was necessary that he be present until it was concluded.

That evening the Valdez firefighters flew back. Alexander, Brady, and I were on site before low tide the next morning and located the small line and had a larger line and buoy ready to attach in its place as soon as the plane was visible. The sky had cleared but the waves seemed worse yet.

We managed to snake it out to the plane—or nearly. I was pulling the end of the line toward the plane (on the other end, near the beach, was the buoy), Alexander and Brady were at the plane. Wading into the surf up to my neck, my lead belts helped keep my feet on the bottom. But the swells would lift me and push me back, then I'd advance quickly between swells; I would feel the undertow about the same time I'd become buoyant. So, after a swell would lift me, the undertow pulled me under, feet first, toward the plane. Being even more buoyant as the water reached my neck, it was a constant battle against the undertow pulling me toward the wreck. I was waiting for Alexander and Brady to attach their short length of line to the plane and bring the other end of their line to me, so we could tie the two heavy lines together. If anything caused the loss of the plane again, the heavy, extra-long line and crab-pot buoy would assure we could find it again.

They finally made the wrap around something on the plane and swam the other end to me. Brady was tying the two ends together. Having made the first loop in the line, he stuck his fingers in to make the second loop when a swell lifted us all up and rolled us beachward. Brady didn't roll. The loop cinched tight around his finger and broke it. Between surges, I worked at tying the knot while Brady and Alexander went back to the plane to search it. Alexander, seaward, grabbed the jagged wreckage, pulled himself down and reached in. He felt a head, and on the second try, felt a shoulder. He could not be certain if it were one person or two. Each attempt he made was a gamble. He was flung blindly into the wreckage with bone-crushing impacts. Brady tried it from the beach side by going down, the undertow tried to roll him under and into the tangle of metal. It

8

was all they could do to hang on to the bit of exposed wreckage, let alone manage to free a body from inside of it. So we all headed for the beach.

The new plan was to lift the plane onto a boat and be done with it. So we flew to Cordova, hopped a floatplane and flew out to the "flats" and landed in the lee of Egg Island and boarded the Enforcer to do the job. The Enforcer was the State Fish & Game boat that was large enough of carry the plane on its deck, and a boom big enough to lift it up out of the water. It patrols the 35-mile wide area known as the Copper River Flats where the depth of the sand bars change yearly and 500 fishermen gillnet for salmon just landward of the pounding breakers coming in from the gulf.

We weighed anchor and headed out the channel toward the gulf, estimating a 4 to 6 hour run back to Katalla. Nearing the channel approach, the Enforcer ran over a fishing boat's gillnet and got it wrapped around the propeller. John Stimson couldn't believe he had not seen the 900-foot-long string of corks and had to stop the Enforcer. Just inside the breakers, we were rolling wildly in the swells as John reversed the propeller hoping to dislodge the net. It didn't work.

I went over the side to cut the webbing out of the "wheel" while Alexander went in and lay back several yards as a safety measure. The stern lifted high up out of the water, then plunged way down. I grabbed the webbing with my left hand and my knife with my right. The rudder was scaring me as it rose and fell past my body. I yelled to have John spin the helm to move it out of my way. Some of the mesh kept hanging up on my inflation tube and yanking me around. I was really getting jittery. With the last two and a half days of ass-busting and then this—there had been a lot of potential for an accident; I was thinking this whole operation was jinxed. I looked back and Alexander's face was showing some mileage, too. He was getting tense. Finally, I said "piss on it" and reached in as far as I could with my left hand and pulled myself down and under the boat, I wrapped my knees tight around the rudder shoe, grabbed the webbing on the propeller blades and started sawing away with my knife. It was hard to hold on, and it was eerie as it would get really dark on the downward plunge and light as I was lifted closer to the surface. But it had to be done or we would have to be towed back to town. Even though I was getting fed up with the entire operation, I wanted to get it over with, and the only way to do that was to stick with it.

It felt like I got most of it out. I surfaced and John reversed the prop to spin out any remaining webbing. Fred lowered the line, and Alexander and I took turns tying our tanks and lead belts to it. Fred's broken finger must have been giving him some grief because he winced each time he grabbed the line to lift our gear up. Then Alexander climbed up the line. When I pulled myself toward the boat with the line, I stuck my feet out to cushion my contact with the hull, and my fin got stuck between the top of the rudder and the hull as it lifted out of the water. My leg then raised high above me, then down into the water again, but not far enough to pull my head under. I pulled up all of the slack on the line, and in a vertical position against the hull, I yanked my foot but couldn't free my fin. Out of desperation, I finally yanked my foot out of the fin and let it go. Climbing back aboard I marveled at how many things had gone wrong. One fin is only good for swimming in circles. In addition, I had used too much air out of that tank.

Anyway, we headed out across the gulf. We had a following sea to the starboard quarter and the Enforcer wallowed the whole way—but it did increase our speed a bit. Poor Alexander knew from the beginning he would get sick. In fact, John Stimson nudged me just as we were leaving to draw my attention to Alexander's chronic, and well-known predisposition to sea-sickness. As predicted, Jim headed straight to the berthing spaces and went to bed. I took him some crackers explaining that's what we always did in the Coast Guard: lots of crackers and a little bit of water (or 7-Up if you've got it). It works great for the flu, too. He was grateful, but not consoled.

As we approached the Katalla coastline several hours later, Alexander was up and bounding out on deck like he was feeling fine. Having spotted our buoy past the mine-field of crab-pot buoys, the Enforcer maneuvered her way through them to a spot as close as the fathometer would allow. Still not close enough to hook the buoy line with a boat hook, we dropped anchor and lowered the Boston Whaler over the side.

Brady and Stimson headed toward the marker to bring it back to the Enforcer where we would wench it aboard. A few minutes later they were back complaining that when they pulled on the line, it had come off of the plane. Attaching it in the first place was a major endeavor and assuring that the knot was fast or that the part of the plane it was attached to was sound was impossible. Now the plane was submerged and the line was off of it. However, the small nylon twine from the original marker was still attached.

The new plan was to dive on it, attach a long poly line from the Enforcer and finish the job. Loading gear onto the Whaler was as much a pain in the ass as everything else we'd done: Getting the vessels close enough to pass equipment but apart enough to prevent ramming together in the rolling sea. It took several passes to get all the gear and Alexander and me into the Whaler.

The four of us and our gear headed for the area of the plane as Alexander and I were gearing up. John was maneuvering over the area as I looked up from adjusting the size of the fin I borrowed and saw, coming in at our starboard beam, a "sneaker." A "sneaker" is an abnormally high wave that old-timers used to call "the seventh wave," assuming they arrive at a predictable sequence. I yelled, "Holy Shit!" Stimson looked up, gunned the engine and swung the bow around to face it. He faced it completely perpendicular rather than at a slight angle.

We were lifted up high in the air, the wave disappeared from under us, and we dropped straight down and slammed hard into the water again.

I couldn't believe that I was saying, "Fuck this plane. I quit!" Then I added, "Maybe if conditions get better, I'll try it again. But for now, I'm through." I told them I was not going to be thrown down onto a wrecked plane, getting impaled, to recover two bodies. I had never before, or since, left before an operation was completed. The others agreed. I oftentimes wonder if they were secretly grateful that I called it quits.

We weighed anchor and made an hour-and-a-half trip to the protection of Controller Bay to anchor up and radioed for a plane for Jim, Fred, and me. As an aside, of the three or four crab boats also anchored there, one was suspected of having a crewmember that had a warrant on him. So, while waiting for the plane, Alexander and Stimson (also a commissioned state public safety officer) put on their Public Safety jackets and took the Boston Whaler over to arrest him. Either he wasn't aboard or the crew hid him when they saw the Whaler coming over. We finally got back about ten o'clock that night

The next day I was contacted by the grandfather of the boys and he desperately wanted the body of the second boy recovered. I explained the conditions at Katalla and informed him that maybe someone else might try it (if they're nuts) but I quit. I was too tired; it was too hard and dangerous. He understood.

I called McAlister in Valdez and thanked him for sending his guys over to help. He said that when they got back to Valdez, they had gone in to see him and told him that they would be glad to dive in the Valdez Arm or the entire Prince William Sound if he wanted, but "Chief Whetsell can take that Gulf of Alaska and shove it up his ass." But, he continued to explain that they were already deciding which of them would come back if their help was still needed.

At any rate, within a day or two, the plane broke up in the storm that had been approaching and the body of Mr. Joy was found. The body of the other son was never recovered.

During that operation, I lost a mask, one fin, and ten pounds. But Brady gained a splint. He sold his diving gear within a week and never dived again.

I initially joined the volunteer fire department to fight fires, but in Alaska, that may be only part of one's responsibilities.

11

Cordova and PWS

I went from Detroit to Cordova, Alaska, in 1965 at the insistence of my employer, the U.S. Coast Guard. If they thought I was having too much fun in Detroit, they were right. In general terms, Cordova rests on the upper-most arc of the Gulf of Alaska. But more specifically, Cordova is located in Southcentral Alaska, on the eastern edge of Prince William Sound (PWS). Prince William Sound encompasses 15,000 square miles of tidewater glaciers, towering snow-capped peaks and calm waterways. Its 3,000 miles of shoreline is surrounded by the Chugach Mountains to the east, west, and north which rise over 13,000 feet. Fifty-mile long Montague Island, twenty-five mile long Hinchinbrook Island and several smaller islands form natural breakwaters between the sound and the stormy Gulf of Alaska. Because the sound was formed by millions of years of glaciations, its shorelines are heavily indented by deep fiords and many smaller bays. The sound has about 40 fjords including the one formed by the vast 35-mile long Columbia Glacier which produced the icebergs that the Exxon Valdez tried avoiding, causing it to run aground on Bligh Reef in 1989.

Prince William Sound exceeds the combined area of Massachusetts, Connecticut, and Rhode Island. And yet fewer than 10,000 people live in the three towns—Whittier, Valdez, and Cordova—and two native villages—Chenega and Tatitlek—situated on the shores of the sound. No roads connect these communities.

Prince William Sound lies to the west of Cordova. To the east, after driving through a cut in the Heney Mountain range, is the 700,000-acre Copper River Delta. Going due east, the marshes of the delta lie on the right side of the road, but to the left are more mountains and glaciers. Still going east, 27 miles away, one crosses the Copper River and sees, straight ahead the Wrangell–St. Elias mountain range. The entire area is described as a "17-million-acre Copper River ecosystem to the east and northeast of PWS, ringed by the world's tallest coastal mountains and supporting some of the most abundant and commercially valuable runs of wild Pacific salmon." Well, you get the idea. Incidentally, the "world's tallest coastal mountain range" mentioned here are the Wrangell–St. Elias mountains. What that really means is that they are the tallest mountains that start right at the edge of the ocean. If you were to continue traveling east (actually, southeast), you would pass by the location of the Katalla (which no longer exists) and on to Yakutat (another small, land-locked fishing town). A lot of people find beauty in that panorama. I ain't one of them.

Cordova's average annual rainfall is much more than the official figure of 120 inches—which is measured at the airport, 13 miles east of Cordova and on the nicer side of the Heney mountain range. Surrounded by mountains on three sides and the Orca Inlet in front, it rains all…the….bloody….time.

Grey skies and rain greet most mornings, mountains disappear into the mist, and distant bays are obscured.

From out in Orca inlet, the town looked like a cluster of wooden boxes had tumbled down the slopes of Eyak Mountain and came to rest at the water's edge. In 1965 Cordova was like stepping back in time to Dodge City: wooden plank sidewalks, mud streets and lots of bars (saloons). There was no TV, and the local radio station was on for 8 hours a day. But most distressing for me was that the only live music in town was a Julius Klein, who sat for hours in the Club Bar and played the accordion. He sure liked the song "Harbor Lights."

In the pitch-black nights nearing winter, what was most striking was walking on the crunching gravel through the rain, past the small houses up on stilts hunkered down against the drizzle, muffled voices coming from behind oily, yellow windows. It created melancholy and inspired whiskey drinking.

But summer—now that was exciting. When the fishing fleet returned to town for the weekend, it was an invasion. Main Street's two-and three-storied wooden buildings—built shoulder to shoulder—bristled with activity. The huge canneries—heavy timber and corrugated sheet steel—perched on trestles of creosoted pylons high above the water roared with work, under clouds of steam and swirls of squawking and diving gulls. All came with the fishermen in wool jackets, hip boots rolled down below the knees, tousled hair, faces red from wind, hands like hams, suspenders, loud voices, the smell of fish and diesel fuel. Bellowing laughter, roaring curses and tumbling wooden chairs, scooting tables, clacking pool balls, 45-gallon garbage cans filling up with empty beer bottles, one after another. And money—my God, the money.

Here's the day that Al Clearwater quit drinking. All the old, rough and tumble, two-fisted drinkers would patronize the Alaskan Bar on Main Street. Al Clearwater was as hard a drinker as any of them. However, he was probably the only one that routinely would drink himself unconscious and have to be hauled home. One Saturday night, the thud of his head hitting the table signaled one of the other patrons to propose an idea. It was such an excellent idea that everyone agreed to help set it up. They volunteered for their assignments.

After phoning the mortician, several of them picked up a casket and brought it back to the bar. Others woke the local florist and got several flower arrangements. A couple of the guys went to Al's place and picked up his suit, dress shoes, dress shirt, and tie.

They set a couple of tables next to the large plate-glass windows that looked out on Main Street and placed the casket on it. Several of them struggled to get

Al dressed up in his suit and tie and they heaved him up and into the casket. Last, they placed the flower arrangements around the casket, and waited.

When he finally woke up, he sat straight up in the casket and started talking to the guys that were standing around the casket. The guys just pretended not to hear him and kept talked amongst themselves. It shook him up so badly, he quit drinking.

Anyway, for quite some time I regretted lugging my tenor sax to Alaska with me, but was glad that I brought my sea bag (duffle bag) of books. It was so heavy I had to stop people from trying to help me by carrying it. I got a lot of use out of those books for the 21 months I was on the buoy tender Sedge. During time off, when we were underway, I read. Even the sax came in handy during the duty nights when I was stuck on the ship in port; I would pace back and forth on the bridge and play. Other than that, my shipmates and I would drink. Some guys, of course, would go hunting or fishing or hiking. I would stare out at the windy, drizzly day and wonder why the hell anyone would do that.

I joined the Coast Guard in '64 because nothing was happening in the world, and I thought search and rescue would be exciting. If I would have waited a couple of years, the government would have provided me lots of excitement in Vietnam. Anyway, the Sedge did a few search and rescue operations, but mostly it was there to keep navigation lights and buoys operating.

I eventually got married to a local girl, and Lou and I settled permanently in Cordova in 1968. I got a job working for the city in 1969 and that's when I joined the volunteer fire department.

Lou's mother and father were life-long Alaskans. Her father, Jack had been a commercial fisherman most of his life. His wife Ida accompanied him on the boat quite often during the fishing season, but not this one time, before Lou was born.

In the mid-'40s, Jack deVille had been gone for some time chugging his way around the sound in search of salmon and had no way of knowing that wife, Ida, had been worrying over the deteriorating condition of their youngest infant son, Bob. Katalla had no hospital, doctors, or even radio communications to summon help from Cordova. The trip from Katalla to Cordova by boat—assuming weather was conducive to the trip—was tedious and very long. Without radar, piloting a boat in the dark when approaching the waters off the Copper River Delta with its sandbars and breakers would have to be a matter of life or death.

On such emergency occasions, the powerful radio at the St. Elias lighthouse on Kayak Island would contact Cordova for help. There were times when this was difficult to arrange. When the weather was bad, a boat from Katalla could not approach the beach line of Kayak Island without being smashed on the rocks. But Ida was becoming desperate. At the kitchen table, she scribbled four notes

14

which read, "Someone sick. Need Plane." She folded each paper and stuffed them into four tobacco cans. She sealed the cans with wax and tape. Leaving her sons in the charge of a couple of teenage neighbor girls, she trekked out into the stormy weather and climbed into a neighbor's boat who then maneuvered it into the deeper water and turned south-southeast. The engine, the reliable "two-bits" "two-bits" "two-bits" sound it made as it dutifully pushed the boat into the bigger and bigger swells, Ida and the neighbor squinted out of the pilot house windows, their eyes fixed on the rotating light of the lighthouse off in the distant darkness.

Reaching the south end of the island they "paced" back and forth directing the beams of spotlights at the lighthouse living quarters until they saw dark figures emerging from the building, which, with flashlights in hands, made their way to the beach. The spotlight on the boat was now directed onto Ida on deck, so the Coast Guardsmen could see she was holding up a can. She then flung it into the water. Then she did the same with the other three cans. Now it was time to wait for the waves to carry the cans to the beach. The Coasties could be seen walking up and down the beach, scanning their flashlight beams on the incoming waves. After a period of time, one of them began waving, and the spotlight revealed that he held up one of the tobacco cans.

The next morning the small plane arrived. It was so small that Ida could only pack for Bob, no clothes for herself. She asked the two Hanson girls what they would like from Cordova as payment for watching the other two boys. "Ice skates," they replied.

In the Cordova hospital, Bob was diagnosed with bronchial pneumonia. He was in the hospital for two months. During that time, Jack had pulled into Cordova and then learned about the situation. After Bob's release from the hospital, Jack, Ida and Bob boarded their boat, the New Josie, and Jack loaded his new prized possession—his new Ford truck—onto the stern of the boat, lashed it down, and headed for Katalla.

In the gulf, in line with the Copper River Delta, the winds tore through the area with such force that crossing was impossible. In fact, turning the vessel around in those swells took timing, nerve, and luck to minimize the time spent sideways to the swells. He was lucky enough to swing the New Josie around to prevent capsizing, but not lucky enough to save his Ford, which was ripped loose of its lashings and washed overboard. That was only the first of many attempts to cross the gulf during the next three months.

When they finally moored up in Katalla, 5 months after leaving, Ida was very apologetic to the Hanson girls and gave them a pair of clip-on ice skates, adjustable in size, so they could share them.

As an aside, in 1993 I went to Wrangell to teach a course to the firefighters there, and spent one evening with Ida's brother, August, who confessed that during that time when Jack and Ida were in Cordova, he (August) and his younger brother were curious about what pool balls were made of. They tightened one in Jack's vice, but it wouldn't break apart. So August hit it with a sledge hammer. The pool ball survived but the vice did not. They were so afraid, they told Jack that Johnny Hanson (the babysitter's older brother) broke it. August went on to describe to me how Jack confronted Johnny in the street that afternoon, and was so angry that Johnny feigned ignorance, that Jack took a swing at him. When he did, he slipped on the ice and fell flat on his back. Now he was in a rage, so he leaped up and beat the shit out of him.

When I got back from Wrangell, I told Jack that August confessed to breaking his vice 50 years ago and wanted to apologize. Johnny was dead now, but Jack felt no remorse for the beating. "Everyone knew Johnny was a thief, too." And Jack got mad all over again.

Incidentally, hurricane-force winds blowing down the Copper River are so common they resulted in two rescue operations I conducted there years later, and later in this book.

It took me a while to learn to appreciate this world and many of the characters that were unique to it. Like the guy everyone called "Wild Bill" (I never heard his last name) who bought an airplane and flew low circles over his house, throwing cans of beer at his chimney. Not to be outdone, Jack and Ida's next door neighbor, Martin Anderson, bought an airplane in Seattle. Never having piloted before, he flew it back to Alaska following the Alcan highway. The deVille kids were all out playing in front of their house when Martin started flying low circles over the mud flats at low tide, when mother Ida ran out and uncharacteristically cussed, "You kids get in the house, that crazy sonofabitch is going to crash that plane!" Within a couple of minutes, he dug a wing into the mud and the plane crumpled in a heap. He stumbled out of the wreckage and, waving his cap in the air, staggered home.

John Wilson, before he got religion and gave up drinking, had a dispute with Junior Thomas on which front-end loader was more powerful. I can't remember which one had the Allis-Chalmers and which had the Euclid. They decided to "settle this like men." Patrons spilled out of the bar to watch the jousting between these giant machines. It seemed like everyone knew about the conflict except the police until they got a call from Gail Steen. She screamed over the phone, "My God—they're trying to kill each other!" The cop arrived to see these giant machines roaring, spewing black exhaust into the air, spinning streams of dirt and rocks with their tires, groaning and twisting and using their buckets to tip the opponent on its side. The cop stopped the contest before a winner could be determined. They went back to drinking.

16

Jon Goerse lived one block away. First, he lived in a house on the slough (Odiak Slough). Since it was in the tidal area, the house was teetering 12 feet up on pylons. In fact, their toilet emptied directly onto the rocks below. One time, he lifted the lid and saw the tide coming in for the second time that day and remarked, "Well, the toilet's flushing again."

This tall, skinny, sixty-something, white-haired Swede and his equally hard-drinking wife, Sandy, would wake up in the mornings and have a contest to decide who would get up first and make the coffee. They would take turns shooting a pistol at the light bulb dangling from the ceiling. The loser had to make the coffee. About once a week, Jon would be at the hardware store, buying a new socket fixture for his bedroom.

He finally moved up to a log house on the road. During an argument, he told Sandy to get out of his house and she reminded him that half of the house was hers. "Okay, fine," he barked, and cranked up his chain saw. He ripped the heavy timber table in half as well as the kitchen counter and cupboards. The place was thick with chainsaw exhaust and wood fibers. He only stopped when Sandy informed him sternly that she was taking the half with the bathroom.

The final straw for the neighbors was one Sunday morning when people came from blocks around to investigate the most horrific racket they'd ever heard. There was Jon, mad as hell at his stubborn old pick-up truck, sawing through the cab with his chain saw.

Jon's behavior was legendary and continued on for years. Even after I got out of the Coast Guard and went to work for the city. Jon was in the Cordova House Bar when Keith Gordeoff began teasing him about something. Keith didn't realize how serious Jon was taking that chiding. Jon made a remark about shooting Keith and walked out. Someone called the cops. The duty officer showed up and patrons said they expected Jon to come back with a gun. The officer borrowed a jacket to disguise his uniform and waited. Jon walked back into the bar and up to Keith, leveled a 9mm Luger at his head, when the officer stepped up and yanked the gun out of his hand.

Later, the officer said he had heard the gun "click," and fortunately it had misfired. I went in to the jail to see Jon, and he swore to me that he had pulled the trigger but it didn't fire.

He never went to trial, but was convinced by the local authorities to move to Texas. After some time had passed, I was chatting with a Public Works co-worker, Bill Bernard, and said I was skeptical about whether the Luger had misfired. The little ramshackle jail behind the City Hall was empty and no cops were around, so Bill and I used a master key we had (to all city facilities) and got the Luger out of the locker and took it to the dump. I cranked a shell into the chamber from the clip, took aim and fired. Blam! Bill and I stared in amazement at each other. "There was nothing wrong with that shell!" Bill remarked. "Let me try." He took the gun and fired off several more shots narrowly missing a can he chose as a target.

"My turn." I took the gun back and fired at the can and missed handily.

We had emptied the clip. "I knew Jon hadn't pulled the trigger, but was too macho to admit it," I reasoned.

Bill said, "We better put this gun back before we get in trouble." So we put the empty gun back in the police evidence locker and never heard about it again.

People can still remember seeing Jon teaching a young boy how to hunt. He would use a couple of sticks and some cord and "make" a bow and arrow. Then, very entertaining to the neighbors who watched out their windows, he and the boy would crawl on their bellies trying to sneak up—unsuccessfully—on seagulls. He was funny when he was sober.

I think Jon accurately described the "old" Cordova best when answering a question posed by a visiting "puker" (a nickname for urbanites who were not commercial fishermen). The puker asked Jon, "What do people do in Cordova?"

Jon answered, "In the summer we fish, fuck, fight, and footrace."

"And what do you do in the winter?"

"We don't fish." Jon said.

Yep, not surprising, brawling was commonplace back then. I was playing the horn in the Cordova House one Saturday night when I saw Keith Gordeoff run down the length of the bar toward the front door and tackle a guy who was standing there, they burst out the door and on to the sidewalk. Now, before I go any further, I don't want people to think that Keith was a bully or was mean. He was just, how shall I say it…rambunctious. He was good-natured and fun-loving. And today he is dignified and respected, and carries a briefcase. Anyway, the crowd spilled out into the street right after Keith and his tackling dummy. I turned around and looked out the plate-glass window behind me. The entire street was alive with punching, wrestling brawlers rolling around in the mud. The most amusing thing I noticed was Keith—shortly after leaving the building—had discretely slithered under a truck and watched the entertaining spectacle he had started. Five minutes later, they were all back inside bellowing with laughter. I went back to playing.

All of that was before television in Cordova.

Salvage Operation at Mile 13

When volunteer fire departments were first organized in this country, starting about 1736 by Ben Franklin, they were mostly salvage organizations. Decent fire suppression was still a long way off. These fire departments would generally rush into burning buildings (or buildings threatened by fires nearby) and haul out valuable contents. Old Ben would have been proud of the Cordova Volunteer Fire Department (CVFD) back when I first became chief in the mid-'70s.

It was early evening in December and the cold snap had been ongoing for about two weeks when we were dispatched to the Federal Aviation Administration (FAA) housing complex at the airport 13 miles out of town. The State Department of Transportation (DOT) fire department was already on scene and had called us for mutual aid. We knew it would be tough since there was no hydrant system out there and if an initial attack by DOT—using their 500 gallons of tank water—was unsuccessful, the use of our tank water (at that time, less than 1,000 gallons) would require a series of surgical and judicious applications. Making it even tougher, we had only two airpacks. That was not unusual for those days, the New York City Fire Department (FDNY) at the same time had only two airpacks per Task Force (two engines and one truck company). They saved them for cellar fires. Their Standard Operating Procedures (SOPs) for tenement fires was that, while the engines were hooking to hydrants and laying out attack lines, two guys would try a quick knockdown of an incipient fire with a 2.5-gallon pump can which one of them would carry. The other guy carried the halligan and pick-head axe. The man with the can and the man with the irons would hyperventilate before charging up the stairway, reach the next floor up, race down to the end of the hallway, break out the window, stick their heads out for a few gulps of fresh air. They would then hold their breaths again until they were at the involved apartment, force the door open and try to extinguish the fire before the hoses were laid and charged. If the incipient fire flashed over and involved the entire apartment, then the hose-humping attack team would use the same method of air supply as the men with the can and irons. They took a beating, and that much smoke inhalation took its toll.

But anyway, by the time we arrived at the Mile 13 airport FAA housing complex, I was surprised to see that not much flame was visible, but there was lots of smoke. DOT had expended its water. Since they were not intended to be a structural fire department, they asked if I would take charge of the event, which I did.

I sent a couple of guys inside to do a size up while I walked around to the back (C-side). The frost was so thick on the ground, it crunched as I walked. Ice

19

crystals floating in the air were dazzling under the street lights. I noted open flame coming out around the chimney, three stories up. The 5-plex, wood-frame apartment building had a full basement, two apartments on the ground floor, two apartments on the second floor and an apartment on the third floor. I talked to some of the occupants who huddled in a bunch, creating a cloud of fog by their breaths. One said after he'd smelled smoke, he saw some fire coming from behind the bathroom wall vent. Others said they had no idea where the fire was.

My interior size-up guys came out with a report. One of them told me all he saw was smoke, and Groff told me that as he crawled up the stairs to the third floor, and placed his hands on the top landing of the stairs, his monkey-grip gloves blistered and his hands were scalded. He was unable to lift his head any higher to look around the top floor. As they explained this, the flames grew around the chimney on the roof.

I gathered the crew into a circle and told them to run three hoses up the stairway and terminate a nozzle on each floor (the third-floor nozzle to be set a couple of steps lower than the landing). While they were laying the lines, I talked with building occupants again. The building's furnace was in the basement, next to the back wall—or "C-side." The pipe-chase for the chimney extended through all the floors adjacent to the bathrooms in the building. That chimney space was also shared by the plumbing and ventilation systems. Each of the five bathrooms had a small window and they were all now breaking out from the heat that was building up in those rooms.

With the lines laid and the crew waiting for direction, I jogged up the stairs, past the firefighters, for a look. The third floor was now so hot, I was afraid that a burst of water to drop the temperature would not drop it far enough or for long enough to allow any salvage work. On the second and first floors, I could hear lots of fire in the rear walls from floor to ceiling. I debated for a moment the likelihood that tearing into the walls and applying water might put the fire out, or only waste the water that could be better used to protect the crew while they did salvage work. When one of the guys yelled up that fire was ripping out the full length of the roof eaves, I knew that a successful attack was impossible with our 800 gallons of water.

Embers were now dropping onto the living room carpets of the apartments. I had the third floor nozzle brought down to the second floor, giving me a nozzle in each apartment on that floor and ordered all portable extinguishers brought up. "Get everything out of these apartments that you can possibly move." The crew erupted into action so fast, it startled me. I never knew how painful it was for them to be patient while I tried to formulate a plan.

I instructed the nozzle man to only use tiny applications of water for protecting the crew. "Don't waste a drop."

Some of the belongings were handed out of the windows to men on ladders that were placed on the front—or A-side—of the building, while armfuls were carried down the stairs by puffing, red-faced firefighters. Fire had flashed over in the third floor apartment blowing out the windows, and now the fire was beginning to eat the upper portions of the C-side walls on the second floor. In one of the second-floor apartments, the crew was lumbering through the room with a full-height, cherry wood-and-glass china cabinet. Sweat was running down their faces as they moved the bulky china cabinet down the stairway, around the narrow landings, across the charged hose lines laid on the steps. When they emerged out the front door, steam billowed off their faces. The owners were jubilant as they saw their prized possession stacking up outside.

Even using the water sparingly, it was depleting really fast. I was mostly staying outside on C-side, watching the fire consume the roof and the third floor. The crew was through salvaging the second floor by the time the flames rolled through both apartments there. It looked so bad from outside, I decided that trying to salvage contents from the first floor would be too dangerous. That's when Stan Shafer walked up to me and said I needed to go talk to one of the occupants. "He keeps trying to crawl through the window of his first floor apartment, and I keep chasing him off. As soon as I turn my back, he's trying to crawl in again."

He was on his tip-toes, looking in the window when I walked up and asked him what he was doing. "My wife's silver setting is right there," he said, pointing in at the chest of drawers a few feet inside the room. "It's been in her family for years. I'd like to be able to get it."

It was just four feet inside. "Hell, easy to do" I assured him. "Give me a boost." Everyone else was busy and I had nothing to do, anyway. It was not uncomfortable at all inside. With the flashlight, visibility was sufficient, it wasn't too hot—even though the floor above was fully involved—and I was breathing comfortably. I handed him the wooden case containing the silverware. He and his wife were very grateful. Embers were starting to drop leisurely from the corners of the ceiling, but they were not presenting any danger. I opened the top drawer of the chest and used my full arm to scoop the items from the top of the chest into the drawer. I then handed the drawer out the window. Then, one at a time, I pulled the other drawers out and handed them out of the window as well. Shit, this was easy. I pulled the drawers out of the end-tables, took the lamps and such and handed them out. Next, I rolled up bedding from the bed into a large cylinder and they went next. To get the mattress out, I had to break the entire window out, but we made it.

I was going to disassemble the bed but I noticed that the embers were showering down now and the carpet was starting to burn and the smoke was getting thick and acrid. It was then that I noticed I'd been screwing around with a mattress while not thinking about their closet with all of their clothes—far more important than a stupid mattress.

"Whetsell, you moron," I admonished myself as I scrambled over to the closet and yanked the door open. My heart exploded as the opened door revealed someone standing right there! I literally grabbed my chest and froze as I stared through the thick smoke at the outline until my watery eyes focused well enough to see that the person was only a tailor's (or sewing) mannequin. I goddamned near had a heart attack. I pitched it out the window, took in a few good breaths of air, and grabbed an armload of clothes. That was the last of my salvage operation, it had become untenable.

I got almost all of their stuff out. I should have been commended by my fellow firefighters for a job well done. But they remembered the rule I had initiated shortly before. That is—anyone seen in the "hot zone" without being fully clothed in protective gear has to buy a case of beer for the department. Shafer noted that I took my helmet off to climb in the window and hadn't put it back on. He couldn't wait to tell everyone. So, instead of a "well done" from my buddies, I got a ration of shit, and they got a case of beer. Actually, they thought the violation was so flagrant, it warranted two cases of beer. They got one case. Screw it, I'm the chief. Rank has its privileges.

The person who really deserved a case of beer was Deputy State Fire Marshal Andre Schalk. He came down to investigate the fire. There wasn't much left to look at: a ground-level pile of debris with a three-storied chimney sticking up in the air. Even before he talked to the building's occupants, he had already formulated a theory. The interviews reinforced his idea. I don't know if every story needs a "point," but this one certainly has one—an important one.

Andre told me that ten days before, an FAA 5-plex had burned to the ground in Cape Yakataga. It was identical to this building, and also built in 1944. It had burned the same way. He theorized that since the temperatures had been around zero, the furnace was pumping away inordinately hard day after day. "These concrete block chimneys and the mortar between them may have deteriorated over the decades and developed pin-hole leaks. The heat escaping through these tiny holes and cracks was slowly pyrolizing (carbonating) the wood surrounding the chimney over the years. Then, the furnaces running non-stop might have generated enough heat through the pin-holes in the chimneys to ignite the pyrolized wood. It only takes a couple hundred degrees to ignite pyrolized wood."

We were standing in the yard behind the remains of the building as he explained this. He looked around and saw that seventy five feet away stood another 5-plex identical to the one that just burned down. "Get a couple of flashlights," he instructed. We went up to the third floor apartment of the building and crawled through an access way into the eaves of the building. We crawled through the eave until we reached the chimney. Sure as hell, the wood near the chimney looked like charcoal. It was fully pyrolized and ready to ignite.

Back outside, Andre told the FAA manager to shut the furnace off, vacate the building and remove all the occupants' personal belongings. He then called his office and instructed that a notice be sent out immediately to the FAA to vacate all such buildings until they were inspected and the problem remedied.

This fire was the first of several I had responded to that were started by relatively low amounts of heat bombarding pyrolized wood. Most of the later calls were due to wood stoves and wood stove chimneys that were installed closer to combustibles than the code allowed. It often takes years for the wood that is exposed to the heat to start slowing breaking down chemically. Once the chemical change is complete, extremely low heat can ignite the wood without warning—in the middle of the night. That's why we have building codes. Incidentally, there is a body count for every code in the books. End of sermon.

CAPE ST. ELIAS LIGHTHOUSE
NEAR OLD KATALLA TOWNSITE, ON TIP OF KAYAK ISLAND, GULF OF ALASKA
Photo: U.S. Coast Guard

RESCUE ON MOUNT ST. ELIAS

On Thursday, May 31, 1984, I was just breaking for lunch when I was toned out. Two medics were needed to accompany the Coast Guard H-3 helicopter to Icy Bay near Yakutat. The call was for an airplane crash; number of persons on board was not known. Erik Havens, 19, a brand new EMT-1 was on call with me. We gathered the medical supplies and our Mustang flotation/exposure suits, and drove to the Coast Guard air station 13 miles away. We were aboard the chopper in 20 minutes.

The pilot was Lt. Rick Bartlet, co-pilot was Lt j.g. Larry Cheek, and also on board were an airman and a radioman.

It took one hour and twenty minutes to make the 200-mile flight to Icy Bay. As we cruised into the area, Rick spotted the fishing boat that radioed in the alarm, and we landed on the beach near where it was anchored. Just before we landed I heard Rick speaking on the intercom asking where Mount St. Elias was. Then he requested a C-130 airplane to that location.

As it turned out, we were not responding to a plane crash, but to a different emergency altogether. While we were enroute to Icy Bay, Rick had radio contact with the fishing vessel Cindy L. The crew of the Cindy L was out on deck that morning when they noticed a plane flying over the area just before 11 a.m. Shortly thereafter, they spotted a flare from the beach and a man waving to attract their attention. Since that part of the country is so remote, and the only sign of life they had seen recently was the plane flying over, they assumed the plane had crashed. They had called in the Mayday before they picked the man up and heard the true story. But while we were enroute, they picked the man up and then relayed his story to the chopper.

The man, Andy Politz, 26, from Tacoma, Washington, a professional climber and a Mount Rainier guide, and his partner, Walter Grove, 46, of Nashville, Tennessee, a 30-year mountain climbing veteran, became the first climbing team to scale the south face of the 18,000 foot Mount St. Elias, the third highest peak in North America. They reached the summit on May 25th and began their descent.

As they descended, Grove began showing signs of cerebral and pulmonary edema (a condition attributed to the decrease in air pressure at higher altitudes). He began having loss of vision, coordination, and had frothy, gurgling breathing as fluids backed up into his lungs. To compound this problem, he was getting frostbite of both feet and his right hand. Temperatures ranged from 5°F to 25°F on the mountain.

24

Finally, Groves was unable to go any farther. Politz managed to locate a knoll at the 6,000-foot level and pitched a tent in the snow near the center. Sheer ice walls formed a semi-circle behind the knoll, and in front was a precipice that plunged down nearly a mile to the glaciers below. The entire area was enveloped in a silent fog.

Grove was bundled into two sleeping bags and placed in the tent. Politz gave him most of the supplies and all the medications: six quarts of water, cold food, and sleeping pills. Politz told him he would go for help and be back in about four days. So, at about noon on Tuesday, the 29th, Politz told Grove to "hang in there," patted him on the shoulder, closed the tent flap, and began his descent.

Without stopping for 41 hours, he crawled down the mountain, then across the treacherous glaciers, finally reaching the coast at about 5 a.m. Thursday.

There, he looked out at the expanse of water and saw nothing, no one. This was Alaska. He lay down right there and went to sleep. At about 11 a.m. the sound of a passing airplane startled him awake. Standing up, he marked an S.O.S. on the beach and began walking. Shortly after, he spotted the crabbing boat Cindy L at anchor, and fired off flares which were spotted by the crew. So, while the captain radioed the Coast Guard, his crew took a dinghy in to the beach to pick up Politz. Two hours later we had Politz in the chopper and were headed for Mount St. Elias, looming 8 miles in the distance.

Mount St. Elias is part of the Wrangell-St. Elias National Park, which stretches north 170 miles from the Gulf of Alaska to encompass 13.2 million acres of the greatest expanse of valleys, deep canyons, and towering mountains in all North America, and includes ten of the highest peaks on the continent.

The snowline on the mountain range reaches down to the 3,000-foot level. Mount Logan, a little farther north, in Canada, is the culminating point of the range, and rises 19,850 feet from the sea. The only thing taller on this part of the globe is Mount McKinley at 20,322 feet. The Himalayas may be higher in altitude, but the base where climbers begin their trek, starts higher up.

Little of the mountain rocks are visible; their rugged valleys are ice-filled and the glaciers flow down from ragged peaks, unite and fan out to form ice fields along the gulf. One of them, the 1,120-square mile Malaspina Glacier, is larger than the state of Rhode Island. The glaciers reach down to the winding, fog-shrouded fjords near Yakutat, the only town in the area.

Staring down while flying over the glacier, I saw the broken and shattered ice field had created jagged crevasses, at the bottom of which lurked ice-blue water. Slipping into a crevasse, especially traveling alone, meant certain death.

As we approached the mountain, the glacier became steeper and finally merged with the mountain itself. Ascending higher up the face of the mountain, fog and mist enveloped the craft. At about the 3,000-foot level we seemed to have worked our way into a chimney, or three-sided vertical trench. There was a wall of ice in front and on both sides of us. The loud popping sounds of our blades reflected back on us. Over my ear phones I heard the pilot of the C-130, which was circling high above the peak, explaining to Rick which way to maneuver the chopper if he needed to pull away. The C-130 pilot must have been using radar for that since he couldn't see us buried in the mist.

Finally, at the 6,000 foot level the wall in front of us rolled back into a wide, gently-mounded snow-covered knoll. There, right near the center was the tent. Desolate looking as it was in that backdrop, we saw the tent flap open and Grove waved at us.

We hovered for several minutes some distance from him, just slightly in from the edge. Rick decided against trying to set down without knowing how firm the snow crust was. If one wheel were to break through the crust, the craft would tilt and the blades would be out of commission. He then had the airman tether himself off, slide open the side door and attach the basket to the boom. He was going to lower Erik and me down so we could tend to Grove and place him in the basket for hoisting aboard. He then decided against that, as well, after noting that the prop-wash was kicking up billows of snow which obscured sight of the basket landing. If he couldn't see the basket throughout the whole operation, he wouldn't use it. Safety first.

He needed time to think this over. He radioed the C-130 that he couldn't do it and would be heading across the Malaspina glacier to Yakutat for fuel. At least Grove knew that Andy had made it out.

Up in the cockpit, while enroute to Yakutat, 60 miles to the southeast, I talked to Rick about using climbers as an alternative. As per his request, the C-130 headed for Juneau, 200 miles to the southeast, for climbers.

Yakutat, population 800, is a windswept, land-locked, isolated community half-way between Cordova and Juneau. It was raining in Yakutat as we dashed from the chopper to the airport lounge. It was now about 4 p.m. there were about half a dozen locals either at the bar or at tables eating. Erik and Andy found a table near the back as I went to the pay phone and made a collect call to the hospital in Cordova. I spoke with Dr. Larry Ermold about Grove's cerebral and pulmonary edema. He told me he was certain the edema would subside at the 6,000-foot elevation and that his primary concern at this point would be the frostbite. At the conclusion of the conversation, I asked him to call the dispatcher and let her know that we were uncertain when we might be back and to let Erik's family and mine know that. I told Rick about my conversation with Dr. Ermold.

Heading back to the table, I knew that Grove must feel less depressed knowing that Andy made it out and found help. He knew efforts were underway to get him off the mountain.

26

We were throwing our packs of medical gear and mustang suits in the corner when a couple of local EMTs and the town's physician's assistant (the town has no doctor) came into the lounge and spoke with us. They gave us some antibiotics and betadine solution. Andy took the antibiotics while I found a waste basket, inserted a new plastic trash bag, filled it with warm water and betadine, and placed Andy's feet in it (they were badly blistered). Andy then commenced to eat about four hundred hamburgers.

While Rick's crew refueled the H-3, Rick spent a lot of time on the phone. Shortly after, the local state trooper came in and got the story. Ordinarily, since the incident occurred on land, this whole operation would have been a trooper operation. But since there is a mutual aid agreement between State Public Safety and the U.S. Coast Guard, there was no conflict with the CG participating—especially since the initial call for help was from a boat. But since the troopers are responsible for search and rescue inland, they would cover the expenses of the operation. He arranged for rooms and said he would cover the food costs for Andy and us.

By 9 p.m. the wind had picked up and rain was pounding the little airport. It flowed in sheets across the tarmac as two C-130s landed. One carried a mountain climbing team from Juneau, and the other had a team from Anchorage (475 miles away from the other direction).

The lounge filled up quickly as about a dozen rescuers burst in the door, dripping and sputtering. Pretty soon the corner containing our gear filled up with packs, bundles of ropes, ice axes, crampons, and coats. As the rescue crews milled around looking for dinners and drinks, the locals began drifting out.

One of the Juneau climbers was Dick Rice, an EMT-3 from the Juneau Fire Department. He recognized me and came over to introduce himself. We talked for a while and I explained Grove's condition to him. I then introduced him to Rick and explained the standard EMS protocol, that since Dick was qualified in more advanced care than I (I was an EMT-2 at the time), that Dick would be in charge of patient care when Grove was reached. Incidentally, Dick Rice, a bit of a climbing celebrity himself, was mentioned in a 1985 National Geographic article about the St. Elias Range.

All the climbers then congregated around Andy and talked about climbing conditions and routes.

Finally, at about 11 p.m. we walked across the parking lot, and Erik, Andy and I grabbed a room for the night.

The weather had improved by 5:30 the next morning. One of the C-130s was flying the summit while Rick was conducting a briefing for everyone in the lounge over breakfast. A small civilian helicopter—Canadian, I think—having heard of the problem, arrived and offered help. The Canadian Mounties (on the other side of the mountain) were also willing to help if needed. Anyway, the small chopper could carry four people and was equipped with skids for landing on snow.

Rick had a two-man climbing team go with the small chopper and two other climbing teams to go aboard the H-3. The plan was that if the small chopper couldn't land on the knoll, the H-3 would drop off the other climbers at some other spot where they could start their climb. They were outfitted with three days' supplies—just in case.

However, the small chopper was able to land and Grove was back at the Yakutat airport by 11 a.m. We transferred him from the chopper into the quiet interior of Yakutat's ambulance for a survey of his condition. An IV was started and gauze pads were placed between each of his fingers and toes. The tips were already black and hard ... a poor prognosis for them.

We all boarded a C-130 and headed for Elmendorf Air Force Base in Anchorage. I asked the pilot to pressurize the plane at as near sea-level pressure as possible to prevent recurrence of the edema. In just a little over an hour we were at Elmendorf and transferred Grove to an ambulance there.

We were given a ride back to Cordova on the plane and arrived around 3 p.m. Friday, 27 hours after we left.

Grove eventually lost several toes and three fingers of his right hand, I imagine the loss of his fingers put an end to his climbing career.

During a brief conversation I had with Andy in our room Thursday night, I asked him about this strange thing he does for a living as well as for recreation, after he mentioned his plans for scaling the west ridge of Everest the next year. I remarked that there must be pretty good money in mountain climbing. He just stared out of the window and muttered, "Nope."

"Well, it must be fun, then," I reasoned.

"Nope. It sucks."

"Well, if you don't do it for money, and you don't do it for fun, why do you do it?"

Without a pause, he replied, "I guess I'm not happy unless I'm miserable."

I guess Andy never tires of being miserable. I saw him recently on the Travel Channel program "The 10 Worst Places To Be" (or something like that), where he was videoed on a 1999 Mount Everest climb, hunkered down in a tent with his crew, tied down trying to avoid being blown off the mountain like pieces of paper. Fifteen years after the Mount St. Elias rescue, the bastard looks the same. I, on the other hand, look 20 years older.

* * * * * * * * * *

When I was a kid, I saw an old movie played on TV, "The Petrified Forest." About the only thing I remember of the movie was a statement made by the star. In a backdrop of dust blowing over some small Southwestern cafe, the man stared off into the distance and explained that man had finally conquered nature. He could cross the oceans in luxury liners, span the continents in the comfort of railroad trains, fly through the air in airplanes. His cities had insulated him against almost anything nature could throw against him. But now, nature had found new weapons called Neuroses and Psychosis. If one conquered nature on the physical field, nature would attack man's spirit with these new weapons.

Perhaps it is like that statement about the professional warrior. If the general cannot find a war to fight, he will wage war upon himself.

Loren Eiseley wrote about our settling the new world, America, in the 17th and 18th centuries. "We toppled instead off the end of time. We, all of us, western man clinging in his little enclaves to the eastern seaboard, had re-entered the stone age. We had starved helplessly in our first winter; Indians had to feed us. Generation by generation we had had to relearn the arts of a vanished era. In order to survive we had had to master what our Paleolithic forebears had taken for granted. The farther we pressed into the forest, the more rank, prestige, and fine garments would dissolve into rags and buckskin. We would be reduced to Elemental man....." (*All the Strange Hours*, 1975, Scribners, NY)

In the end the mind rejects the hewn stone. Instead it asks release for new casts at eternity, new opportunities to confine the uncapturable and elusive gods. It's important that creature comforts don't dull our minds to somnolence.

Hell, if today's 17-year-old city dweller can no longer run away to sea—there are no more cabin-boy positions on whaling ships—and if he's tired of creating adventures in his neighborhood streets and alleys, perhaps his last alternative to fight off his screaming frustration is to go to the mountain with Politz. There he can face down the Ice Age, his original adversary, one-on-one, and in silence and cold, play the final game of winner-take-all. The Elemental Man.

WRANGELL-ST. ELIAS MOUNTAIN RANGE
SCENE OF RESCUE OPERATIONS
Photo: National Park Service

COSTELLO AND THE DEKABRIST

In the Coast Guard, duty on a 180-foot buoy tender in Alaska was different than it was on other buoy tenders I was on. One of the interesting chores we had to do was chase away Russians who were fishing in American waters. Usually it didn't amount to much. A commercial airliner flying across the Gulf of Alaska would radio that they saw the Russian fishing fleet fishing inside the 3-mile limit (that was before Congress lengthened it to 200 miles), we would tear out of Cordova, and go streaking across the gulf at a terrifying 12 knots. As soon as they saw we were coming on their radar, they would scoot back out and wave at us as we approached. Occasionally, we would go out to pick up an injured crew member. We would lower a lifeboat which would go alongside the ship and have the injured person lowered to the boat, get him aboard our cutter Sedge, and haul him back to town. But one of our missions lasted a little longer and was more interesting.

* * * * * * * * * *

It was Thanksgiving morning when we got underway to head toward Yakutat and escort a Russian fishing factory ship, the Dekabrist, into Yakutat Bay for repairs. Here's what happened: When a new ship is sent from Vladivostok to fish as part of the fleet, it is years before it returns. The crews are usually gone for a year at a time and look forward to heading home. So, when one of their trawlers—an SRT—got word it was to return home, they were so excited, they full-throttled out of there weaving through the fleet; they rammed and punctured the Dekabrist. The Dekabrist was given permission to anchor in protected waters to make repairs, but we had to go with them.

The Coast Guard crew got underway that morning and everyone was pretty grumpy because no one had Thanksgiving dinner yet. Wives were pissed, too. The cooks on the Sedge had only prepared enough for the duty section and were trying to prepare more food. I was on the bridge for the four-hour trip through the sound and was being relieved. I took a bearing on the Hinchinbrook Light House which was directly to our port beam (90 degrees to our left). I figured that as usual, by the time I was to come back on the bridge, four hours later, we should be two-thirds of the way to Kayak Island and our rendezvous with the Dekabrist. I hit the sack. Four hours later I walk back up on the bridge. The ship was diving and soaring through the worst swells I had experienced thus far in the gulf. Even as we crested a swell, I squinted ahead but could not see the St. Elias light house on the tip of Kayak Island. I looked out the port side and couldn't believe I was looking at the Hinchinbrook light just slightly aft of our beam. We had been pounding into the sea for four hours and hardly made any headway at all. What made matters worse was that the crew was so mad about being called away from home before Thanksgiving dinner that they gorged themselves on the ship. Now, with the pitch and roll, a lot of them were getting sick. There had been a rush for the saltine crackers and 7-Up which settles the stomach.

We had a new anchor winch cover that was lashed down tightly over the anchor winch, but it was torn to shreds since we were taking green water over the bow. (It's not remarkable to have waves crest over a bow—white water—but plunging your bow into swells and watch it literally surface again like a submarine is a lot more impressive.) To get a fix on our position and mark it on the chart—to record the range (or distance) from Hinchinbrook Island—I locked my knees on the sides of the radar and hugged it like a humping bear. Then I stepped out on to the wing of the bridge to get a compass bearing on the light. That was an adventure. Anyway, it took us a total of sixteen hours to meet up with the Dekabrist. It finally calmed down as we, the Dekabrist, and another accompanying SRT made it to Yakutat and dropped anchor.

Our captain went over to the Dekabrist to say hello and to tell them they couldn't go ashore or take pictures or anything. Of what? We all laughed about that. The Old Man came back and up on the bridge bemoaned the fact that he'd been invited for dinner with their captain. So he got all dolled up in his class-A uniform and had one of the boatswain mates ("bos'n") and a seaman run him over there in the life boat.

It was late that evening when the bos'n radioed back that he wanted a crew of guys to come over to the Dekabrist, and he wanted them armed. We couldn't believe what we were hearing. Not only were the U.S. and the Soviet Union involved in a cold war, with emphasis on "cold", these Russians were fishermen, not military. What had happened was really funny.

When the boarding party got to the ship, they were informed that the seaman on the captain's lifeboat was missing. After the captain had gone to their captain's room for dinner, our seaman—an Italian kid, named Costello, from New York City, got into a shoving match with one of the Russians in a passageway. Now, when our captain was ready to leave, Costello couldn't be found anywhere on the Dekabrist. The armed boarding party's search widened to the SRT tied alongside the Dekabrist. Down on the mess deck of the SRT, the lights were out as the crew was watching a movie. They turned on the lights and there was Costello sitting on the deck in the middle of all the Russians watching Marlon Brando in "On The Waterfront" with Russian subtitles.

During the next few weeks, the Russians were welding the side of the ship and we were bored to death. Then we started trading stuff. The Russians were rationed short strips of toilet paper, so for a full roll of American toilet paper we were getting loaves of Russian bread. A pack of American cigarettes was exchanged for several packs of perfectly awful Russian cigarettes and magazines. One magazine I got was pretty enlightening to me. I couldn't read it but I recognized some photos in it because I had seen them before. They were of the scene of the shooting of black civil rights leader Medgar Evers.

But the coup de grace was Playboy magazines. A Playboy magazine was worth big, brass belt buckles with the USSR stars on it, lots of cigarettes, and beaver hats. Now I know why they needed that toilet paper.

The captain kept urging them to get the repairs done and we managed to get back to Cordova just in time for Christmas.

FIRE ABOARD THE ALASKAN SWEDE

One Saturday night in July, at just past midnight, dispatch received a series of 911 calls reporting a boat on fire in the harbor. Approaching the harbor, I noticed a fishing vessel Alaskan Swede, in flames and adrift several hundred feet away from the dock. It was a clear, warm night and we had to weave our apparatus through the Saturday-night crowd of fishermen who had poured out of the waterfront bars. It had been a long, long time since a burning boat had been cut loose from its moorings and set adrift to protect adjacent boats, but someone had done it this time.

I had Engine 4 lay a line from a corner hydrant out on to the dock. The crowd was unruly and really getting in the way. Out on the water, there were several jitneys carrying shouting fishermen, buzzing around the Swede. Flames boiled out of cabin and wheelhouse of the steel vessel. I shouted through the bullhorn, trying to get someone's attention down there, hoping that one of the jitneys would push the boat over to us so we could make an attack. I also asked dispatch to contact the Coast Guard cutter Sweetbrier for their assistance. During the mayhem, two of the jitneys collided and one fisherman fell into the water followed by shouts and maneuvering of boats trying to pick him up. Behind me there was cussing and shoving going on between fishermen and the firefighters who were trying to advance lines to the edge of the dock. It was Saturday night in Dodge City. I radioed for the duty police officers to help us clear people from the dock.

Finally one of the jitneys got hold of the Swede and pushed it over to us. It came in stern first. Flames were blowing out of the cabin and leaping high into the sky. The fishermen were pointing and shouting to assure that I noticed that on the stern were barrels of diesel fuel and a thousand-gallon tank of gasoline. The Alaskan Swede was a "tender" who would pick up the day's catch from fishermen out in the sound and resupply them with food and fuel. The tank of gasoline was so far aft, I didn't feel there was any danger yet. On the dock, we were high above the vessel. One of the nozzlemen played some water on the fire through the open cabin door while we tried to figure out how to get airpack men on to the boat. My brain was frantically scrambling for any outrageous solution. There was no ladder going down the face of the dock where the boat was laying, but if we could get a stern line tied securely enough to a pylon and somehow keep the bow out away from the dock we were standing on, we might be able to get an attack team aboard and knock the fire down. But before we could do that, the fishermen who were afraid of fuel exploding, pulled away and the Swede drifted out again.

As this happened, the boat's owner, Jon Branshaw, came up and with a shaking voice told me he thought one of his crewmembers might still be on the boat. I was just sending a team of airpack men down to the floats to hitch a ride with a jitney out to the Swede when I saw the Sweetbrier's small boat entering the harbor. "Are you sure?" I asked.

"I can't be sure, but I think he might."

The Swede was in tow and heading for the Sweetbrier. Obviously, no one would survive in a fire like that unless, by some miracle, the man could drop down a hatch into the engine room. Even then, I can't imagine an oxygen supply adequate enough to keep him alive. Nevertheless, I could feel a boiling rage welling up inside me. For years, we had notified fishermen that if a moored boat caught fire, leave it tied up and move the adjacent boats away if they were endangered.

Anyway, we repositioned our apparatus down on the Coast Guard dock a block away. I informed Lieutenant Commander Ernie Blanchard, the captain of the Sweetbrier, that there may be a victim on the vessel.

The Sweetbrier tied the Swede alongside the buoy deck and Coasties were advancing their 11/2-inch handlines as we were stretching our handlines across their deck. The Coasties started on the stern of the Swede and worked forward toward the cabin. We boarded her bow, took out a forward window and started cooling the interior. The Coasties had oxygen breathing masks (OBAs) but didn't have them on, so when they reached the main door, they just crouched down and played water inside. They were generating some steam, so Loyd Belgarde spun the wheel on the forward forecastle hatch cover, pulled it off and set it aside. He didn't wait long enough for the heat to vent, and just dropped down into the hatch. A second later he was scrambling back out, scalded. We took out a couple more windows.

Then Rex Goatcher dropped down into the hatch. To help with the search, I sent a couple of airpack men into the rear door where the Coasties were working, but it was still so hot in the galley part of the cabin, they could not advance very far. Goatcher's hand came back out of the forward hatch waving for help. A couple of guys leaned over and reached down into the opening and lifted the fisherman's body out.

When I turned around to motion the medics over, I saw that Dr. Osborn was there. He examined the body and determined not to bother attempting to resuscitate. The body was transported to the mortuary.

After I did a Point of Origin investigation on the vessel, I had it towed back into the harbor and tied alongside the harbormaster dock. I asked the harbormaster to keep an eye on it while I went back to the station for the video camera and a 2½-gallon water extinguisher. While the fire crews were getting the apparatus and equipment back in service, police officer Fred Brady and I went back to the Swede. While I walked through the Swede narrating the fire cause investigation, Fred ran the camera.

Because of the burn patterns, the warping of metal structural members, then the examination of the inside of the galley stove, I determined that it started as a result of a faulty temperature control device in the stove. But what really took my attention was the forward hatch cover over the forecastle. The hatch cover, which had been placed back over the opening, could not be opened from the inside.

On the stern, the hatch cover over a rear hatch was missing and had apparently been removed during overhaul. So we set a pallet over that hole for safety. I was really stewing over that forward hatch cover and checked with the Coast Guard regarding regulations for boats. Is there a regulation requiring two means of egress from the berthing area? No. But when I mentioned this issue to the boat owner, Jon Branshaw, during a phone conversation, he told me that when he had the boat overhauled in the shipyards recently, he had that hatch cover "fixed" so it could be opened from the inside. I told him that I looked at it and it could not be opened from the inside. He told me that the hatch cover I was looking at had come from the stern. He explained that when he was looking over the boat after the fire (apparently when I had left to get the video camera), fearing that someone might fall into the open hatch on the bow, he removed the cover from the one in the stern and put it up forward. I didn't believe him.

I asked Loyd what he did with the hatch cover when he removed it during the fire fight and he told me he just set it down on the deck. Branshaw insisted that it must have gotten kicked overboard during the action.

Finally, a couple of our department divers went down and searched the bottom near the Coast Guard dock and they retrieved the hatch cover. What Branshaw said was true. A handle was welded to it which allowed it to be opened from the inside. The victim had been on the bunk right below the hatch, an arm's length away. He must have been overcome by smoke in his sleep.

* * * * * * * * * *

Incidentally, years later I realized a much better tactic was readily available to me which would have simplified that job immensely. To my knowledge, the tactic is virtually never used because its effectiveness is never considered.... it's too far "out of the box." To see what it is, you will have to read the chapter "Uncommon Approaches to Fire Attacks."

THE SEA LARK

The boat harbor is the centerpiece of the life and economy of Cordova. After doubling its size in the '70s, the harbor has slips for about nine hundred boats. During the peak of the fishing season, many of the slips have two boats tied abeam of one another, and it's safe to say that we may have a thousand boats down there. From another perspective, the assessed value of the boats in the harbor always, coincidentally, equals the assessed taxable value of the community itself. In one report I wrote about our fire protection many years ago, the taxable value of the community was $100,000,000 and so was the value of the boats in the harbor. An aspect of this often overlooked is that those boats represented hundreds of independent businesses upon which the rest of the community depended. But the city's investment in fire protection system for the harbor amounted to hundreds of dollars, not thousands. This became very clear one chilly Monday morning.

Andy Foreman was a grizzled and crusty old fisherman of the old school. What I mean by the "old school" is that his boat, the Sea Lark was a small seiner by today's standards. The old wooden-plank, 30-footer required annual caulking and constant attention to keep afloat. Andy often neglected to do that. To keep the moisture at bay inside the cabin of the boat, he had suspended a simple light bulb from the overhead and kept it on. The Sea Lark had taken on a lot of water and Andy had not visited his boat during the weekend to pump her out. Finally, she'd taken on all the water she could support before her decks were awash, her mooring lines snapped, and her cabin began slipping into the water. As her hull sank, the gasoline in the fuel tanks started bubbling out of the tank vents and spreading across the surface of the water and the surrounding boats. This went unnoticed until the last eighteen inches of the cabin was all that remained above the water, then the hot light bulb in the cabin struck the cold surface of water, burst, and ignited the gas.

In my office, two blocks away, I could hear the "whooomf!" I walked to my window and saw above the roofs of the buildings across the street the column of smoke curling black. The 911 phones in dispatch started ringing before I walked out of the office to don my turnouts.

I got there before the engines and saw the water burning and engulfing at least three other boats in the flames. I radioed my report to the responders and headed down the ramp for a closer look. I met a ballsy bunch of fishermen carrying portable extinguishers and heading for the fire. The professional thing would have been to order them back. So naturally, I grabbed an extinguisher and went with them. You have no idea how relieved I was—and astonished—to see the gasoline on the water was actually going out before we reached it. Gasoline burns quickly, and there was just a thin layer floating. However, the fishing boats Silver Spray, and Natalie, and the puker boat (pleasure craft) Northern Comfort were burning.

The crew had arrived and was lugging our "portable" pump (178 pounds) down the dock, along with cans of foam and other supplies. The boats were quickly extinguished, and we were flowing a blanket of foam onto the water when I noticed that the foam was being pushed away by gasoline that continued to boil up from the bottom. It was then that I realized what had happened. I ordered everyone—especially civilians—out of the area and hoped the blanket of foam would offer protection against another ignition. There was no way of knowing how long the gas would continue to come up and spread. The air was thick with the fumes, and the whole place was a bomb. Yes, it certainly was a Monday.

I had dispatch contact the electric company and also inform the harbor-master to shut down all electric power to the harbor. The cops would evacuate the other floats. EMS Captain, Don Endicott, suggested that he dive down and plug the fuel tank vents with an underwater putty called "monkey shit," then attached large hoisting straps to the boat for lifting. Diver Mike Carr suited up to go with him. We contacted Ken Simpson, owner/skipper of the Jet Belle—a vessel big enough and equipped to lift the Sea Lark off the bottom.

Don said not to worry about him and Mike. They would enter the water upwind of the gas, and go down. If the thing lights off again, they would not surface until they were well clear of (and upwind of) the fire. We all backed off and waited for the power to the harbor to be shut down.

As soon as that was done, Don and Mike dove down and stuffed monkey shit into the vents and the leak stopped. We spread another layer of foam on the water. The Jet Belle hoisted the boat to the surface, chugged out of the harbor and headed down to the ocean dock where the city crane lifted her out.

Several hours later, Andy mean-dered down to the harbor and flipped out because he couldn't find his boat.

CVFD DIVER DON ENDICOTT
DIVING ON THE SEA LARK
Photo: Cordova Times

37

Bristol Monarch

The National Fire Protection Association (NFPA) has developed many terrific standards over the years—even though I get mad at them from time to time. One set of standards I have recently embraced is NFPA 1405 for "Land-based Fire Fighters Who May Respond to Shipboard Fires." Before they developed that, we used to just make shit up as we went. The Alaska Fire Chief's Association paid some attention to the problem many years ago after the Prinsendam burned and sank in the Gulf of Alaska.

But the fire aboard the Bristol Monarch was definitely make-it-up-as-you-go. One April, as most Aprils, the herring fleet was in town poised for the opening of the season. The larger vessels—many worn, weathered, and ladened with deck loads of supplies—bobbed, bored at anchor in the bay. At night, with all the lights, it looked like another city out there. The processing vessels, which bought the herring and processed and boxed up the fish out there on the fishing grounds, gambled their entire investments on a short season. So did their crewmembers.

Oftentimes, most of the crews would have come in to town on skiffs, leaving only a couple of people aboard the anchored vessels. That was the case the day the processing ship Bristol Monarch had a fire in the forward lazarette.

I grabbed a portable radio that had the marine frequencies on it—particularly channel 16—and held that in one hand and my other portable radio for department conversations in the other. I radioed the Monarch and asked if they had power to weigh anchor and get underway. If not, I would have sought a way to have them towed to the dock rather than shuttle firefighters and equipment out there. Our conversation was interrupted by the U.S. Coast Guard Marine Safety Office in Valdez who could clearly hear our transmissions because of the series of repeater sites established throughout Prince William Sound. They asked us to get off of the areas' main channel and switch to "conversation" channel. I explained that the ship was on fire and we would be getting off that channel as soon as they were tied up to our dock. No problem.

As the ship slowly made its way toward us, we saw an increase in the smoke from the forward portion of the ship. We set up a good water supply, and even though the hydrant capacity and pressure was very adequate for the small flows needed for ship fires—and especially since operating off hydrant pressure would be more than adequate considering we would be stretching handlines down into the ship, increasing the nozzle pressure—still, I wanted the control we get by running the water through one of our pumpers. In addition, should we need to use

38

foam, we would have that hose line stretched out as well. I had all the airpacks and all spare tanks brought up and placed in one spot.

Tied alongside the dock, the smoke didn't seem too threatening, and one of the guys went below with a vessel crewmember for a size up. Back up, he explained the situation: The interior access to the forward lazarette is through a very narrow and low passageway two decks down. But fortunately, the air on the first deck below (Deck 1) was fine and the conditions two decks down (Deck 2) were not too bad. Also, there was an adequately large space on Deck 1 right where the ladder (stairway) passes through it for a good staging area. I know how odd it is to have a major operation going on above a fire floor, but with metal decks, good air, and easy to evacuate if needed, it was the most convenient place.

The metal deck directly above the fire was not alarmingly hot—no blistering or scorching of paint. The smoke coming out of the limited openings was being squeezed out under tremendous pressure. It appeared to be still an incipient fire. Backdrafts only occur in fires having gone through the full, free-burning phase long enough to deplete oxygen in the compartment. What I am not sure of is this: If a compartment is so airtight and intentionally shut down to exclude air, how much of the contents have to be consumed—or can be consumed—before you create a backdraft situation? Does the watertight/airtight construction of compartments of ships create backdraft conditions even when free burning has not occurred? I really don't think so, but the threat of it kept me nervous. I should say that the unignited volatile vapors and stored heat would certainly ignite spontaneously when opening the room, but that there would not be enough volume for a devastating explosion since all the combustibles have not been consumed.

I couldn't help but remember an engine room fire aboard the vessel Salmo Point which happened my first year in the fire department. We had only two airpacks then, which no one ever wore because no one knew how. Having just left the Coast Guard, I had some training wearing OBAs (oxygen breathing apparatus), so I donned an airpack and descended the ladder to the engine room to find the incipient fire. The space between the engine and other machinery was so small, I could barely wedge myself through it. It was pitch black, so I followed the sound of the fire and would take my glove off from time to time, hold my hand up and try to feel where the heat was coming from. Suddenly, everything got light as Chief Jack Dinneen had found a way to lift the large hatch cover from the deck above. The heat and smoke disappeared, and it was easy to knock the fire down. I just wished I could overcome the common problem of being unable to ventilate a fire aboard a ship.

Anyway, with the spare air tanks now stacked on Deck 1 and the lines deployed to Deck 2 for attack and back-up, and the other lines stretched to the other compartments adjacent to the lazarette (to keep bulkheads cool enough to prevent fire spread by conducted heat to combustibles in them) we were ready to do something. About this time, some frantic man came running through the ship shouting, and pointing. A couple of the guys told me that whoever it was, he needed to be removed. I never bothered to ask him who he was, but explained that he was interfering with the crew and he would have to leave. I interrupted his protest by telling him he'd be thrown off the ship if he didn't leave. I didn't know that he was the captain of the vessel, nor did I know back then that a ship is different (legally) than a building. The master of a ship is exactly that. Fire-fighters are invited and can be un-invited just as easily. Apparently, he didn't know that either, because he left.

I knew we would go through lots of air, so we called Mile 13 (airport fire department) for extra tanks. Bob Pudwill took the nozzle, and Mark Kirko was right behind him. A crewman from the ship led them down the ladder to Deck 2. It was black down there and Kirko's flashlight beam skimmed across the corners of the landing they were on. The crewman positioned them in front of the nar-row companionway that led to the lazarette, turned around, and left. Two other firefighters positioned themselves there to feed the hose to the attack team as they advanced. The companionway was only 24 inches wide, and as they advanced down the 18 feet toward the hatch, Pudwill remarked, "If something goes wrong, we'll never be able to turn around in here with the airpacks on. We'll have to run backwards." Mark agreed it would be impossible.

Anyway, I still wanted to be able to ventilate the compartment before opening the hatch but I didn't know how. A couple of the guys who reconned the area came up with an outlandish idea. To direct the smoke from a hatch in the overhead of the lazarette and force the smoke straight out to the main deck and open air, they suggested using the ship's supply of cardboard boxes lashed together with duct tape into a very long tube right through Deck 1 and to the main deck, at which point the electric smoke ejector would be sitting to suck the smoke with great force. I asked them what would keep the cardboard from catching fire? "Well," they said after a pause, "you gotta be fast." How do you argue with logic like that?

I told them to round up everyone that didn't have something better to do and start making the smoke tube. While they were doing that, others were frantically hauling empty air tanks up the ladder and full ones back down to the attack team, back-up team, and exposure teams. The crews on the cubicle's perimeter were on air. Pickup trucks carrying the tanks were speeding to and from the station, filling empty tanks. It didn't take long before the ventilation team said they were ready.

In the meantime, Pudwill and Kirko moved up to the door. It was so black down there, Pudwill didn't see a CO_2 extinguisher laying on the deck right in front of him and kicked it. When those things go off, they sound like a dragon's roar. Pudwill, who was one of the biggest guys in the department, and who predicted he would never be able to turn around in a space that narrow, turned around, and he and Mark were lens to lens—for about one second. In about two seconds they were both back out of the passageway. They were a little embarrassed after the flashlight revealed what had happened.

Everyone else was still poised to start the operation and I was still waiting for word from them to start the ventilation. They went back down and Mark touched the steel door, then opened the three dogs on the hinge-side of the door. As they are designed to do, the door seam opened up a couple of inches and stopped. There wasn't much pressure behind the door, so they radioed to ventilate. The ventilation team opened the overhead hatch and started the fan. Hot smoke billowed out to the main deck and it only took a few moments for the cardboard—particularly at the joining edges—to develop red, glowing embers then open flame. At the same time, the attack team entered and sprayed water, the steam shot up the tube (as well as back down the companionway), and extinguished most of the cardboard. The rest of the cardboard was extinguished with small squirts of water and boot-stomping. In a few seconds it was over.

It was a beautiful afternoon, and when I got back to the station, the crewmembers of the Bristol Monarch, armed with stiff bristled brushes were scrubbing our hoses out on the apron of the station and re-packing the hose beds with clean dry hoses. I walked back to the pool table and saw 30 cases of beer sitting there, a check for $2,500, and a box of tan baseball caps embroidered with a picture of the ship and the name "Bristol Monarch." That box was ripped apart and the caps distributed within a blink of an eye. The captain of the ship was grateful that he would be back in business by next day, which turned out to be just two days before the season was opened.

As an aside, I remember the following morning truckie, Dana Smyke, complained about his contact lenses. He had been crawling all around through that vessel, constantly checking perimeters and working on the ventilation system, and had gone through six airpack tanks. He had been up all night trying to get his contact lenses out. The airpack air was so dry, his contacts stuck to his eyeballs and he went through an entire bottle of Visene to get them out. Members need to know that. I learned later that some departments limit a firefighter to two tanks of air per incident.

THE LESSONS LEARNED WERE:

Backdraft potential—extremely diminished if a compartment has not gone through an adequate free-burning phase first. One still must be careful around small openings adjacent to the burning room. I learned that suppression operations are the options of the ship's captain and the vessel owner, and that local fire fighters are resources only. I also learned that cardboard—though combustible—is still useful. But the most important thing I learned was that I really had no clue about fighting ship fires. Of course, I had firefighting training in the Coast Guard, but it had nothing to do with tactics and strategy. "Winging it" is a good way to get somebody killed if the fire is bigger than what I dealt with.

That reminded me of what fire service icon Chief Ronny Coleman said, explaining why he wrote his first book on tactics and strategy. Initially, what caused him to write his book was back when he was a captain, asleep at his station, he woke up in the middle of the night and couldn't go back to sleep. He got up and looked out the window at the nearly-deserted street. A large fuel truck went lumbering by, and he imagined it involved in an accident and catching fire. "I was a captain, responsible for the scene and all my people, and it occurred to me that I don't know what the hell I'm doing." He continued, "Routine training in the fire service seldom addresses the needs of the decision maker." He was absolutely right.

How little I knew about commanding a shipboard fire was revealed to me years later. It took a number of years after, for me to develop a plan for ship fires. Our vessel fires were always on small vessels. We had not yet been impacted by the cruise ship industry.

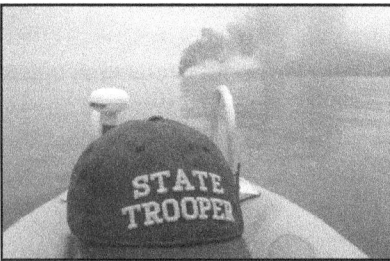

TROOPER BOAT ON WAY TO BOAT FIRE
Photo: Mike Hicks

BOAT FIRE ON PRINCE WILLIAM SOUND
Photo: Mike Hicks

MARINE FIREFIGHTING SYMPOSIUM

The Alyeska Pipeline Company hired fire protection specialists Hildebrand and Noll Associates from Port Republic, Maryland, to assess the resources and expertise in the sound for dealing with either a massive fire at the tank farm or on one of the oil super tankers plying the waters of the Prince William Sound or Gulf of Alaska.

After visiting the communities of the sound and making their report, they were contracted by Alyeska to conduct a large, week-long marine training symposium in Valdez and invited firefighters from the region to take part.

They asked me to teach a couple portions of the training: Vessel and Water Safety and Fire Ground Operations and Logistics. I was co-teaching with Bill Burket and Bob Rumens, Maritime Incident Response Team from Hampton Roads, Virginia. The entire teaching staff was very impressive since it included Coast Guard officers from the state and from Washington, D.C.; representatives from the Pilots Association; Marine firefighters from Portland, Oregon, and Seattle, Washington; two chief officers from FDNY; and lots of people from the oil industry and marine shipping. I had to crack the books to prepare, as well as correspond with some of these folks. It was then that I first read NFPA 1405, the standard for structural firefighters who get involved in shipboard fires. It was one of the best standards I ever read, and decided that shipboard fires was my new love. By comparison, structure fires can be a bit boring and not that much of a challenge.

Establishing a water supply for a ship fire at berth is not very complicated when you realize how small most of the compartments on ships are. Water for fire attack is relatively minimal, but distributing the rest of it to surround the six sides of the cubicle takes planning.

For safety, besides assuring that other boats in the area are assigned to stand off and watch for anyone falling off the ship, trying to survive such a fall was something I wanted to address. I remembered from Coast Guard training that if a helmeted seaman were abandoning ship by leaping off the main deck, he either must remove his helmet or hold it down tight on his head. When the brim of the helmet hits the water, it will pull sharply up and the chin strap may cause your neck to snap. The other thing I wondered was how fast would you sink to the bottom wearing a full set of turnouts and an airpack which weighs 30 pounds? So, I got one of our really old airpacks that was so far out of compliance we could no longer use it, put on my turnouts and walked over to the swimming pool. Robbie Mattson went with me. I tied a loop on both ends of a line. Robbie held one end and I grabbed the other. I figured that if I was so heavy on the

43

bottom and couldn't shed my stuff fast enough, I could pull myself up using the line. I turned on my air tank, cinched my mask up tight, and jumped into the water. I bobbed on the surface like a cork. Oddly though, as soon as my regulator was submerged, it went into free-flow. Air blew into my mask with such force, it pushed my mask out away from my face. But still, no water could come in.

Robbie helped drag me out. I said that maybe I had so much air trapped in my bunkers, it floated me. But now that I was totally soaked, I might sink. I jumped in again and just bobbed on the surface. In subsequent tries, I discarded the safety line, and tried jumping from higher up. Sinking was out of the question. So I put that in my safety lecture.

We held the last day drill on the Arco Juneau. Even though it is one smallest of Arco's fleet, it is 883 feet long (about three city blocks) and half a block wide. When loaded, the bottom of its keel is 52 feet below the water, and the bridge towering above the ship is 12 stories above the waterline.

Prior to the drill, several of us firefighters crawled all over and through that behemoth. There was an aft stairway (ladder way) that led from the main deck—behind the superstructure—to the shaft alley seven stories down. At the bottom, we stooped over and went forward through the shaft alley to the lower-most portions of the engine room. I thought that under the right circumstances, that might be an innovative approached to engine-room fires on very large ships.

For monitoring the temperatures of steel decks or bulkheads, mariners often use welder's chalk to mark the area they are monitoring and watch for color changes. Later, when developing my plan for Cordova, I wanted to use thermometers often seen on the wood-stove chimneys or in ovens. They come with magnetic backs and can simply be stuck to a deck or bulkhead, and the temperature can be read regularly to detect the advance of a fire from one compartment to another. I also toyed with the idea of a thermo-couple attached to a thin rod to be inserted through a hole in the bulkhead made by a welder to assess the effectiveness of a fixed CO^2 system discharged into the area.

Anyway, besides reading NFPA 1405, I read every detail of the Coast Guard's "Prince William Sound Fire Protection Plan." Later, when in Juneau, I discussed the plan with the Coast Guard Captain of the Port for Southeast, Alaska, Captain Dave Ely from the Juneau Coast Guard base. I mentioned that I was writing our (Cordova's) SOPs based on 1405. He was doing the same thing down there.

Ely retired from the Coast Guard shortly after and came to Cordova to compare his plan with mine. He did something really neat. He took all the steps of 1405 and wrote it like a firefighting checklist and put it on a computer disk. Our two plans looked very similar. The neat thing about his disk was that it was a generic template for a plan. Anyone could take the disk, fill in the blanks with local information and be ready to fight fire. Every chance I got, I promoted Ely's plan. He wasn't making any money from it and, in fact, gave a stack of disks to Alaska's State Fire Service Training Division in Juneau to distribute to anyone who wanted one.

Some time after that, when the cruise ship Clipper Odyssey docked in Cordova and the passengers disembarked for the day, I talked the skipper into having a combined drill with his crew and mine. Of course, the drill was primarily for firefighters rather than officers. We did lay some supply lines and connect to their international shore connection, but mostly it was firefighters crawling around and advancing hoses. No simulated stripping combustibles out of adjacent compartments or setting up perimeters, farting around with the HVAC system, or planning a dewatering operation. Like I said, most training is directed at the basic level, and very little comprehensive training for command officers.

My fire plan checklist was in a 3-ring binder and it was a pain-in-the-ass to read, turn pages, and write notes on, while using my radio. My solution was ingenious. Afterwards, on my computer, I reformatted the pages to be 4 inches wide and 8 inches long. I printed the plan out on write-in-the-rain paper and made a spiral binder out of it. What was ingenious was that I put the spiral binder on the bottom of the pages rather than on the top. I slipped an elastic cord through the spiral binding and hung it around my neck. It was upside-down. When I lifted it up in front of me it was right side-up; I could read it, jot down notes (for example, when was the last time the draft marks and inclinometer were read and what were those figures), then drop it and my hands were free. It was perfect.

Funny thing about that drill on the Clipper Odyssey: The dispatcher announced it was a drill when she first toned it out. Plus, everyone was expecting it. But we neglected to announce that it was a drill during subsequent radio transmissions afterwards. I forgot that the hospital had a radio at the nurse's station. Dr. Elizabeth Turgeon was walking by the nurse's station after we were well into our drill and heard that we were pulling "burn victims" off the ship. She sprung into action and ordered a mass recall of all hospital staff to prepare for these "victims."

She was livid. I guess it had something to do with paying all that overtime.

Incidentally, our Mass Casualty Incident checklists (Field Operations Guides, or FOGs) for years had been on clipboards; one for each position—Incident Commander (IC), Triage, Treatment, Transport, Equipment/Supplies, and Staging. I reformatted all those into the 4-by 8-inch hang-upside-down-around-your-neck FOGs. And they are very popular. As an aside, I wrote an assessment and evaluation checklist for disaster drills the same way and have used that when I travel to other communities to conduct disaster drills for them and evaluate their proficiency. That way, I almost never have to set my coffee cup down.

I must confess that I became so focused on marine tactics and strategy, played so many scenarios over in my mind that I began to crave an incident of that magnitude. Sick, huh? Hey, don't look at me with disdain. You'd do the same thing.

Those tactics are so different from structural firefighting, thinking about them perked up my days. In fact, in preparing for a bid on a contract (which we did not get) to teach firefighting to crews of off-shore oil drilling rigs out of Sakhalin, Russia, I was drawn by the similarity in tactics to that of marine firefighting. I truly believe that fire protection specialists and text book authors should take a hard look at the similarities and promote those similarities in developing effective tactics for off-shore oil rigs.

THE TANKER ARCO-JUNEAU WAS SIMILAR TO THE EXXON-VALDEZ SHOWN HERE AGROUND ON BLIGH REEF, PRINCE WILLIAM SOUND. THE EXXON-VALDEZ WILL BE DISCUSSED LATER IN THE BOOK.

Photo: Wickipedia

THE CHUGACH CANNERY FIRE

(AND THE OTHER MAKE-IT-UP-AS-YOU-GO STRUCTURE FIRES)

Structure fires were nearly a daily occurrence in our small town in those days. But most of them were simple house fires that, because of devouring NFPA books, soon became like a walk in the park for us.

In NFPA's 1974 Fire Protection Handbook, they suggested an approach to fires like these—one position attacks—that I initiated as soon as I became chief in 1975. Our first-in engine (with its 1,000-gallon tank), would drive straight to the front of the building (side A), and firefighters would make an immediate attack (or exposure protection, if needed), using just the tank water. Our second-in engine would lay a supply line from the attack engine to a hydrant, hook up, and pump water back to the first engine. A third company was saddled with doing only truckie work: laddering, forcible entry, ventilation, lighting, etc. But water supply for attacks was the focus of our attention in the beginning.

We drilled over and over again to make the operation so timely that the supply engine got water to the attack engine before the attack engine's water tank was empty. We practiced the various types of hose lays that we might need to facilitate this operation. We stopped losing firefights immediately. Of course, many years later when Large Diameter Hose (LDH) became available, our newly purchased 5-inch supply line negated the need for a second engine to tie in to the hydrant and actually pump the water back to the attack engine. But we still used the second-in engine to lay the line.

Fires of nearly conflagration size occurred, of course, in Cordova's early history. But they also occurred in recent history. In 1963 several downtown blocks were razed (just one year before the massive earthquake and tidal surge). In 1968 the Ocean Dock, Standard Oil facility, and part of Parks Cannery went up in a blaze. In 1969 another downtown block burned down. Shortly after that, I joined the fire department.

* * * * * * * * * *

My two biggest fires were both at the same cannery complex, which is the oldest complex in the area.

Chugach Cannery was previously known as the New England Fish Company. Before that, it was called Orca when it was built at the turn of the century as a whaling station. Labor had been provided by Chinese workers who came equipped with small bottles of opium to help them through the arduous labor in the rain and mist at this isolated location. Divers still find these bottles right off

47

the dock of the cannery. Orca was built on the edge of the inlet before the town of Cordova was established. When Cordova was built, it was separated from Orca by three miles of roadless, rugged shoreline. For most of the existence of Cordova, cannery workers traveled from Cordova to the cannery by a launch, stayed the week in the large, multi-storied wood-frame bunkhouses, ate in the mess hall and spent time off around bonfires on the rocky beaches. After a week of long hours on the sliming lines, they would board the launches for a trip back to town for a weekend of R & R.

Eventually, funding was acquired to punch a road along the beach line to the cannery and the trip to and from town became a short 6-or 7-minute drive down the gravel road. The road was brutally rough, very winding with sheer rock cliffs on one side and boulders plunging straight down to the water on the other. Over the years it had been the scene of several mishaps of inebriates inadequately navigating the curves and tumbled into the bay.

The cannery had it's own water supply for years from Crater Lake which was located on the mountain high above the cannery. But hydrants were never installed for firefighting for our first conflagration-size fire there. Naturally, the main challenge was water supply that night.

It was well after midnight when my pager went off. I must have been in the deepest stages of sleep, because I kept falling against the wall while trying to get dressed. More irritating, was that the dispatcher would never say what the call was. I only heard a "beep," then silence. I fumbled for the phone and called her to learn the fire was at the Chugach cannery. I blurted out, "Page it out again," and hung up. All I heard a few moments later was the beeping of the pager and no voice.

The darkened streets were deserted and all the windows black as I tore through town. I was hoping that the dispatcher would fix whatever problem there was with the paging system, because all I heard were beeps every few minutes. As soon as I crested the hill on First Street, I saw the sky north of town was glowing red, and I was alone. Even in my agitated state, I knew the sun did not rise in the north, so this was not going to be a walk in the park. A few minutes later, the pager sounded and I heard the voice of Assistant Chief Robert Varnam announcing the fire's location. The beeping awakened him and he drove to city hall to investigate the problem, and did the page himself. Something I should have done.

I was halfway down the connecting road when I saw the actual flames licking against the black sky in the distance. Both my legs began to shake. And you have no idea how heartening it is to hear "Engine 4 responding with 6 (persons) aboard." Then I heard another engine was out, then another, then the truck company. Now, I was a bad-ass leading a cavalry charge. To me, there is no sight in the world as beautiful or awe-inspiring as black-clad, helmeted firefighters dismounting from fire engines, ready to kick ass.

I had Engine 4 draft salt water from the bay. The tide, of course, was receding and the slope was so gradual that a half-foot drop in the tide would move the edge of the water several yards further away from the engine. As the rate of the drop increased (which is normal), the engine had to stop pumping and drive forward more often. This had the engineer, Dan Jager, hopping faster and faster, from managing the gages on the panel to losing prime, shutting down, driving forward, and getting prime again. I'm sure this was aggravated by my constant bellowing over the radio about the water supply.

Finally, almost in a rage, Dan leaped into the cab of the engine and drove straight into the water, leaped out of the cab chest-deep into frigid water. For many hours after the fire, we were draining and flushing everything in that rig to clean out every drop of rust-causing salt water. Nevertheless, Dan established a water supply that would last us awhile.

The sporadic water supply from Engine 4 would have caused all sorts of problems for the guys on the hoses had it not been for a second engine drafting water from a creek. During rainy days (and this night was rainy) torrents of water poured down the mountainside and through huge culverts under the road and into the bay. That engine company used large sheets of plywood to block the culverts and create a "pond" to draft from. More reliable than a receding tide, it had its own set of problems: The creek became a pond instantly and water could not be drafted out fast enough to prevent torrents from spilling over the road and begin washing it out.

In addition, gravel was churned up as it was drawn up into the suction hoses and they had to be continually disconnected, propped upright, and beat upon to dislodge the gravel. Most of the time, the troubles of the engines supplying water alternated and they were never both shut down at the same time. They managed to keep the tanks on the attack engines full. Of course, the engineers on the attack engines were constantly alternating water supply to attack lines between their tank water and relay water. Seventeen-year-old Gayle Groff and Reverend Gary Davidson were studiously walking laps around their engine, opening and closing valves and bouncing their feet on supply lines to test for pressure, then radioing to the attack teams any time that supply lines went limp and tank water was low. This went on for hours.

The mutual aid DOT crash/fire engine managed to make it to the scene and start pumping before its drive engine froze up. There would be no moving of this monster airport crash truck. This was in the mid-'70s, before anyone was using large diameter hose for water supply. These were the days of street spaghetti.

We didn't extinguish the fires in the buildings, but we held it back mostly with handlines like WWI trench troops—hunkered face-down and refusing to be pushed back. We only had one deluge gun (master stream device). For the other sides of the fire, crews lay prone on the wooden dock and created walls of water to hold back the flames. I would stoop over and move from team to team, trying to sound calm and confident. We never felt scorching heat in those positions because of the hose streams we crouched behind. With every swirling change in wind direction, black oily smoke would roll over us. We were blinded and could barely breathe. "Breathe shallowly....off the top of your lungs...pant lightly," I would instruct when I could talk. Occasionally, I would lay flat and press my mouth flat down on the planking, trying to be patient while waiting for the smoke to lift. A couple of times, it took so long I felt panic at not being able to take a breath. I wasn't alone. Firefighter Rich Wilson leaped up to run away, completely blind as to which direction to run. Varnam managed to grab the hem of his coat and sling him back to the ground. When the smoke lifted that time, Varnam told him to take a break.

Back in my normal position I watched the wind pick up intensity and flames curled over the handline crews, but they couldn't feel it. The crew looked like surfers in the curl of a wave. But fire brands were blowing on to Engine 1. Gary Davidson never stopped his routine of switching valves between filling his tank, supplying handlines from relayed water, back to using tank water, then refilling the tank whenever the drafting engines pumped it. Gayle, on the other hand, scurried around atop the hose bed stomping out small flames and attaching a short length of fire house for their own protection. I was thinking about pulling everybody back. About this time I saw Davidson stop momentarily and look straight down at the ground. A short distance away, Richard Harding was doing the same thing. I was dumbfounded. Then it dawned on me, they were praying. My first impulse was to yell at them to pray on their own time, we were in trouble here. Now, I know you aren't going to believe this, but the whole damned fire department and half the town saw this. The wind switched direction—as quickly as throwing a switch—and blew like a sonofabitch out toward the bay. It blew so hard the other way that the flames looked like a torch.

Eventually, it was over. First, under control. Then, over completely. Now, we still hadn't pulled out the last of our equipment out until 32 hours after the alarm because of overhaul and point of origin investigation. The body of the firefight was six hours long, and the overhaul took a full day.

Gary Davison, Robert Varnam and Richard Harding were ministers. Some of the guys joked that we should rename ourselves the "Cordova Friar Department." Naturally, someone else piped up "Oh, Holy smoke!" Anyway, Davidson, Varnam, and Harding insisted that the attendance roster for that call include "God." Fine. I wrote His name down. In fact, when writing out compensation checks at the end of the year for fires attended by members, I—yes, I really did—wrote a check to "God" for $5.00. I gave it to Harding saying, "Here, you mail it."

Actually, that fire was a water-supply challenge. The second one we had there about ten years later was a challenge of tactics and strategy. By that time we had a full supply of large diameter hose (LDH) for our supply lines on our engines, more master stream devices, and large flow nozzles for our larger handlines and hydrants.

Coincidentally, the week before this second large fire at Chugach cannery, we had run a series of drills based on an objective: "Flow more water, in less time, with fewer people," which previewed the exact initial attack we used on the fire. In fact, it worked so well that two professional fire investigators from Kirkland, Washington, remarked that they had never seen such a "stop" in their long careers.

* * * * * * * * * *

The drill that prepared us for this second fire used a forward hose lay from a hydrant to a building, laying out a 5-inch hose which can supply more than the rated pumping capacity of the engine itself. While the LDH is being connected to a suction inlet on the pumper by the engineer (1), one firefighter stretches out the "bomb line" (2). The bomb line (sometimes described as water cannon) is set on the ground and is supplied by one 3-inch hose. The 150 feet of 3-inch hose is preconnected to a discharge port and can flow about 900 gallons per minute. It takes about 20 seconds to stretch it out. While that is being done, two other firefighters are stretching out a handline capable of 200 gpm (3 and 4). If needed, when the hydrant man returns to the engine, he can help stretch a second handline (5). Counting the officer (6), this evolution takes six firefighters and can flow a total of 1,300 gpm in one minute, and fifteen seconds (time starting when the engine stops at the hydrant half a block away). A 5-man crew can flow 1,100 gpm in the same period of time. By contrast, most paid departments, running 3-man or 4-man crews usually flow 200 gpm as an initial attack. That is sufficient for 2 or 3 rooms of a house, certainly insufficient for a large commercial or industrial building.

Our turnout time on this second fire wasn't too much better than the previous one. Anytime after 10 p.m. adds a minute to our response time, because some members are in bed and must get dressed. So, 4 minutes to get the apparatus rolling and 6 minutes to the cannery site automatically means a well-involved fire. Of course, we'd made numerous calls to that complex in the interim but they were small fires extinguished by "civilians" before we arrived. I had hoped that might be the case here. But, the responding police officers informed me that would not be the case. And as I saw the orange reflections on the clouds above, I knew they were right.

Ten in the evening was still early enough to attract public attention, and word spreads pretty fast around town. Two young men were trotting down Orca Road to see the fire; I stopped and asked if they were sober and they said yes. I told them to hop in if they would help me when I got there.

The cannery complex starts on the pylon-supported dock, which also supports—closer to the beach—a huge warehouse full of boats and forklifts. In the second story of the warehouse were chicken-wire and plywood cubicles containing personal gear of the fishermen. The warehouse was connected via a suspended walkway from the upper story to the huge complex of cannery buildings. The heavy timber, steel "I" beam, and corrugated-tin warehouse was fully involved, and the flames were reaching through the enclosed walkway heading toward the cannery.

After you locate the enemy, you cut off all his avenues of escape, then move in and kill him. The challenge is to find sufficient resources to do that. Having heard that Engine 2 was first out, I radioed for them to make a forward lay on the upper road and stop in front of the cannery.

I would have Engine 4 lay in on the lower road just above the water line. As soon as E-2 stopped, I had the hitchhikers pull the bomb line out of the rear compartment while I disconnected the 5-inch line from the hose bed and attached it to the suction port. As soon as the engineer engaged the pump and took his place at the pump panel, I told him to charge the bomb line, then honk his horn for water (when the man at the hydrant hears a 5-second blast on the horn, he knows the suction hose is attached and ready for water). The hitchhikers and I positioned the deluge gun (water cannon) and directed it up toward the walkway. As soon as the tremendous stream of water hit the walkway, the fire stopped advancing. Three of the firefighters from E-2 stretched a handline into the cannery, up a flight of stairs and to the elevated walkway.

To say the least, they were surprised to see two cannery workers lying prone on the floor, spraying water with an old cannery fire hose, and had been holding the fire back until the bomb line took it over. Engine 2's guys told the cannery workers, "We got it," and they gratefully left. But after a couple of minutes, E-2's guys saw they were not needed there, the bomb line could hold it. But later, I would need them back there when I moved the bomb line in a maneuver I'm so proud of I can't help telling people about it.

Anyway, shortly afterwards, E-4's bomb line was deployed and the handlines were stretched and manned and I asked the plant manager about hazardous materials that might be inside the warehouse. He mentioned LPG (propane) tanks lashed to the wall inside at the A/D corner. The truckies (T-5) ripped a large hole in that corner and we manned a handline in the hole to keep them cool.

Meanwhile, to our left, the fire burned through the dock on the (B side), igniting the creosote pylons underneath. With no wind, the fire under the dock mushroomed in both direction—toward the outer end of the dock and toward us. Realizing that with the fire approaching us from underneath, we would lose it all. I told T-5 to plan to cut the dock in half and create a fire break if we couldn't hold it back. I had dispatch call Whitestone Logging Company and request loggers and chainsaws.

In the meantime, the harbormaster crew showed up with the extra portable pumps I'd asked for. While the truck company crew was crawling around the dock, planning how to cut it in two, firefighters from both engine companies who were not already committed were sent under the dock. Lt. Kirko placed the guys under the dock where they stood waist-to chest-deep in the water and used the handlines supplied by the portable pumps to fight the creosote-pylon fire and hold it back from crawling up to under our feet.

Later, my son, Jason told me how eerie it was under there when boats and forklifts would fall through the dock and splash in the water below, creating wakes that would sometimes slurp up to their chins. He also said that, oddly enough, they became used to the cold and no longer noticed it. I was able to change that approach after some time, when the tide receded a bit and we took control of the warehouse. In the meantime, we had to rotate crews under there to prevent hypothermia.

Taking control of the structure fire was ingenious, if I say so myself. I have recounted the tactic many times in tactics and strategy courses around the state. The fire investigators from Washington State said they'd never seen an offensive maneuver on a fire that big before. With the entire structure burning, there was really nothing inside to save, but the only way to truly defend the rest of the complex with our limited manpower was to put the fire out, not wait for it to go out.

The fire was still roaring, downstairs and upstairs. You have to recapture the downstairs first, of course. I had T-5 cut a hole in A wall, low and near the D corner, just left of the LPG tanks. I moved E-2's bomb line away from the elevated walkway to the hole cut by T-5. In place of the bomb line, we used handlines to hold the fire back in the elevated walkway. The tip of the bomb line protruded through the wall and the stream was able to reach the far wall (C-side). The canon swept up and down and blackened the fire down in that portion of the downstairs. Continuing the up-and-down sweeping, this wall of water kept the fire at bay. I then had T-5 cut another hole in the A wall a few yards farther down, and place E-4's bomb line there. It did the same maneuver, and we had doubled the size of our recaptured territory and were flowing nearly 2,000 gallons per minute.

We slid open one of the big doors on D wall and directed handline streams on the backside of some of the boats involved. With that done, 25 percent of the downstairs was recaptured. Now, with E-4's stream holding back the fire, T-5 cut another hole, even further down A wall and E-2's bomb line was lifted and pulled down to that hole. We now had recaptured nearly half of the downstairs. Now the 3-inch lines supplying the deluge guns were becoming "braided." It took several minutes to remedy that, but we continued working our way down A wall, until the entire downstairs was extinguished (except for minor, non-threatening contents fire). Then all it took was one person and minimal work with the stream to keep it out.

With that done, we went after the upstairs. It was a contents fire mostly, emitting occasional small explosions we assumed were fuel containers owned by the fishermen-renters of the storage compartments. The walkways upstairs ran east and west, one adjacent to A wall and one adjacent to C wall. The compartments occupied the entire center strip of the upstairs. Visibility was zero up there. In fact, had we been able to see upstairs I would not have sent Kirko's and Jason's teams up there. We were lucky that no one got hurt.

As those teams commenced the slow march down the A side and C side hallways toward the B wall, stopping at each compartment along the way, they were groping in the blind. It was slow, tedious, backbreaking work, because each compartment was crammed full of gear. The entire time, we watched for any rekindling downstairs and under the dock. It took forever.

While they were inching their way forward, police officer Fred Brady reminded me that the fire was creeping west down the dock and the ice house at the end of the dock was burning. The entire dock would soon be lost if something weren't done. Fred wanted a firefighter, a boat, and water supply for the dock.

I'm not sure who arranged it, but a local fisherman who had been cruising up and down the bay watching the fire, volunteered to shuttle Brady and a firefighter out to the face of the dock. For a water supply, someone lashed the end of a 3-inch hose to the boat. The same person(s) lashed a series of crab pot buoys to the hose to keep it afloat. Fred and the firefighter slowly cruised out to the face on the dock.

They attached a smaller (manageable) hose to the supply line and climbed up the steel ladder while the fisherman, who had tied his boat to pylon, fed the hose to them. Fred picked up some dunnage there, held it tight against the planking on the dock, aligned with the space between the planks, tilted the dunnage at an angle, and had the nozzleman spray back and forth along the dunnage. The water was deflected down the spaces and onto the burning pylon caps. They systematically worked their way back up the dock toward the origin of the fire.

Occasionally, they were in line with a dock ladder (leading over the side of the dock) were able to descend a few feet and direct the stream of water directly on to the fire underneath. Slowly they made their way back to a close proximity to the ice house. The fire was now only on the top of the dock. The ice house was a gonner, and there was nothing one handline could do except keep the fire from spreading. That's where they stayed, holding the fire in check, until the ice house fire had burned itself out.

I'm not sure how long they were on that job, but it seemed like hours later they returned justifiably proud of themselves. With some embarrassment, I'll have to admit that during that whole time I'd only looked out in that direction a couple of times and wondered how it was going. I was so preoccupied with my bomb line offensive attack idea, everything else was surreal.

By daybreak you could see some exhausted, but visibly proud—to the point of being giddy—firefighters bobbing as they walked, muttering, "We're bad. Oh yeah, we're bad."

The other thing that you could see as the darkness lifted was the upstairs of the warehouse. For the first time, we saw twisted steel "I" beams that had twisted in the intense gasoline-and creosote-fueled fire, had racked, and fallen in. They were barely held in place by stretched bolts or welded seams. Of course, the attack teams could feel that they were worming their way through "I" beams, but seeing in the daylight, how precarious it was, made their eyes grow wide. Mine, too.

Well, for the second time, we'd stopped a massive fire at that location. The first innovations were in the water supplies and the teamwork in the relaying of an unreliable source, and the second was going offensive in what almost any other department would have considered a defensive-mode fire. But, I want to give credit for the idea of using a bomb line to Juneau's Fire Chief Al Judson, who used it on a hotel fire. What made my use different was in recapturing (and holding) areas an increment at a time.

CHUGACH CANNERY FIRE
Photo: CVFD

NORTH PACIFIC PROCESSORS CANNERY FIRE

Back in '75 when I became chief, I would visualize the perfect fire department conducting a perfect fire fighting operation; a fire fought by tactics that were conveyed throughout the membership of crack troops nearly by telepathy. Well, that vision materialized at the North Pacific Processors fire.

At 6:14 one chilly October morning in 2001, we got toned out to "an unconfirmed fire at North Pacific Processors." Since I was already up, day-trading on the stock market with my computer, it didn't take me long to be bunkered up and on my way. I didn't have much anxiety about this because this cannery, the town's biggest employer and single biggest economic resource, was sprinklered. I tried to look through the night sky in the direction of the cannery for any sign of smoke or fire but saw nothing. I pulled in to the deserted parking lot and was met by Don Roemhildt, the brother of the plant manager, Ken Roemhildt.

We entered and walked through the plant toward the C-side, or the dockside. The place was so spacious and deserted, every sound we made echoed. The last room we passed through before walking out on the dock above the bay was the "egg room" where salmon eggs were extracted and processed. It was enormous and empty, and there was a light layer of smoke near the ceiling 30 feet above. I could hear no water running from a sprinkler head and certainly no sprinkler system alarm. I could hear the radio transmissions of the responders as they left the fire station.

We opened the back door and stepped out onto the dock. Laying to in the calm water a short distance away, was the SERVs vessel Crystal Sea who had been cruising by and called in the fire. I climbed an exterior stairway to get nearer the roof and from this C-side vantage point, I could see flames emitting around a vent pipe on the edge of the roof. I raced back down and popped my head inside the egg room, then back outside again, and estimated the concealed space between the interior ceiling and the roof was about three feet. Also inside, I could see that embers were beginning to fall from the ceiling. I stood motionless for a moment, listening, and could hear that the fire had spread through most of the ceiling assembly in the rear segment, about 25 percent of the cannery.

Hearing the responding apparatus pulling up out front, I ran through the cannery to talk to them, making sure to be walking when they could see me. Running is something we don't allow. Not only is it a stumbling hazard, it makes you look frantic—which means you are not in control. I yelled that it was in the ceiling assembly and to stretch in some lines. One of the medics came up carrying my bag of turnouts from my rig, and she heaved them up onto the loading dock where I stood.

Acting Company Officer Robbie Mattson had his crew stretch a 5-inch (supply line) into the cannery about two-thirds of the way back toward the egg room and attached a 4-way manifold to it. He instinctively knew that it would be a large-flow fire. Next came the bomb line. The 150-foot long 3-inch hose was coupled to the manifold and stretched into the egg room. Also, three handlines were attached: One went into the egg room, one went to the center of the main cannery facing the box room (the next room to the right), heretofore uninvolved, and one went over by D-wall facing the tank room (the room furthest right).

Several cannery workers arrived to help with the firefight. Normally this would not be allowed, but we had good visibility in there and they would be very helpful. Assistant Safety Officer Loyd Belgarde informed me that a technician had arrived to shut down the ammonia system and bleed off the pressure. A few moments later, Belgarde informed me that the electrical power to the involved area was shut down but that they would keep power to the rest of the cannery—to give us light—as long as possible. One officer remarked about the "civilians" (cannery workers) wandering around the hot zone and that we should evacuate them. I replied, "Not yet." Before long, one or more firefighters were scrambling around handing out hard hats to the cannery workers that didn't have them.

By now, the roof was sagging in the egg room and was burning furiously at the ceiling level in the box room and the tank room. I was hoping an opening would appear so we could play water into it with our bomb line, but it remained pretty much in tact as it bulged downward. The ceiling space the entire breadth of the cannery was roaring and I wanted to locate a place to "draw a line in the sand" and make a stand. When one walks into the cannery from the street (A-side) and walks through the building to the opposite wall—and the dock—one is walking west.

I had firefighter Kyle Marshal crank up the chain saw and had a pallet placed on the forks of a forklift for him to stand on. The next room back from the egg room—east of the egg room—had a ceiling lower than the ones in the egg room, box room, or tank room. That's where I had the fork lift hoist Kyle up where he cut a strip out of the ceiling so we could see into the concealed space. While Kyle was cutting the ceiling, Loyd was working the bomb line and some others were using a handline trying to get streams of water into the ceiling assembly of the egg room. That was just busy work and everybody knew it.

Funny, it was about that time the dispatcher radioed me, but because of the noise, I couldn't understand her and stepped into a compressor room and closed the door. She said she had on the phone a Tom Wells, HAZMAT specialist from the Anchorage Fire Department. Wells wanted to know if we wanted a 3-man "assessment team" to come down. I asked "What does he want to assess?"

A moment later, she replied, "The ammonia system." I had her thank him and explain that we had the ammonia shut down. I went back out, but a few minutes later she called me back. I stepped back into the compressor room where I could hear her say that Tom was on the line again and he wanted to be sure I knew that "it was 'free'." I almost felt guilty by saying no again.

Incidentally, the next day when I talked to Tom on the phone, he explained that he'd received a call from the Alaska Department of Environmental Conservation (DEC) wanting him to go to Cordova to assure there would not be a release of ammonia. They had heard of the fire and even before the fire department was in full operation, they were calling for outside expertise.

Anyway, I walked up to Loyd and the others and tapped them on the shoulders and told them to pull out. They backed out and shortly after, the roof came down in the egg room. We lost a handline under the collapsed roof and had to get another one. Looking up, we could see the ceiling assembly flashover through the hole that Kyle cut, and directed the stream from the bomb line into it. If one were to consider the exterior rear wall (C-wall) as the first wall, and we had set up at the second wall to reach the egg room, box room and tank room, we were now on the third wall back and the roof assembly became fully involved between wall two and wall three.

This was the last wall of defense. If the fire breached this wall, it would walk through the rest of the cannery unstoppable. I'd never realized before how personally I took these firefights. This was the first time I ever thought I would lose a fight. Since we were set up down on the ground floor (Division 1), Deputy Chief Kirko and Robbie said they were going to set up defensive lines upstairs and become Division 2. Robbie had people stretch a second 5-inch line through the cannery to D-side and up a flight of stairs. With a manifold, he supplied numerous handlines the full breath of the cannery. Visibility was fine and the air was very breathable.

Truck Company Lieutenant Dana Smyke had his company ladder A-side and drag their stuff up on to the roof. Kirko stepped outside and had the crew from Engine Company 2 hoist the other bomb line up onto the roof, and went up to direct operations up there. We only had two chain saws, so the unflappable EMT Captain Joanie Behrends had dispatch call the U.S. Forest Service for more chain saws.

In essence, there was no Incident Commander (or Chief) running this operation. I was running Division 1, Robbie was running Division 2 and Kirko was running Division 3. No one was outside running the whole show. Everyone was working the fire.

As the roof began giving way where it was attached to the third wall, there was a space about one foot wide tearing open and burning like hell. This separation broke the 3-inch sprinkler main and water spewed uselessly out of it. I had my crew direct their streams straight up to play water on the third wall to keep it from igniting. I instructed them to remain under the steel I-Beam because the roof would eventually come down.

I had our Explorers (the Explorer program, sponsored by the Boy Scouts for older teenagers and who were released from school for this fire) pack in another shoulder load of hose and couple it to our manifold. Lt. Mike Hicks, asked them to help stretch the line toward D-wall so he could take it up a short flight of stairs near the tank room. After a few minutes, he asked Belgarde to run that crew while he went off to do something. Belgarde could not tell that the crew were Explorers and advanced further into the fire. For him on the catwalk above the stairs, the fire was now down to his level. It wasn't until later that we realized we had these kids in the hot zone. Anyway, Division 1 was stalwartly holding the fire at the third wall.

I heard radio transmissions between Divisions 2 and 3 and some talk about doing a "trench cut." That's where a long, narrow part of the roof is cut from one side of a building to the other to create a fire break. The bomb line on the roof directed its stream over the truckies to land between them and the flames. With this protective barrier in place, Dana and his crew started the cut. When it was completed, no flames were coming out. Kirko asked them to move closer to the fire and make another cut. They did. The first cut was very tidy and straight. The closer cut turned out to be very erratic....more oblong than rectangular, indicative of cutting frantically and dangerously close to the fire. Flames blew out of this cut. Division 3 drew their line in the sand.

Down below them, Robbie also had the crew do a trench cut on the ceiling right below Divisions 3's cut. Then they entered three office spaces whose west walls were the third wall. They ripped off the walls and ceilings on the C-side of those rooms and were able to see and stop the fire in that wall.

What was most odd, was that no matter how many times you walked through that cannery, the way it was laid out, you were never really sure of where you were in relation to other floors. It was purely luck that the wall we were protecting on the first floor happened to be the same wall that Division 2, above us, chose to make a stand. And coincidentally, the same place the roof crew made their second cut.

On the ground floor, when the roof finally came down between Wall 2 and Wall 3, we were able to totally drench the third wall all the way to the top of the building and see the other two divisions right above us. We waved at one another. We still had a lot of fire going, but we had sliced it off from the main part of the cannery; its bunkhouse, offices, cafeteria, and primary canning equipment worth many millions of dollars.

Extinguishing the collapsed roof was tedious but not very interesting. However, the box room was three stories tall and full of pallets of collapsed cardboard boxes banded together and covered with plastic shrink-wrap. There were also stacks of large plastic "totes" 6 feet long, 4 feet wide and 4 feet deep, stacked inside one another like papercups. These stacks also reached up three stories high. This place was still burning. Dana's guys from Division 3 came down to play water on it from the doorway—accessed by a catwalk—on A-side of the room. But they could not reach the fire or the C-side of the room.

Robbie and I walked out on to the dock above the water and he suggested we cut a hole through the wall of the box room and send in a team. One of the cannery workers cut a small hole and we looked in. There was a steel scaffolding that had to be crossed and ascended to be able to reach the fire.

"Boy, that's gonna take balls," I said.

"Where's Brandon?" Robbie asked.

Brandon Doig walked up, his tiny Superman action-figure wedged into the rubber band around his helmet. He, Oscar Delpino, and Tracy Whitcomb donned airpacks and went in. They humped that handline in through all the legs and braces of the scaffolding and when they got next to the pallets of boxes, started their climb up the scaffolding to the top. We could keep an eye on them because they were clearly black silhouettes against the orange backdrop of the fire. It was grungy, sweaty work and they went through a pile of air tanks, but eventually they knocked the fire out.

We stuck around for several more hours during overhaul and worked with the cannery workers using forklifts and a borrowed backhoe inside to rip walls apart and the contents of the box room.

The eventual tally on that fire was $12 million. That was just a fraction of what was saved. The cannery was rebuilt and running when the fishing season began the next spring. Subrogation by the cannery's insurance company focused on the fact that the inside of the roof assembly was not sprinklered. The question ultimately was whether a retrofitted sprinkler system in the year it was done required coverage in that concealed space. It sure as hell does now.

At the beginning of this story, I mentioned that I had always pictured the perfect fire department as one that needed practically no supervision; everyone would do the right thing automatically. I was elated because I just saw that. No one sought direction from me. Guys that I had known for decades, who used to look to me for direction, advice, and reassurance during fearful events, breezed by me like I didn't exist. I went from being elated to being saddened. I was no longer needed. I was starting to suffer from mental "empty nest syndrome" when your kids grow up and no longer need you. They just pretend to. You become a figure-head. It was time for me to plan stepping down.

CONCEALED SPACE FIRES #1

Of course, the North Pacific Processor's fire was difficult because it was a fire in a concealed space. But it wasn't a surprise. Over the 28 years that I was chief in Cordova, I had been surprised by concealed space fires that either caused firefighter injuries or near misses. I realized years ago that I needed to pay closer attention to these things, but I couldn't find current texts or courses with detailed coverage on tactics for combating concealed space fires. So I wrote my own to insert in my tactics and strategy classes. Here are several incidents I use as examples:

A few days before Christmas in the late '70s, in the wee hours of the morning a family was awakened by strange sounds coming from downstairs. Downstairs was the wife's home business, a very large single room furnished with shelves of inventory. The florescent lights for the room were an integral part of the drop ceiling, which was suspended about two feet below the floor of the second story. To enter the store, one would walk into the main door from Pipe Street, into a hallway which led to an open stairway leading upstairs to their kitchen and living room. Or, before reaching the stairs, one could turn left through a door and into the store.

Smoke filled their home, and the family of four, dressed only in their underwear, dashed down the stairway, ran across the ice-covered street—slipping and sliding—to the neighbor's house to call the fire department.

Dick Groff sat in the jump seat and donned the airpack while Engine 4 drove to the house. Not having heard a report of conditions yet, he saw no reason for putting on the mask just yet, so why not insert a pinch of snoose into his cheek? On arrival, the downstairs was fully involved, so Groff and Gary Davidson stretched the handline into the main door, kicked open the door to the store, swished the nozzle around and blacked the fire out in about ten seconds. They looked out to the street and indicated to me that they would be going upstairs to check for extension. They donned the masks and started stretching the hose up the flight of stairs.

The steam from the downstairs attack was still billowing, but it was obvious to me that it must be out. While Engine 2 was laying in from a hydrant from the next corner down the street, and the truck company guys were setting up lights and such, the rest of the guys from Engine Company 4 were casually stretching out more handlines…just in case. I ambled around the building to see what—if anything—might be happening in back.

What nobody knew, or noticed, was that all the fire downstairs had been extinguished except the fire between the drop ceiling and the floor above. The full width and breadth of the room, and two feet deep was still roaring—but undetected. It only took a few minutes, about the same length of time it took for Groff and Davidson to make their way upstairs, through the kitchen and into the hallway leading to the bedrooms, for the fire to re-ignite downstairs, roll into the downstairs hallway and up the stairs.

Just about the time Groff was thinking about lifting his mask just a bit to spit out the snoose, the sound of the fire made them turn around to see the kitchen light up. They swung around and scrambled toward the fire, applying water. Other firefighters from Engine 4 picked up the second line, swung the nozzle like a sledge hammer, breaking out the large window of the store and shot water into the room blackening it down again. The guys from Engine 2, seeing what was happening, grabbed a line from their engine and raced up a back exterior stairway on C side which led directly to the kitchen, assuming Groff and Davidson needed help. They kicked in the door and saw Groff and Davidson now at the top of the stairway, fighting their way down. The E-2 crew advanced inside in case there was trouble, which there wasn't.

I met Groff and Davidson as they exited. I tried talking to them about not going upstairs until they are assured that the downstairs is completely secured, but I couldn't keep a train of thought while watching Groff spitting snoose, grimacing, and gargling with water that he was slurping from his nozzle, and complaining he felt sick from swallowing tobacco juice. The others from E-4, by the window, were watching and laughing, and just over their shoulders I could see the downstairs ignite again. They directed a stream into the window again, then shouted that they saw a ceiling tile fall which solved the mystery of the re-ignitions. The guys from Truck 5, armed with pike poles and ceiling hooks accompanied the engine guys in. They pulled the ceiling and extinguished the fire.

Even though we routinely stretched a back-up line for working fires, since manpower is not really a problem, we decided that on multi-story buildings, we would have a back-up team near enough to be called upon. Also, if the deputy chief is not on scene, someone else would cover the back of the building. Presumably, if I were at the A/B corner—scrutinizing the A side and B side—the deputy chief would be at the C/D corner watching those sides. However, that early in my career as a command officer, I hadn't dealt with concealed space fires enough to make them a part of my thinking in residential fires.

CONCEALED SPACE FIRES #2

Years later, on a residential fire, I was confronted by a really odd one. It was a sunny, mid-morning when I rolled up on a one-story house fully involved downstairs with fire boiling out of the living room windows on A side and out of the kitchen windows on C side. Engine 4 rolled up to A side and knocked the fire down with tank water applied through the living room window, then around through the kitchen window on C side. The guys put on their masks, entered, hit a couple of small spot fires, and vented the smoke out using their streams.

62

While this was happening, others hand-dragged the supply line back to the nearest hydrant about 75 feet back. Engine 2 just staged and Truck 5 moved up for overhaul. A neighbor was in the yard with a video camera.

The fire was not only out, the broken-out windows downstairs allowed enough clear air so that guys were strolling around through the rooms waiting for the overhaul to be complete so we could go "back in service."

Lt. Kirko took two young firefighters—my 18-year-old son, Jason, and 19-year-old Robbie Mattson—and squeezed up the steep, narrow stairway to the upstairs. Three guys in airpacks, dragging a hose up the out-of-compliance, afterthought-of-a-stairway was a chore in itself. They quickly scanned the full length of the "finished attic," converted to bedrooms, and saw nothing. I was standing in the yard below the gable window on B side, when Kirko stuck his head out of the window and tapped the palm of his hand on top of his helmet, indicating it was really hot in there, then pointed up to the roof. He wanted the roof ventilated.

All of us outside were perplexed. We saw some smoke curling around the roof, but that was not unusual. People were walking around inside the downstairs almost as though the fire had occurred a week ago, and yet Kirko was complaining about the heat upstairs and wanted the roof ventilated. He disappeared back into the room and the crew was laddering the building when I saw—simultaneously—Kirko, Jason, and Robbie running out the front door while a ball of orange flame belched out the gable window. The team was yanking their masks off and I saw blood running down my son's face.

The other firefighters knocked the fire down again while Kirko explained what happened. "There was nothing upstairs. No fire. Not much smoke. But it started getting really hot, so I asked you to ventilate, then went back to the guys at the hose. We were at the top landing of the stairs, when suddenly the temperature rose so fast, we could almost hear it. It was searing. Then the room got hazy…or vaporous looking. The vapors started turning orange, so all three of us dove headfirst down the stairway. We just tumbled and wound up in a heap on the floor." Besides Jason's bleeding head, they all complained about body pains from wearing airpacks while tumbling down the stairs.

Here's what happened: These types of houses are not commonly found in suburbia USA. In coastal Alaska, because of the heavy snow accumulation, roofs are peaked very high. That creates a lot of space upstairs that most people would not let go to waste. Occupants retrofit some barely manageable stairway inside somewhere. Then convert the upstairs space into a couple of bedrooms. Windows are placed in the gables. In this case, the roof ridge ran from B side to D side and the roof sloped to the bearing walls which were A side and C side. Upstairs, this creates one, long vertically triangular room.

Up there, the triangle is modified. A "ceiling" of two-to-four feet wide, running the full length of the room, is placed for attaching lights. That creates a triangular-shaped concealed space up top. The sloping roof then meets the floor creating unusable space on each side of the room, so, short walls are erected, called "knee walls." Now you have created triangular-shaped, concealed spaces in the eaves (which is sometimes used for storage). Now you have three concealed spaces surrounding the bedroom(s) space upstairs. The wall coverings (in older houses) are often plywood and not sheetrock. The eave spaces are "connected" to the ceiling triangle by the roof rafters. Got it?

When I arrived at the house, the fire was ripping out of the windows on A side and C side. These flames were impinging on the roof overhangs and, unbeknown to us, burned through the soffits (the horizontal, underside of eaves) and had ignited items stored in the eaves on both sides. The hidden flames then traveled up between the rafters to the ceiling triangle and combusted the wood there. So the upstairs bedroom was totally surrounded by unseen fire…just like an oven. When the bedroom heated up to its ignition temperature, it ignited spontaneously—a flashover.

In hindsight, the soffits outside and the knee-walls inside should have been pulled immediately with axes and halligans and streams of water played into them. But, at the time, it seemed normal to Kirko to make a sweep of the room first, before starting overhaul. It would seem normal to anyone.

Try this: Anytime that the heat is too intense for this type of interior operation, or the construction materials and limited space make it too tough, ladder the gables and cut three holes—one just below the peak, and the other two at the two triangles where the roof meets the bearing walls. Do it on both ends. Play water into each of the three concealed spaces. If needed, do it from each end. Then, the final overhaul can be completed in comfort inside.

We experimented with other ideas such as driving a piercing nozzle up into the eaves from down stairs but no one can swing an 8-pound sledge hammer up, and drive a large spike through 1 inch of wood. Cutting holes for it with a chain saw worked, but if one were going to do that, it's easier cutting the gable walls and play water in horizontally.

When I was teaching strategy and tactics in Juneau, I told this story and they informed me that they had an identical problem less than a week before. Making gable cuts is an idea I have been championing for fires in large warehouses that have cocklofts.

CONCEALED SPACE FIRES #3

This single-story home was several miles out Power Creek Road, a winding, dangerously narrow road that twisted above the edge of Eyak Lake. It was the site of numerous fatal or near-fatal accidents where cars would plunge down the

64

vertical drop-off into the water below. When I arrived that morning and saw the light-colored lazy smoke enveloping the house, I figured the most dangerous part of this operation would be the drive out there. That type of smoke characterized an incipient stage of fire, requiring we search out the little bastard and stomp it out. No one was home.

It was a windless day, so the smoke just curled around the house at ground level. While we were stretching out hose and establishing a draft water supply, we tried looking in through doors and windows but couldn't locate—from those vantage points—the seat of the fire. This was before the availability of gas-powered positive pressure fans. The standard approach was vertical ventilation by cutting holes in roofs. This approach is completely ineffective in pre-flashover fires when the smoke is lazy and too cool to lift through the roof. Electric smoke ejectors simply didn't have the capacity to make the effort worthwhile.

Oh, well, back to the old way—go in and crawl around blindly until you find it. Gundlach took the nozzle and crouched on the ground in front of the front door. The door was a few inches above the ground level. There was no indication that there was any space below the floor, yet when he stepped into the door, he got suddenly shorter.

From his waist down, he was into the space under the house. A quick burst of flames shot up around him, and we yanked him out of the hole. After we cooled him down with water from the hose, the only place he was burned was a thin strip of flesh on his right cheek that was exposed between the edge of his mask and the ear flaps of his helmet.

On the opposite side of the house, C side, facing the lake, a sundeck stretched out from the main floor of the house. Below that was an entryway to the space below the house, but no one had been able to see it through all the smoke. It turned out to be a full basement complete with furnishing including a refrigerator. Like many concealed space fires, once you locate them, extinguishment becomes routine. And after we found the entrance, this also became routine. A quick knockdown and overhaul. The only thing remarkable thing that happened during the overhaul was that one of the guys noticed the refrigerator door had sprung open a bit and there were smudge-covered bottles of beer in there. He took one out to Gundlach who was in the ambulance getting his burn flushed with water. There was a lot of beer in there, so we drank it. It was warm.

The medics thought it would be prudent to have Gundlach stop at the hospital ER to check his burn. Heading back into town, we stopped at the ER to see how he was doing. We just crammed into the small emergency room while the doc was checking him out. The doc stopped for a moment, looked up, sniffed the air and asked, "You guys been drinking?"

"No," we chorused in unison.

Gundlach was fine and the burn didn't leave a scar, just a slight discoloration that you barely notice today. But since nomex hoods were just appearing on the market at that time, I ordered a bunch.

Hey—it was only one goddamned beer (each), okay?

NIOSH CONCEALED SPACE FIRE FATALITY

Anyway, that firefighter fatality report that the National Institute for Occupational Safety and Health (NIOSH) composed, following the death of an engine company captain in the spring of 1998, explains clearly how dangerous these fires are. Less than 25 minutes after the 17-year veteran captain had entered the building, he was found dead. This warehouse fire was fought by 5 engine companies, 3 truck companies, 1 rescue company, and a battalion chief. There were 30 firefighters on site.

The warehouse, which prepared and stored dog treats, was tall, but single storied. It was 110 feet long and 59 feet wide. The operation took place a little after 2 in the morning. During the substantial time it took for firefighters to gain entry to the building (8 or 9 minutes), truckies made it to the top of the wood-truss roof. On the way up, they reported to the IC that they could see fire in the ceiling area. And on the roof, they saw lots of grayish-brown smoke emitting from a roof vent.

When they approached the center of the roof, they saw fire coming from the roof vent. The ventilation team opened up a 4- by 4-foot hole and were immediately driven back by heavy fire and heat. It was a full 5 minutes after that, that the initial attack team advanced inside with handlines, followed by a second attack team which included the victim, and a third handline was advanced inside also. About 15 feet inside the door, the smoke conditions were so bad that visibility was zero, but there was very little heat. A minute later, three firefighters entered with pike poles to see if they could pull the ceilings.

Inside the building, 30 or 40 feet, the attack teams could still not find the fire. The guys with the pike poles could not feel a ceiling above them. All the firefighters inside could hear the saws of the ventilation teams on the roof and knew ventilation was taking place, but the heat and smoke level was intensifying as they advanced into the structure.

At this time, the truckies reported to the IC that they were getting "real good fire out of the roof and are getting off." The captains on the attack teams radioed to the IC that they couldn't find the fire in those black conditions, yet they were continuing to advance. When the attack teams came together, they decided that the conditions, bad as they were, were deteriorating and a retreat was called for. The captains ordered their crews to follow the hose lines out, but the hoses were in such a tangle on the floor, it became a problem. The teams could feel the heat coming down from above, and there was no visibility. Most of them made it out. One firefighter got separated, but his buddies re-entered and got him out. The captain (victim) did not exit. Six minutes after they started their retreat, part of the truss roof collapsed and blocked the entryway.

Even though a Rapid Intervention Team was assembled, and forced entry was underway at the opposite wall, entry became impossible and the volume of fire was unaffected by the water being applied.

The minute details of this report made it very long and it took considerable effort for the reader to visualize everything that was occurring at the time. Following the report, NIOSH made several remedial recommendations. However, I think they missed the primary problem here: The inability to recognize this as a concealed space fire.

In a nutshell, when truckies on the roof reported lots of fire up in the ceiling area and smoke and flames boiling out of the ventilation hole, where did it all come from if the smoke and heat did not lift from the warehouse? Everyone in this business knows that when a building is significantly involved in fire, after cutting a vent hole in the roof, the smoke and heat goes spiraling out of the hole and the heat inside the building dissipates and the smoke lifts, making an interior attack easier. If the heat and smoke don't dissipate, there is a full ceiling, or membrane, between the ground floor and the roof. The fire and smoke coming out of the hole is from a compartment above…a cockloft or some other concealed space.

Having recognized this, a better approach to reach the body of the fire—particularly a large building with a truss roof—would have been by cutting large holes on the gable walls of the building—just below the roof line—on each end, and taking turns applying water from each end until steam generation stops. In the meantime, the crew downstairs could be making entryway cuts on opposite walls and set up for vertical forced ventilation. And even though teams eventually would have to be sent in downstairs, ceiling overhaul should be done when visibility improves enough to ascertain the stability of the trusses. If roof stability can't be guaranteed, and you can't wait until daylight to overhaul, screw it, drown the place.

In the tactics and strategy course I mentioned earlier, on the last day, I divided the students up into 3-person teams. I sent one team to the following scenario I developed: The Baranoff Hotel in Juneau was built at the turn of the century and over the years had been the site of numerous remarkable fires which the Juneau FD somehow, miraculously, beat. One ingenious attack commanded by Chief Al Judson will be described later under a different topic. Anyway, while I was walking around town the night before the course, I noticed that the distance between the rows of windows on the 6th floor and 7th floor was greater than the distance between other floors in the eight-story building

I asked the local firefighter who accompanied me if the last two floors (7th and 8th) were additions to the original building, and he said yes. Going inside, as was expected, the ceiling height on the 6th floor was the same as the others. This meant there must be at least a 3- or 4-foot concealed space above the 6th floor ceiling. I asked the hotel manager where the access was to this space and he told me the closet of room 623.

I described this scenario to that team: They had responded to the hotel and were told by the simulated firefighters inside that the fire, which originated in room 623, was out. But, shortly afterwards, the team on floor seven—"Division 7"—reported that they could not clear the smoke from their area and that it was getting thicker. In fact, Division 7 went on to say that the heat was increasing rapidly on the 7th floor. The 3-person student team was to develop a strategy for a problem that initially made no sense to them. What I hoped was that the emphasis I had put on concealed space fires in my lecture would prod them into taking a closer look at this building and look for indications that a concealed space might be the source of the problem. I expected them to go inside and question the hotel manager the same way I had. Then, send an attack team (simulated) to room 623. And don't forget to take the attic ladder. Pull Division 7 back to a stairway to stage on the landing of the 6th floor. They could fine-tune their strategy with some tactics like preventing a backdraft before opening the closet hatch. They did, and passed the test expertly.

CONCEALED SPACE BACKDRAFTS

One additional word of caution: I read of a fireground fatality which resulted when the attack team pulled the ceiling—a drop (or suspended) ceiling—and that area backdrafted as soon as the air got to it. This was a basement fire.

Another case in point: Dick Groff tells the story of when he was firefighter in Ketchikan, and the oddest thing happened to him after the department extinguished a well advanced fire in a 3-story, wood-framed building, which had been a nursing home. During overhaul, the firefighters had removed their masks and hung them on their airpack harnesses. Axes, halligan tools and pike poles were resounding throughout all floors of the buildings. Dick explained that all the windows and doors were opened, the air was clear, the whole thing was over. Dick was working alone, dragging a hose with him as he walked through the place ripping parts of the rooms apart and wetting down any suspected embers. He noticed a small amount of smoke emitting from a grate on the wall that was covering duct work. It was about head height.

Dick used the pick-head axe to pry the grate off the wall. He stuffed the nozzle down into the duct a few inches and cranked back on the bale. The explosion blew him across the room, where a moment later he was laying against the opposite wall, dazed, and watching the nozzle of his hose flipping around the room. The entire place was full of smoke. He heard the other firefighters frantically dropping down prone as they—and he—grappled with donning their masks. So, there is another example of well…not really a concealed space fire, but a backdraft in a small concealed space. Watch your ass.

I could go on and on and on, but here is a better idea: Observe, think, and study. This is serious shit.

FEEDING HOSE
TO THE NOZZLE CREW
Photo: Joan Jackson

DIVISION 2, NORTH PACIFIC PROCESSORS FIRE
Photo: CVFD

DIVISION 3, PROTECTING THE ROOF TRENCH CUTTERS
Photo: CVFD

69

Unique Approaches to Fire Attacks

The credit for our successful offensive attack on a huge cannery goes to the fire chief of Juneau's combination department, Al Judson. The Baranoff Hotel has had numerous memorable fires during it's long history. It had been a six-story, wood-frame, non-sprinklered downtown hotel. One of the lesser known stories had nothing to do with tactics, but did exemplify firefighter ingenuity. Firefighter Sandy (I didn't catch his last name) was assigned to the bucket on an elevated platform on the main street in front of the hotel to combat a 4th-floor fire.

The platform was elevated to the 4th-floor window level, waiting for something to show. The engine company guys were walking up to the 4th-floor. Sandy was fidgeting uncomfortably and informed his partner, Al Judson, that he needed to take a dump. "Well, hell," Al said, "nothing's happening here yet. We probably have time; lower this thing, you run across the street and take a shit." Sandy started lowering the platform when a ball of fire blew out of a window right in front of them. They opened the nozzle and started to work. Finally, Sandy could stand no more. He removed his helmet and sat it upside-down on the platform behind him, pulled down his pants and defecated into it. He wiped himself with his glove, then rinsed his glove off in the stream of water.

Al was completely prepared to tell everyone about it later, but when they were back on the street, several others became aware of the incident when they saw Sandy washing out the inside of his helmet with a fire hose. And just in case there might have been someone who hadn't heard about it, at annual banquet, the crew presented him with a helmet fitted with legs so it would sit upside-down. They attached a toilet seat to the brim.

Anyway, back to the big one. It was late one evening, back in the '70s, before the Baranoff was sprinklered, that a man who'd just been fired from his job working in the kitchen, returned to the hotel and spilled a large amount of kerosene on the kitchen floor and lit it. By the time the engines rolled up on A side, the kitchen, the bar, the dining room, and the front lobby were ripping. Al did the usual positioning of the engines and trucks but saw that he was completely out-gunned. There was no way he could man enough handlines with his available personnel to meet the required fire flow. The rooms of this wood-framed, non-sprinklered hotel were filled to capacity. Occupants were spilling out of every available exit. Many were standing at the windows of the hotel rooms above the fire, hoping to hell that the IC knew his job. He did.

Al told his crew to take the deluge gun (monitor) off of the top of the engine and set it on the sidewalk. Next he told them to couple the two 2½-inch hose lines to the intakes. He then told them to take the monitor into the front doors. They took it up to the front doors and set it down on the sidewalk. "I said 'Take it in the front doors,'" he bellowed. He directed two hose teams to march alongside the monitor team and use the wide fog patterns to protect and keep the monitor team cool. It was touchy, but they got the monitor inside the doors, sat it down, and backed out. Al signaled to have the 2½-inch lines charged. The entire downstairs exploded into steam and the fire blacked out.

Al looked at the attack teams holding their handlines and yelled, "Go!" They charged inside, hitting spot fires as they went. In a few minutes it was done. The rest of the occupants of the hotel walked out relieved to be alive.

I'm not sure where the nickname "bomb line" came from, but that's what we have always called the monitors that we keep preconnected with a single 3-inch hose and store in the rear compartments of our two front line engines. Of course, we removed the doors from the rear compartments. Three of our engines have top-mounted monitors and two of them also have bomb lines as "extras." The hose which is preconnected to them are double-rolled "donut rolls" of hose, so that when the monitor is carried away from the tailboard, the hose unrolls easily and without twists. One person can deploy it 150 feet away, but two people make it easier. The devices are for large-flow fires in large commercial or industrial buildings. Don't use it on somebody's house; you'll look like an idiot.

STRAIGHT STREAMS ON ROLLOVERS

I was reading a book once on how to improve one's writing by thinking in metaphors. With that fresh in my mind, I responded to a fire call and, as most of us do, I was looking over the situation, formulated a plan, and was deploying the crew. Pretty routine. But a metaphor occurred to me as I did this. When a new recruit takes a certified Firefighter-1 course, one of the first topics he studies is Fire Behavior. He doesn't study Building Construction until much, much later in the course. However, in the medical field—EMT or Paramedic—the segment "Anatomy and Physiology" (a single course) comes early in that field of study. The Behavior of Fire—metaphorically speaking—is physiology. Building construction, metaphorically speaking—is Anatomy. The physiology of what is occurring within a patient's body is the focus of the medic. The "body" that the fire officer is examining is the building within which this chemical chain reaction (fire) is occurring. How fire behaves is really only relevant to the structure that contains it. Otherwise, what's the point?

It dawned on me that I watched the fire differently than the general public, certainly, but maybe differently than the line firefighter. I watched it like a mongoose watches a cobra. I did it to anticipate where it would move to strike next. Confident that I knew its plan, I unleashed the dogs. Oddly enough, at the call following that one, the mongoose/cobra metaphor was so intentionally before me that my confidence, (which has always been high) became higher

After Loyd Leyman wrote about his experiments using fog pattern for fire attacks, the fire service uniformly over the following years purchased combination nozzles so we could take advantage of the technology. As a result, solid streams became a part of the past. Straight stream use became very limited, also. Except for overhaul of burning mattresses and the like, no one bothered to use them any more if they could afford the combination nozzles. As an aside, one method still in use somewhat is to direct two separate streams into an upper story window, but have the streams collide as they enter, the smaller droplets strike the ceiling and sprinkle down on the fire, on the lower contents, imitating a sprinkler system. This is primarily to drop the temperature upstairs to allow attack lines to make it down the hall toward the room of origin.

Anyway, Chief Tom Opie of North Slope Borough told me of an experiment he conducted using straight streams. In a large Quonset hut he had set a fire in the far back wall and let the fire build from its incipient phase, heating the ceiling, and watched the rollover progress toward the A side where his nozzleman sat waiting. When the leading edge of the flame was about mid-room, the nozzleman applied a straight stream, sweeping from left to right the entire half-circumference of the building and the flame front retreated back to the spot of origin. The visibility remained good and the thermal balance within the room was not upset. He then allowed the fire to rebuild and used the straight stream again repeatedly, always with the same result. No steam was generated, and consequently the temperature did not drop dramatically, but most important, firefighters could still see the length of the building and eventually advanced through the room and extinguished the fire with short bursts of water directly on the small incipient fire.

I had included this in my tactics lectures before I had the opportunity to try it out myself. When we (CVFD) finally got an old house to burn, I learned a couple of other things about this technique which we now use as standard practices. The non-sheetrocked, wood-frame house had a large living room about 15 by 30. The enclosed entryway ("arctic entry") was on A side. There was an old couch that sat along the inside of A wall. Some of the firefighter students sat there to watch the demonstration

The nozzleman, Robbie Mattson, dragged his hose past the couch potatoes and squatted down in the middle of the living room facing C wall. The remainder of the firefighters crouched behind him. Dana Smyke manned the nozzle of the back-up line in the arctic entry. We started a fire against C wall. Before long, smoke covered the entire ceiling and mushroomed down halfway to the floor. The temperature rose steadily but remained tenable. Even with the heavy smoke one could easily see the wall of the fire on the ceiling leisurely making its way from C side toward A side the full 15-foot width of the ceiling. As planned, when the flames reached about midway, just above Robbie's head, he directed a straight stream from left to right like a trench. The flames stopped their march immediately and retreated back to the fire's origin. Visibility remained unchanged.

Then I asked Robbie to try a 20-degree pattern next. The fire rebuilt and Robbie used a wider pattern. Some steam was generated, the fire retreated, but the smoke level dropped to about one foot off the floor. One was still able to lie flat on the floor and see the length of the room. Not too bad. The next time Robbie used the recommended, short-burst, 30-degree pattern. The fire almost went out but the entire room went black.

I got the video camera out and asked that the fire be allowed to rekindle and to use the straight stream again. The room gradually heated up again and the wall of flame walked toward the A side again. When it passed over Robbie's head, Dana poised to douse the flames but I told him to wait, figuring Robbie must be trying a new technique—I could see he was looking straight up at the ceiling. The fire rolled all the way to the A wall, banked down and ignited the couch that the student firefighters were sitting on. They excitedly rolled off the couch and onto the floor.

The heat level shot straight up to scorching levels when Robbie cranked back on the bale of the nozzle. Again, he sliced the ceiling down the center and the flames parted in the center like the Red Sea and retreated in both directions. The visibility remained excellent. He then advanced to the seat of the fire and extinguished it with short bursts of straight streams. The camera didn't pick up any of it. Outside, we all discussed what we saw.

It was an accident that Robbie let the fire get past him. Looking up, the brim of his helmet and the mask of his airpack obscured his vision, and he couldn't see how far back behind him the fire had gone. It was the ruckus caused by the guys rolling off the couch that prodded him into action. But what we actually learned was that it was not necessary to apply water in front to the edge of the fire to have the technique work. Right in the center of the wall of fire on a ceiling works just as well. The advantage is that if a crew is not yet ready to enter a building, and wants to use this pattern, it will work from almost any vantage point. With the new "2-in/2-out" regulation, a nozzleman from a doorway can stop the advance of a rollover while waiting for other firefighters to get a back-up line stretched and manned. And since the purpose of this stream selection is to maintain visibility inside by not upsetting the thermal balance, when the rest of the attack team is ready, they have only an incipient fire to deal with and deal with it easily. They don't have to use PPV.

It is critical that attack teams understand and can predict the difference between a "rollover" and a "flashover." An incipient fire transitions to a free-burning (fully involved fire) by either a rollover or a flashover. The straight stream will not stop a flashover, and being caught in a flashover is a death sentence for firefighters. Flashovers occur when the room is uniformly heated up by the incipient fire. The radiant heat from the incipient fire uniformly heats up most of the contents of the room to their kindling temperature. The ceiling, wall panelings, furniture, and so forth become heated to the point the ignitable vapors are emitted from them all.

When their ignition temperature is reached, these gasses ignite in unison. Oftentimes a ball of fire erupts with enough force to belch an orange flame out of a doorway or down a hallway. Test fires have shown, through the use of thermal couples, that the temperature momentarily spikes at 3,200°F before dropping down to the steady ceiling temperatures of 1,200°F. The quick heat absorption of a 30-degree fog pattern is needed to drop the temperature before this occurs. No one survives a flashover. This phenomenon can also produce a lot of force.

Firefighter Whitethorn, Anchorage Fire Department, was killed in a super-market fire in a flashover of backdraft proportions. The event defies all reasoning. The supermarket was a much larger area than one associates with a flashover, because it's hard to imagine uniform heat build-up in such a large space. But, according to the report I read, the flashover was so violent, it threw Whitethorn against a shelf so hard that the impact severed his aorta, killing him instantly.

Nevertheless, most of us have seen flashovers with frightening impacts. A rollover is much gentler. The incipient fire heats the ceiling area just ahead of the advancing flames, so the flames spread across the ceiling like the tide coming in. Actually, it's sort of pretty. The more open an area is, the more air currents available, the more spacious the room is, the more likely the transition will be via a rollover as opposed to a flashover. I emphasized to my crew that if in doubt, use the 30-degree pattern. Only use the straight stream if it is obviously a rollover.

Point in fact: We went downstairs in the same building and tried it again. We set the fire in a small bedroom, approximately 10 by 10 feet. The attack team waited in the adjoining living room. Back-up was at the front door. The bedroom, with only one opening—its doorway—produced a flashover that blew a ball of flame out into the living room. The living room, in turn, was big enough (with an opened exterior door behind) and had enough air exchange that the ceiling began a rollover.

The crew sliced the ceiling with a straight stream and the fire retreated back to the doorway to the bedroom but had no affect on the fire in the bedroom. The top of the doorway to the bedroom (the header) was at least 18 inches below the ceiling. It required a second application of water (30 degrees) to extinguish the bedroom. So, the second lesson we learned during this drill was: separate rooms divided by a doorway with a header require two separate applications of water—a straight stream for the rollover and the standard recommended fog stream for a post-flashover, free-burning room. Even with that, the visibility in the living room remained pretty good and was not terribly affected by the steam and cooling of the bedroom. I heard someone refer to the tactic as "penciling."

I don't want to leave this topic without explaining the flashover that killed Whitethorn. You need to know this. The heat from the source fire naturally gathered at the ceiling. Also, naturally, you would not expect heat to accumulate in a ceiling as large as a supermarket. One would expect a rollover. However, the ceiling was held up with large, solid "I" beams which created 15 foot wide, long "ponds" in the ceiling to "catch" the heat. The heat built up uniformly in one of the "ponds" until the ignition temperature of the combustibles was reached. That huge "pond" of gasses that was 3 feet deep, 15 feet wide, and stretched the width of the building was a bomb. Those "I" beams are more stable than bar-truss beams, but they car create "ponds" for capturing pre-flashover gasses. The next time you go shopping, look up.

* * * * * * * * * *

At a conference of the Alaska Fire Chief's Association, Glennallen's Fire Chief Rocky Ansel and I were talking about these not-so-traditional approaches to firefighting one evening, when he told me how Glennallen uses piercing nozzles. Vehicles that have fires in the engine compartments are difficult to knock down because it is often difficult to open the hoods with fire blowing out in all directions. Generally, we beat the shit out of the hoods in a time-consuming operation of unlatching them. Rocky's people use piercing nozzles. They will rest the spike on the top of the front wheel. They move it a couple of inches aft of that spot, then they punch it through the plastic cowling and into the engine compartment. Crank back the bale and that's it. Then they can take their time opening the hood.

He also uses one in conjunction with the thermal heat sensor (Fire Finder) in fires under trailer houses. Lots of trailer occupants use heat tapes on their water pipes under their trailers. There is usually 1.5 to 2 feet of space under a trailer, and even with wooden skirting around the trailer, extremely cold temperatures can freeze those pipes—consequently, electrical heat tapes. Those tapes have a habit of catching fire. Rocky simplifies their operations by walking around the skirting with the Fire Finder until he finds the hottest area. They drive the piercing nozzle through the skirting and start applying water. After a few seconds of that, they rip the skirting off and finish the job and overhaul. He told me that his little department has nine of those bastards. We only have one and seldom use it. That will change.

By the way, Anchorage Fire Department Captain Jim Kenshalo invented a neat way to open car hoods when the regular release level under the dash can't be used. It is a long, steel rod that has a tuning-fork-looking thing on the end that you slide through the grill, and a "T" handle on your end. You slide it in to straddle the fork around the latch cable, grab the "T" handle, and rotate it. The fork grabs the cable, pulls it back, releasing the latch. He made one for me.

WHEELED EXTINGUISHERS ON STRUCTURE FIRES

I decided one time to try to develop fire attack procedures for Alaska's village fire departments: They had few resources—money, manpower or training. The new emphasis on discouraging firefighters from making interior attacks unless they had breathing apparatus and the proper training to use them, made conventional interior attacks not recommended.

That complicated things considerably for rural fire departments. In the '70s when oil started flowing from the North Slope, and barrels of grant money started flowing across the state, many small Native villages were getting fire engines. And some training was provided for many of them as well. But what happened afterward was very discouraging. Often the engines weren't housed, and were seen years later parked outside, buried up to their hubs in mud, getting rusty, and completely useless. Often, villages would use the vehicles for other purposes and they would fall into disrepair. Training the firefighters initially was done well, but as those people moved away, newer residents didn't have a clue.

But there were villages that never even got that much. They have a couple of portable pumps and some hose and appliances, and that's about it. No turnouts either. We often refer to them as the "Carhartt fire departments." When the Carhartts respond to a fire, they have to set up draft with the portable pumps, and in winter they might have to cut holes in the ice on the river or pond to get to the water. Then they must lay out enough hose to reach the building. The time required for this usually means a complete loss of the building and contents.

* * * * * * * * * *

Kodiak's Fire Chief Mike Dolph once told me of the mathematical formula for calculating how much ABC dry chemical is needed to put out a fire in a cubical...a 3-dimensional fire. This could be used in the villages. The rating printed on the side of portable fire extinguishers refers to the size of a 2-dimensional fire that can be extinguished. Square feet....flat surfaces. Here's the formula: The cubic feet of the building, or room (length x width x height), multiplied by .0385 will tell you how many pounds of dry chemical are required to extinguish a fully involved cubicle. After that, I actually read that formula in NFPA's Fire Protection Handbook.

Anyway, let's say the living room of a house is fully involved and the size of the living room is 12- by 14- by 8-foot ceiling. That's 1,344 cubic feet (ft^3). Multiply that by .0385 and you get 51.7 lbs. So, if applied properly, 52 pounds of ABC dry chemical should knock that fire out. In other words, three Carhartts armed with a standard 20-pound extinguisher each, and schooled in the proper use of them, could station themselves at windows and doorways of this room and knock the fire down. Of course we know that the heat is not dissipated nor the air breathable, and that spot fires still need to be hit with water from up close, but the fire is out and will not travel to the rest of the building or other buildings. Again, that was just a formula that most people I talked to had never heard of. It needed testing.

Back in Cordova, we finally got an old cabin on the outskirts of town that we could burn. The living room was 12 x 12 feet with an 8-foot ceiling (1152 ft^3); the kitchen was 12 x 10 feet with an 8-foot ceiling (960 ft^3) and the bedroom was 12 x 10 feet with a 9-foot ceiling (1180 ft^3); for a total of 3,192 cubic feet. Multiply that by .0385 and we would need 123 pounds of agent. Our harbor is supplied with numerous large extinguishers on wheeled carts. They are 150 pounders and are configured with a large tank filled with agent and another tank filled with compressed nitrogen to propel the agent. Most harbors, I imagine, have these extinguishers and so do industrial plants of all sorts. They cost less than $3000, have no moving parts, require only minor maintenance, little training, and are easy to refill. They also take up a lot less storage space than a fire engine. Anyway, we borrowed one and took it to the drill site.

Looking in the picture window of the cabin, through the living room, one could see the full kitchen just beyond. The bedroom, on the other hand, was off to the side of the kitchen and could not be seen from there. One would have to walk into the kitchen, turn left and go through the bedroom doorway. We surmised that we would have to apply most of the agent through the living room window, knocking the fire down in that room and most of it in the kitchen, then move the 50-foot hose around to the other side of the house and hit the bedroom through its window.

We used diesel fuel to get the fire evenly started in all the rooms at the same time and it went almost immediately into the post-flashover, free-burning phase. Fire was blowing out of every opening in the building. The nozzleman walked up to the living room window and shot the powder in. Worth mentioning here, since we are all conditioned to spraying the ceiling first, was that with dry chemical, you start at the base of the fire. So he sprayed down low first, and swept the stream back and forth as he moved it up the walls to the ceiling. From our vantage point, the fire was out in about 5 seconds. He then moved to the other side of the house to apply the chemical through the bedroom window, but discovered that there was no fire there. We were surprised and pleased to see that the agent actually turned a corner and went into the third room. Also, there was powder left over, so the formula was pretty damned accurate. We video taped it and sent it to State Fire Service Training.

Incidentally, one can use that formula in reverse. How many pounds of agent do you have on hand? Take that number and divide it by .0385 and it will tell you how many cubic feet of space can be knocked down with it.

Anyway, put a couple of large, wheeled extinguishers in any remote village, and provide some training in their use. They can be wheeled almost anywhere. If there is a lot of snow or mud in the area, a couple of guys can place them on the back of a 4-wheeler and get them to the scene. Knock the fire down instantly with them, while the rest of the Carhartts are establishing a water supply with portable pumps to knock down the spot fires afterwards. Hell, they could even use pressurized water extinguishers, if there are enough of them, and they are easy to refill. Use smoke ejector fans to clean out the air inside for the folks to finish off the overhaul.

* * * * * * * * * *

If you read the account of our firefight aboard the fishing vessel (F/V) Alaskan Swede, you will recall that the initial problem was our inability to get firefighters aboard the vessel. Another vessel could have ferried an attack team to the Swede which was adrift in the harbor. Since the firefighters would have been without hoses, I could only think of having the Swede pushed back to the dock. But then, there was no way to tie her up because the fishermen in the jitneys were afraid to get too close to the roaring fire.

Unfortunately I don't have the dimensions of the Swede's cabin and pilot house, but I'm estimating that the cabin and berthing space was about 2,900 cubic feet. I estimate the pilot house as being about 1450 cubic feet. So, 112 pounds of ABC dry chemical would have snuffed the fire completely out in the cabin and berthing space. The pilot house would have easily been knocked out with 56 pounds of powder…actually much, much less because of the sparse combustible furnishings up there. At most, a total of 168 pounds would have snuffed the entire fire out like blowing out a match.

Two guys can easily load a wheeled 150-pound extinguisher aboard a jitney. They are available all around the harbor. Since they have 50-foot hose and nozzle attached, one could leave the unit in the jitney, stretch the hose out and blast it through the doorway or a window. Then a couple of 20-pounders would have knocked out the pilot house. All gone. With the fire out, the vessel could be tied up again at the dock, then commence cooling with water and overhaul.

I never thought of that because that tactic is not "normal." It wasn't until I started experimenting with dry chemical on larger structure fires that I noticed with astonishment how effective it is.

A final statement about dry chem on structure fires. Fire Flow calculations are based on the rate at which water must be applied to a free-burning fire to absorb the BTUs being generated by that much space and that much fuel. Think about the tetrahedron for a minute. Dry chemical does not remove any of the three sides of the fire triangle, but it does attack the fourth side of the tetrahedron. It stops the chemical chain reaction of the fire. The fire is instantly snuffed out. There is no longer any propagation of flame. BTUs are no longer being generated. All you have left in front of you is a hot room. Water applied at any rate will absorb the temperature and drop it. You don't have to worry about the rate at which you are applying it, because no BTUs are being generated. Get it? That means that after the fire is knocked down, the fine mist of the "Bazooka," or any pressurized source of water, would work.

DROWNING IN THE OMNIPOTENT WRITTEN WORD

Dick Groff used to work for the U.S. Forest Service. I remember the day he decided to retire. He was disgusted by the way the organization had changed during his career. "We used to have an official policy manual that was about an inch thick. Now, the series of binders takes up an entire shelf and looks like a set of encyclopedia."

That is also the single, biggest change in running a fire department, too. Written policies and procedures were the exception, now the rule.

You've heard that common sense may not be all that common. One thing that you can take to the bank is this: people (and organizations) find comfort in order, even if that order is established by procedures meant to deal with routine events. Written procedures remove the need for reasoning. That's okay, because they create uniformity of actions from one person to another, one era to another. Drawback. Once written, a procedure becomes a commandment regardless of the original intent. Yep, Dr. Frankenstein, you created it, then it turns to control you.

One winter evening, just before dark, five high-school age boys were making short work out of a case of beer and driving out Power Creek Road. Because of the seemingly endless rain, the gravel surface was not covered with snow or ice. In fact, the ice on the lake was getting too thin for skaters. Chucky Davis maneuvered the "muscle car" leisurely around the tight bends of the winding, narrow, oftentimes one-lane road that had Eyak Lake on its right and steep rising woods on the left. The rear windows were rolled down, but they only went down halfway, about twelve inches. The front windows were rolled up.

Just beyond the last house on the road, and not noticing that he had been speeding up, he lost control of the car, flipped off the road and landed upside down in the lake. Miraculously, the four passengers, submerged in the pitch-black and freezing water, all squeezed out the back windows. The front passenger, too. Chucky never made it out. Two of them ran back to the house to call for help while the other two chest-deep in water, tried reefing on the doors to get at Chucky.

Don Endicott, Stan Shafer, and I rolled out of the fire station just a few minutes later. I was driving only as fast as could safely be done with a rig that size on such a treacherous road. It took us at least 15 minutes to get there. It also took that long for the car passengers to free the car door, since the roof of the car pinched down on it. Don and Stan were able to drag Chucky out while I pulled the gurney out to the ground.

Heading back, while they were doing CPR in the back, I radioed the dispatcher and instructed her to call the hospital (at that time, the hospital did not have a radio) and inform them that we had a cold-water-drowning victim and that we were doing CPR and that our ETA would be about 15 minutes. Then Don yelled up that they had an IV started and injected 50 cc's of Sodium Bicarb (the protocol back then for any cardiac arrest). I radioed that to dispatch to relay to the hospital. I radioed again when the EOA was inserted. Each time, the dispatcher would radio back to me that she had relayed the information to the hospital's duty nurse.

I finally pulled up to the hospital; we quickly pulled the gurney out and never stopped CPR as we rolled it down the hallway and into the emergency room. Don and Stan were exhausted from doing CPR while bracing themselves against the hard curves that I was navigating. They were soaked from jumping into the lake. Chucky was lifeless, covered with twigs and gravel, and dripping puddles on the ER floor. Except for us, the ER was empty. A moment later, the duty nurse stepped in, took a look and said, "I guess I better call the doctor." We were speechless.

I took over chest compressions and Don followed the nurse out to the nurse's station. "Didn't the dispatcher tell you that we had a drowning and were doing CPR?"

"Yes," she replied.

"Why didn't you call the doctor so he and the staff would be waiting for us when we got here?"

"It's the doctor's policy that when someone comes into the ER, the duty nurse does a patient assessment before calling him."

Don could hardly believe his ears. "Does that make sense to you in this circumstance? You knew we were doing CPR."

"I know," she said, "but that's the policy."

Chucky didn't make it and probably would not have anyway.

The "policy" was changed right after that. Actually, it wasn't changed, it was lengthened with a series of "excepts" to cover every conceivable non-routine set of circumstances so that—Lord forbid—some unforeseen event might not be covered by the written directive.

As the years passed, the inter-operability of the hospital staff and EMS division became more fluid. Part of that was due to having a physician sponsor from the hospital. When our crew qualified to start IVs, and later, for advanced cardiac care, we were required by regulation to have a physician sponsor. Before long, the EMTs and hospital staff felt like a large family.

Still, there was a clear line between their jurisdiction and ours. That line was crossed numerous times to everyone's satisfaction.

THE ICE AUGER AMPUTATION

Throughout the fishing season, the boats in the fishing fleet and the canneries use mammoth amounts of crushed ice. Each cannery has its own ice house and ice crusher for its use and for the boats that fish for that cannery.

John Watts, 20 years old, carrying the short shovel, tentatively maneuvered his way into the ice house's cubicle. In the center of the cubicle stood the auger that, when turning, drives the ice chips down through the hold in the floor.

Periodically, someone has to manually shovel the residual ice away from the walls and into the hole where the auger pulls it down to the room below. To facilitate this, the auger is run so that the shovels full of ice hit it and are sucked down below. This athletic, young summer employee, earning money for college, spread his legs—spanning the width of the cubicle—and wedged them tight against opposite walls, leaned forward and scooped ice toward the turning auger. The deeper he got, the harder the pack was. He was working up a sweat in that frosty room as he slammed the shovel down harder to shatter the ice pack. Then his foot slid and he slipped backwards, his legs flung forward toward the auger. In disbelief, he saw—too late to prevent it—his foot wedge between the blade of the auger and the hole and was drawn down into it.

The pain was excruciating as he did the splits and his leg was crushed and ripped and drawn into the auger before the machine was shut down. He was horrified to see he was ensnared up to his groin, bleeding profusely, his leg shredded.

I doubt that a minute had passed between the time his screams alerted fellow workers and the fire department was toned out. The rescue truck and ambulance responded. The rescue truck crew was responsible for disentanglement or extrication; the ambulance crew responsible for patient care. The truck company lieutenant and his crew were talking to employees about whether the auger's rotation could be reversed or if the machine would have to be dismantled. It either could not be reversed or none of the employees present knew how to do that. The crew erupted in a scramble for tools and equipment.

Meanwhile, day-shift medic Gayle Groff, the only one small enough to do it, climbed up to the cubicle and crammed herself inside. Watts was chilling rapidly. His panic initially kept his blood pressure and level of consciousness peaked up. But now the bleeding was taking its toll. His mental status was starting to dull and his speech slur. His respirations were faster but more shallow. He had EMS training as a member of the ski patrol and understood what his symptoms meant. He discussed these with Gayle. Attempting to slow this decline, Gayle set up to start an IV. He was so cold, so shocky, and his blood pressure so low, his veins were flat. She inserted the needle into his arm, hoping to strike a vein but didn't hit it. She hoped the oxygen would slow his slide into shock. She positioned herself under him as much as she could to insulate him from the ice. Blankets over him were welcomed.

Gayle got a briefing from Lt-5 and didn't like the news. This was not going to be quick. The medics did something that hadn't been done before: Called for the doctor on shift—Dr. Osborn—to come to the scene. They believed that maybe the only way to save the boy's life would be to amputate his leg before he bled to death. Even the last of all options—a tourniquet—was impossible to apply. In fact, amputating was easier to say than to accomplish for the same reason: getting to the leg, then cinching off the blood supply above that. His butt and lower abdomen were all that was accessible above the auger.

Dr. Osborn was equally unsuccessful in starting an IV while the truck company was feverishly attacking the machine. They finally dismantled enough of the equipment around the auger to finally reach its shaft. The cannery's oxygen/acetylene cutting torch was wheeled up to the site and their equipment maintenance foreman lit if off. Sparks flew as onlookers froze in place, watching. Finally the auger was off and John was dragged out of the icehouse, still conscious.

There was nothing left of his leg. Thanks to the cold, his bleeding was slowed enough to prevent him from bleeding to death. Today, he is still an avid skier, even on a prosthetic. In fact, he is an instructor in the use of prosthetic ski attachments.

DROWNING AND THE SMALL TOWN CURSE

There is one primary advantage to providing emergency services in a small town: You know everybody. There is one primary disadvantage to providing emergency services in a small town: You know everybody.

It was raining the night a carload of high-schoolers, driven by a senior girl, was heading into town at a high rate of speed. About four miles out, she lost control on a curve near Mavis Island, shot across the road's shoulder, became airborne, flew over a float plane that was moored in the lake, and landed right-side up in the water. No one was significantly injured by the impact, and the kids climbed out of the windows. They made their way toward the bank. The water was only chin deep. The first one to the bank looked back to see everyone was out including David Bruce who was the last to exit the car.

We pulled up with the ambulance, the rescue truck, and an engine. I was in Squad 6 (Chevy Suburban). I had pulled my diving suit on before I took off, and had someone zip me up as I climbed out of the Suburban. Then we heard that David Bruce didn't make it to the beach. The medics were examining the kids and wrapping blankets around them as I clipped on my belt of lead weights and jumped into the water. I yelled back to have someone ready my tank and mask in case I didn't find him right away. It was a short distance to search. I plunged down and felt around the bottom, grabbed a breath of air and plunged again.

A feeling of dread swept over me because I hadn't located him in such a small search area. I had just about decided to don my tank and mask when I remembered that I hadn't groped under the vehicle. I went down the edge of the vehicle and reached underneath and felt his head. I pulled him out and up by his hair and headed for the beach. Other members jumped in to help me.

On the beach, we began CPR, stopping periodically to ascend the steep bank until we reached the road. The ambulance was gone. It was used to take the other kids to the hospital. We loaded David into the aisle of the old, beat-up rescue truck and headed for town. The space on the floor was so narrow, I positioned myself at David's head to perform mouth-to-mouth, and another member straddled him to do chest compression. We had an oxygen unit there, but it was not equipped to do forced ventilations. I strapped the nasal canula to myself and inhaled oxygen through my nose then blew air into David. After a few ventilations, I realized that the most I was getting into me was about 4 liters per minute. I was absorbing some of that, and what oxygen I was blowing into him was not what he needed. So I attached a larger oxygen tube to the tank and stuck the open end of the tube in my mouth so I could inhale 12 – 15 liters per minute in, which was 100 percent oxygen. After absorbing about 4 percent, I was blowing 96 percent oxygen into the patient.

In those days, it was standard protocol to start an IV line for injections of Sodium Bicarb on anyone in cardiac arrest. But all the advance care equipment was in the ambulance. We had no suction unit, no Esophageal Obturator tube, nor Endotracheal tubes. We just prayed he wouldn't start vomiting.

At the hospital, the staff met us with a gurney, and we wheeled him into the ER, still doing CPR. But, busy as we were, I noticed that the lobby was filling up with people. Parents and schoolmates milled around and looked alarmed when they saw that CPR was being done. In the ER, we went through the entire routine, but the patient's heart was Asystole (flatlined). Defibrillating was not an option. That would have been okay, and expected, if his body temperature were really low, but it wasn't. So Asystole was an ominous sign. Finally, the doc called off the treatment.

We headed for the door, and the doc peered out and down the hall to his left and saw the crowd of people there, including the boy's parents. I looked over his shoulder and saw their expressions were saying things like, "Kids, they'll be the death of us," whimsical embarrassment, "What are ya gonna do?" Obviously, they had come in after we rolled through the building doing CPR. The doc exhaled, steeled himself, and walked out. I froze at the door and didn't want to walk out, especially still wearing my dive suit. People always looked at me like they were looking for answers. I don't have words for these situations. I heard David's mother's voice, "He died!"

I peeked out and saw my chance to escape by turning right and out a secondary exit, which put me in a walkway between two buildings. I walked to the front of the building, which was jammed with vehicles and, avoiding people, made it to the rescue truck and ducked inside and sat in the dark.

I suppose the crew inside was waiting for me to come by and officially excuse them from the operation. I couldn't do it. I couldn't go back inside and, frankly, resented being expected to be tougher than everybody else.

Too many times over the years I avoided meeting and returning eye contact with people after incidents like this. One time I was walking into the post office and the elderly wife of a man we unsuccessfully tried to revive at their house was walking out. She saw me, shook, put her hand up to her mouth, and started bawling. I went back to my car and drove off.

THE SHOOTOUT AT THE EPISCOPAL CHURCH

When I first got to Cordova as a civilian, I was a garbage collector. Our most dreaded stop was at the state built and managed low-cost housing complex. It was named "Eyak Manor." It consisted of four, two-story buildings with four apartments in each. To be a resident in one of these sixteen 2-story apartments, applicants had to qualify by showing their level of poverty. The four buildings sat in a semi-circle around the parking area. In the middle of this parking area was the "garbage shed." It was like a gazebo: a sturdy roof, a concrete floor, and chicken-wire walls. Inside, garbage cans lined up along three sides, each one was fitted with a lid. For the first few months the complex was tidy and clean. Parking in front of the buildings was orderly.

As time wore on, the place took on a different look. The parking spaces became a storage lot for vehicles in disrepair. They would be up on blocks with tires or wheels removed; hoods were propped up and engine parts scattered on the ground. In these cases, spare vehicles were parked there, and parking spaces became scarce. The vehicles in disrepair would sit there, untouched, for months at a time. The porches of the buildings became collecting spots for old couches, wrecked bicycles, boxes, old TVs…you name it. But worst of all, my partner and I would hate walking into the garbage shed. People had stopped walking in and placing their garbage in a can and replacing the lid. They would open the door, fling the garbage into the building, and close the door. Even those who wanted to use the cans would refuse to wade into the building to find a can because of the mess and the rats. The place would be heaped waist-deep in garbage, and upon hearing us enter, the rats would start moving around and the whole place rustled. After a couple of fires, some assaults, and one murder, the state decided to set some standards.

They announced that residents were allowed one vehicle per family—or unit—and it better be usable. There was an outcry. The state held firm. It was the state's property, and if someone wanted to reside there, they must follow the rules.

"What am I supposed to do with my truck? I'm waiting for parts so I can repair it," one could almost hear.

"Put it somewhere else. Don't try to make your problem my problem," was the type of answer the state gave.

The vehicles did not move, so the state gave a deadline, after which the state would have them towed away and the cost billed to the owner. Eventually they were gone.

The state also told residents to get rid of everything on the porches. Another outcry from residents who complained they didn't know what to do with all that stuff. "Just get it off the porches. And, by the way, don't drag a bunch of clutter into the apartments, either." Eventually that stuff was gone. The state tore down the garbage shed and placed one garbage can in front of each unit. That way, each person was accountable for his space. Of course, back then, we picked up garbage twice a week.

That was thirty years ago. The place is clean, neat, and tidy to this day. The state continually either repaints or applies new siding. They paved the parking lot, put in sidewalks and so forth.

I guess the point of the story is that standards of behavior must be clearly explained, then compliance insisted upon. This oppressive authoritarianism is oftentimes unpleasant, and can easily border on being stupid and anal. On the other hand, it can be similar to the way parents raise decent kids. Kids—or adults who have a childlike inability to establish their own standards—must be either "raised" by authority figures or we all must suffer from their behavior.

I don't think one can fairly place the blame of the Eyak Manor problem on the "poverty" level of the residents, either. I chuckled when I read about the time actor Richard Burton was going to sue a magazine for saying that he had been raised in the slums of Wales. He was from a small, impoverished coal mining Welsh town. A friend of his calmed him down by having Burton affirm that no, electricity was not available around the clock, just during the times it was needed most; and no, water was not plumbed directly into each building—residents had to go to the city well, in the center of town, and tote their water home. This, the friend explained to Burton, is poverty by our standards. Burton retorted that there is a difference between living a simple, modest life, and living in the slums. Slums are dirty and dangerous. His town was tidy and meticulous. Peer pressure kept it that way. "Quaint" was a better term for the town. Slums—it's not a lack of affluence; it's a lack of standards.

* * * * * * * * * *

Mike Donovan's (not his real name) licentious girlfriend took a liking to her brother's friend, Gary Kavisto. Gary thought that in this civilized world, women should be able to change their minds about who they wanted to be with, and even though Donovan made himself quite clear that he would not tolerate two-timing and viewed it as a personal affront, Gary could not comprehend it.

Gary's crowd of chain-smoking, boozing slugs had no standards of behavior at all. And Gary should have known that Donovan had the world's shortest fuse since he had detonated a number of times in the past. For example, as a mechanic, having completed repairs on a customer's truck, he presented the bill to the customer. The hapless customer questioned the amount owed, and things suddenly went black on him. When he regained consciousness, Donovan had left. The customer did not suffer a broken jaw from the punch, but he had been knocked out so proficiently he hadn't even seen the punch coming.

Well, Donovan had made it clear to Kavisto and to Kavisto's friends that Gary should look elsewhere for a girlfriend.

The evening of the shooting, Donovan walked into apartment #3 of Eyak Manor to see Kavisto laying on the couch and several friends hanging out in the living room. Donovan pulled a 9mm pistol and opened fire. Gary held up his hands as though to block the bullets. Donovan shot him five times, including once in the head, turned and walked back out.

Gary Davidson and I responded to the page for the ambulance. Enroute, we heard a police officer on scene radio "expedite" to us. When we entered the apartment, I saw nothing but our patient. I learned there had been others in the room, but the Titanic could have been moored in the living room and I wouldn't have noticed it. The officer said Kavisto had faint respirations. "I wet my lips and put them next to his nose and I could feel his breath." Of course, those aren't effective respirations, but it indicated that his heart was still beating. We ventilated him with a bag-valve-mask. His heart beat was strong.

We lifted him onto the gurney and blood ran out of his head like a faucet. In those days, we were not IV qualified, so I stuffed compresses into the exit hole in the back of his head as we wheeled the gurney out to the ambulance. The other bullet holes weren't bleeding that profusely. We loaded him so fast and Gary sped off so quickly, it was only then that I noticed we hadn't latched the doors well enough and they sprung open and flung wide with each corner we made. In that moving vehicle, I struggled with the saturated packing in his head wound and couldn't stop the bleeding. I was slipping and sliding in the blood while squeezing the bag-valve-mask.

In the emergency room, Dr. Larry Ermold automatically went through his routine but kept shaking his head no, knowing that the wounds were mortal. He did remark that Kavisto would make a great heart donor, because his heart was beating like an athlete although he must be brain-dead. Then Ermold stopped, looked up and asked, "Did they get the shooter?"

"Not that I know of," I replied.

"Jesus, we are thirty steps from the front door. The ambulance is right out front and the lights are on in here." (The frosted ER window was on the front of the building right next to the parked ambulance.) "How do you know a nut like that isn't going to walk right in here to finish the job? How about getting a cop down here."

When I phoned dispatch and asked if we could get a cop down here, she put me on hold, then informed me that they were all out looking for the shooter. So

I called two new fire department members that I expected would be very level-headed and mature, to stand guard at the hospital: Reverends Robert Varnam and Richard Harding. They had come from California to run the Baptist church here. They still had that urban flare in clothing and wore overcoats. I asked if they owned guns, and they said they had a shotgun and a pistol. I explained the situation and they agreed to stand guard in the lobby of the hospital. When they arrived, I told them that I didn't know who the shooter was, so just turn out the lights in the lobby and wait there. They each sat down, guns in their laps, and waited while I returned to the emergency room.

The dispatcher never contacted me to tell me she had contacted Wendell Jones to come down and guard the hospital. Wendell was a Fish and Wildlife protection officer under the State Department of Public Safety. And as such, he was a commissioned state trooper. He quickly threw on a pair of blue jeans and a jacket and drove to the hospital. With his pistol in hand, he walked into the darkened lobby. Varnam and Harding sprung up and pointed their weapons at him. In reaction, Wendell pointed his gun at them. Everyone froze.

"Who are you?" Wendell blurted out.

"Who are you?" Varnam and Harding replied.

The three of them stood frozen in place waiting. Eventually they got it worked out and lowered their weapons. Harding stuck his head into the emergency room, explained that there was a cop in the lobby now and that he and Varnam were going home, and then added, "When this is over, we gotta talk."

Eventually, the doctor declared Kavisto dead and we left to go hose out the ambulance. We could hardly stand up in there because the floor was so slippery with blood.

When the phone rang the next morning, I could hardly move to answer it. I had thrown my back out lifting the patient last night and hadn't noticed. The dispatcher phoned to say that the police officers had Donovan cornered in the basement of the Episcopal church (right next door to the hospital), and since they were thinking of lobbing tear gas in, they wanted a couple of airpacks standing by.

My back was so painfully stiff, I could barely move and got my wife to agree to drive me to the station. By the time I'd placed a couple of airpacks in my vehicle, I could move a little better. At the church I laid out the BAs and joined Stan Shafer behind a patrol car parked in the street at the front entrance of the church (A side). Stan had been called out to stand by with the medical gear. I was waiting for a couple of the cops to come over for a briefing on how to wear a BA. Then Police Chief Bagron asked if the fire department could establish a traffic-free perimeter (vehicle and pedestrian) around the area. Shortly after I asked dispatch to tone out the department, I was radioing company officers and assigning them intersections to block off.

From where Shafer and I were sitting, our backs against the tires of a patrol car, we looked up at the hospital a few feet away and saw a crowd of hospital workers looking down from the second-story window of the breakroom.

Before the fire department took up positions at intersections, Stan and I saw the hood of a car creeping up from the ally on C side of the hospital. In it were Roy "Flapping Eagle" Anderson and several other men, all armed with rifles. The cops, staring intently at the church, did not notice. I waved to them and said, "Flap, goddamnit, go home." He didn't even look at me. Groff walked over to the car, making sure his coat was open so his gun would show and told Flap that the cops didn't need his help. Flap left.

The area was a bit tense. Between local PD and State Public Safety, the church was completely surrounded. One officer was with Shafer and me and leaning across the hood of the car, his gun pointed up in the air. Bagron, armed with a shotgun, was against the door frame of the front doors. Lt. Tom Page was with him. To enter the church basement, one walked into the church's main doors on A side, walked through to the rear (C side) and down a narrow stairway. One would walk a few steps down to a landing, turn, and take a few more steps to be in the basement which was the church's office. The basement did not extend the entire length of the building. Adding a little natural light to the basement were several windows on C side which were so high from within, that the only one accessible was the one at the landing. Outside of C Side was where the rest of the cops were, standing behind trees. There was no access or windows to the basement from A, B, or D sides. During the search through town, Tom Page had quietly walked into the church, crept through to the back stairway and looked down the stairs. He couldn't see Donovan, but he could hear him making odd "animalistic sounds." Donovan wouldn't show himself as Page instructed, so Page summoned the rest of PD who surrounded the place. And there everyone stayed well into mid-morning.

A couple of the cops came over and I showed them how to don the airpacks. They left them on the ground and returned to their positions. Page, carrying a tear gas canister, crept into the church again. Shortly after, I heard a rustling noise and Page came running full-speed back out the front door yelling, "Here he comes." I heard a volley of shots from behind the church and Bagron stepped into the doorway and shot several times toward the rear of the church. Shouts from the cops on C side signaled that they had Donovan and it was over. He had fired numerous shots out the window from the landing, hoisted himself out the window for a final shoot-out, but dropped his gun as he went through the window and the cops were on him before he could retrieve it. He was unscathed except for some shotgun pellets in his ass.

Bagron said that since the threat had been removed, it would be safe for Shafer and me to go get the canister. Sometimes tear gas canisters start fires in buildings. As Shafer and I entered the building, wearing the BAs, I summoned an engine to stand by and also to set up a fan. Bagron was right; the canister had started the carpet on fire, but we stomped it out. My back felt better after exercising it a bit.

Working with Cops
and Miscellaneous Gun Calls

Some firefighters I've met found it odd that we (CVFD) worked with local cops so often. It never occurred to me that it might be seen as unusual in other places.

There was another time when we set up a perimeter for the cops on a barricaded gunman, when Officer Dan—sparkling with bravado—stood in the middle of the street to talk the guy out. The guy started shooting at him, and Officer Dan, dodging bullets, leaped around in the street like an Irish dance troupe—funniest goddamned thing I ever saw.

Speaking of Officer Dan reminds me of the time he hijacked a SAR call from us because he thought those things were exciting, I guess. Dispatch was notified that 19-year-old Cynthia Cain had been in a climbing party ascending the west side of Mount Eccles, next to town, when she fell about 900 feet. Rather than paging out the fire department, Officer Dan contacted Chisum Flying Service for a helicopter. He grabbed hold of fire department medic, Janice Farnes, some equipment, and choppered up to Mount Eccles. By this time, we heard about it and gathered in the parking lot in front of the fire station and watched. The chopper could only off-load Officer Dan, Farnes, and the equipment by touching the tip of one of the skids to a rock, and the two rescuers stepped out and scrambled onto the rocks. From our distant point of observation, we saw something yellow fall out of the chopper, fall and flap several hundred feet straight down. We all held our breaths fearing it was Farnes or Officer Dan falling. It was only a blanket.

Since the U.S. Army 172nd Brigade from Anchorage's Fort Richardson was in the area training near Sheridan Glacier, they sent a chopper capable of sling-loading Cain out after she had been stabilized by Farns. What was ridiculous about Officer Dan taking that SAR call was that he had nothing to contribute. He could not perform patient care, he was not a climber, nor did he have enough gear to lash the party to the mountainside if the weather had turned or if airlift out would have been delayed.

Later, back in the building, one of the firefighters said to him, "Next time there's a barricaded gunman, we're going to get there first and shoot the place full of holes." Everyone snickered. "But, first," the firefighter continued, "you have to teach us that little dance you do when being shot at." The whole place erupted.

A middle-aged local guy, well known around town, was pretty stressed-out over some legal issue, pulled a .45 and took the magistrate hostage in the court house. The cops found it hard to surround the place with only four guys. So, we set up a perimeter and were given the added duty of watching windows and doors with binoculars and report to the police captain any movement we saw (chief was out of town). We didn't see any.

What was comical about the situation was that when the state trooper SWAT team flew into town and assembled at city hall for a tactical briefing, I started laughing. It was the middle of winter, the whole place was covered in snow, and these guys all wore green/brown camouflage. But the funniest part was that they painted their faces with green and brown greasepaint...so's they'd blend real well with the goddamn snow. Finally the gunman just gave up.

Police Chief Yerrick once asked us to create a diversion so his officers could charge into a hotel room and grab couple of guys. Hell, we can do diversions.

The bad guys were in a room in the Alaskan Hotel with a window on the north side overlooking Browning Street. The cops snuck into the building and crept up the stairway and down the hall. I'm just making this up. I assume they crept down the hallway—I don't know because I was outside in the fire engine.

I had two other guys with me that night when I drove down Browning, the fire engine's lights flashing, and came to a halt right under the windows of the Alaskan Hotel. I gave several long blasts on the air horn, had the two guys stretch out a couple lengths of fire hose and drag it down an alley. Everybody was watching, including the two bad guys. They were standing at the window of their room when the cops went in and got them.

* * * * * * * * * *

When Kevin Clayton became police chief, practical jokes abounded. He would put fingerprint ink on Captain Ed Weibl's phone ear piece, then call him. "Ed, I need to talk to you, can you come into the office?" Ed would walk into Clayton's office. Clayton would start to talk to him, stop, stare, and ask, "What the hell have you got on your ear?" Weibl would walk into the bathroom and see a circle of black around his ear. Clayton just shook his head and mutter, "That goddamn Jason." (Whetsell—my son).

Kevin would take two raw eggs out of the refrigerator, go into the apparatus room and put them into Jason's turnout boots. Ready for the next call.

One of the guys bought a vial or Morning Breeze at a novelty shop in An-chorage. One drop on a piece of paper smelled like the raunchiest fart ever. So once, when Chief Clayton was interviewing someone in his office, our guy put two drops of Morning Breeze on a sheet of paper and stealthily slid it under the door to Clayton's office. Several of us sat in my office and snickered. Later that day, Clayton remarked how his interviewee kept farting during their meeting.

More recently, one of the cops "modified" the keys of the computer keyboard the officers use for writing reports. He pulled all the keys off of the little metal stubs and put them back on out of order. So, when a "hunt-and-peck" typist pecks a "b" (for example), a "k" appears on the screen. It didn't bother the department's only touch-typist, but the rest were livid.

* * * * * * * * * *

Here is a problem familiar to many volunteer fire departments: Aside from the rift between volunteer and career firefighters (much less pronounced now than in the past) there oftentimes exists a condescending attitude of law enforcement officers toward volunteer firefighters and medics. It was not unusual, when meeting a new state trooper assigned to Cordova, to see an expression of disdain on his face. The same was true, occasionally, when a new police chief was hired. It usually took a period of time for them to respect our professionalism and skill level. But occasionally, we gave them good reasons for hesitating to use us in their operations. There is an inherent difference between our two groups.

Generally, cops look good. Their mannerisms in public are professional and carefully rehearsed. Firefighters are more boisterous and undisciplined, and volunteer firefighters are more likely to blunder during calls that are not routine calls. We had a couple of blunders that really made us look bad.

When the ambulance crew responded to a gunshot victim on the Coast Guard Cutter Sweetbrier, they were led to the berthing area. There, the seaman, who had committed suicide by shooting himself in the head, dropped the gun next to himself on the bunk. The cops were just arriving when one of the medics picked the gun up and set it on a nearby bunk. It was obvious that the head wound was fatal and that no treatment would be administered. The medic just moved the gun impulsively and unnecessarily. The cops were pissed because it complicated their investigation. They had a right to be pissed. Even though they never said it, I could almost hear them thinking, "Volunteers have no self-control."

Afterwards, we talked about our actions at crimes scenes. We even wrote down several SOPs. I thought that would be enough. But when the troopers asked us to search Hawkins Island for a guy who left town acting despondent and suicidal, he told our searchers that if they discover a body, do not disturb the scene. Just call the troopers and wait there. But as soon as one of our veteran members, a cool-headed, never-blundering officer, found the body, he picked up the rifle and cranked the bolt back several times to empty the cartridges. It was something he had done a hundred times in the Forest Service, flying in and out of remote locations. He did it out of impulse. As soon as he did it, he realized how badly he just screwed up. Now, the state troopers were pissed off.

I have always been very sensitive to any demeaning remarks about volunteer firefighters and especially my crew. But when the cops are right....they're right. There was no way I could defend those actions. I just apologized and then reiterated our SOPs to the crew. On the other hand, we had worked with local PD and AST so often in the past, they had confidence in our professionalism. Besides, I had witnessed them screw up enough that they couldn't say too much about us. Like the night a good friend, Frank, was shot.

* * * * * * * * * *

It was a rowdy weekend night when I heard the ambulance crew get toned out. I wasn't on call but I went because it was a shooting on the fishing boat of a friend of mine. When I arrived, the ambulance was already at the harbor and a couple of cop cars were also there. Down on the floats, a crowd of fishermen gathered and were milling around. Frank was a popular fisherman and was inside the boat with a bullet in his head. The crowd was upset, many of them had been drinking. One cop had a young female by the arm and was talking to her, but she was belligerent and drunk. It was Frank's girlfriend. The other cop, a female, new to law enforcement, had a pistol in her hand, and as I walked by her she turned and whispered to me, "I don't know how to get the clip out of this thing. Do you know how?" She was never cut out to be a cop, eventually realized that and resigned, but at this moment she was stuck and didn't know what to do. She was supposed to be handling the crowd that was almost snarling at Frank's girlfriend, who was bellowing, "He shot himself. He just shot himself."

I walked right past the officer, stepped into the cabin of the boat and saw Joanie Behrends, our EMS team leader, looking flustered. She looked glad to see me. Walking forward through the galley, one takes a step down into the berthing area in the bow. Frank was on his knees at the step, facing Joanie, who squatted down talking to him. He stared straight down at the deck, wouldn't look up, was shirtless and his thick, beefy shoulders were slick with sweat and blood. His arms stretched out at a 45 degree angle from his body as he clutched the door frame in front of him. Joanie was trying to convince him to accept treatment. He was frozen in place and every muscle tensed, probably trying to brace himself against the pain, I don't know. But he wouldn't speak or budge.

Joanie pointed her finger at her own head, indicating that the bullet hole was right behind his right ear. "Let's go, Frank," I directed. Nothing. "Come on," I said, putting my hand on his slippery shoulder, and giving a little pull. My hand slipped off.

I nodded to the rest of the crew to lay the backboard on the deck right behind me. I put both my hands on his shoulders far enough back that my fingertips curled into his arm pits, and reefed back. My hands slipped off, and I fell backwards into the galley and against the table, knocking things over. I got so pissed off, I grabbed him again around his chest and lunged backwards lifting him off the deck and we both crashed flat on the backboard. I slid out from under him and told the crew to strap him down, face-down on the board. The only sounds he made as we carried him out on deck and off the boat were moans and growls. The crowd followed us up to the ambulance and kept making remarks to and about Frank's girlfriend who was ahead of us and being taken to the police car.

The female officer kept the pistol concealed and looked uneasy and confused about what, if anything, she should do.

By the way, Frank survived like a man who had suffered a massive stroke. He could never remember what happened that night on the boat, and the cops never could prove anything one way or the other.

* * * * * * * * * *

On that incident, I was about half in-compliance with my own standards about not responding to a scene where you were not required to be. But, I could easily justify my responding to a call on the hillside behind Morpac Cannery. The ambulance was toned out for a shooting, and they, in return, had the truck company toned out to help them. I heard that the police were on scene, so I went. If two separate agencies are going to be responding to the same incident, oftentimes a command post comprised of the senior officer from each agency can keep things orderly between the two. In this case, it was really unnecessary, but my curiosity was killing me.

We get a lot of transient unemployables in Cordova during the fishing season. They hope to get jobs on fishing boats or in the canneries. They barely have enough money to get to town, and certainly not enough to leave. They often crawl up into the woods out of sight, making ramshackle shelters out of pallets or plastic tarps. Sometimes you can find clusters of them, dripping miserably in the rain, the occupants sooty from squatting close to small campfires. A lot of these folks are decent kids off on an adventure before returning to cookie-cutter suburbia and a lifetime of wishing they were Indiana Jones. But occasionally, you will run into drunks and screwballs like the two involved in this incident.

The young drunk, barely in touch with reality, had stumbled up the winding path to the "camp" that night. Kerosene lanterns lit the place up, displaying an old mattress, seat cushions, and the clutter of fire wood, Visqueen tents and such. He lay face down on the mattress and lapsed into sleep. The screwball, not even remotely in touch with reality, returned to the camp, told the drunk to get off of his mattress. The drunk was out cold. The screwball kicked the drunk a couple of times to roust him but got no reaction. Now, the screwball got mad, aimed a shotgun at the drunk's back and fired. Screwball was a few yards back so the pellets hit the drunk in the butt and groin. That woke him up.

97

I don't know who witnessed it or who called 911, but the cops had the screw-ball in the back of their car when I pulled up. He was making faces at me like a little kid does: contorting his face and sticking his tongue out. Creepy.

The truck company provided the manpower for the ambulance crew and packed the patient down the trail out of the woods. Some stayed at the camp a few minutes longer to pick up the paper wrappings and such left by the medics. "What's this?" one of the firefighters asked, picking up a small object out of the dirt. "Christ!" he shouted, flinging the object away in revulsion. It was the drunk's testicles.

They told me afterwards that the scene up there with the kerosene lamps and the shadows they cast all around reminded them of eerie Pink Floyd music.

I learned that later the drunk was outfitted with a colostomy bag and moved to Anchorage. I was in Anchorage a few months later and visited an old friend, Anchorage Fire Department paramedic (now chief of their EMS), Frank Nolan. He said, "I met a fellow Cordovan a few nights ago." I asked who it was but he could not remember the guy's name. He recounted, "We got toned out for a public assist just a couple of blocks away. There was a guy standing on the sidewalk, covered with tattoos, so drunk he was weaving back and forth. He said he was from Cordova. We sat him down in the ambulance and the stink just mushroomed throughout the rig. I looked him over and he was covered with shit from head to toe. He had been drinking for about three days and neglected to wear his colostomy bag. It took us hours to get that smell out of the ambulance."

I replied, "If you know where he is now, I could mail him his testicles."

AIR MEDEVACS

I've heard a lot of EMS services refer to Air Medevacs as Search and Rescue operations. They really aren't the same. An Air Medevac is just a long ambulance run without an ambulance. You use an airplane instead. Of course, that alone can make it interesting. Mostly, there is only one medic on board for this very time-consuming operation.

A department medic traveling to a fishing boat or a Native village or some island out in the sound by small airplane or leased helicopter may burn up as little as an hour or several hours. But what makes these operations a challenge—besides just the threat of bad weather—is that the medic has no access to any other medical equipment other than what he/she packs and loads into the aircraft. Compounding this, for over 20 years, there was no means of communicating between the aircraft and the fire department or hospital. If trouble was anticipated, another medic was placed at the office building of the air taxi service to listen to the airborne medic on the radio, then relay information to the department, the hospital, the state troopers, or whomever. But, for the most part, the airborne medic was on his/her own.

Being on your own can be traumatic if you doubt your own skills. I was on call when summoned to a boat out in the Gulf of Alaska for a 19-year-old male who suffered a sharp, debilitating pain in his chest, not a result of trauma. It sounded like a classic (although rare) spontaneous pnuemothorax. The treatment for this "blown and collapsing lung" is to insert a large-bore IV needle into the intercostals space (between the ribs) and into the sac surrounding the lung (pleural sac) to expel the air. In the leased helicopter, flying through the crappy weather about 150 feet above the water—because of the fog—the pilot thought I looked sick because of the flight. He didn't know that I was in near panic at the prospect of trying to do this procedure. Luckily for the patient, and me, he only had pleurisy.

I almost gave up making these runs after I flew out to a bay in the sound to pick another young man who "was sick." We landed the float plane and taxied up to the boat where he was crewing. He was standing on deck looking perfectly fine. The captain tossed me the guy's bag of clothes, marched him over to the plane and boosted him inside. I was trying to ask questions and take notes. I said to the captain, "You radioed that he was sick. What's wrong with him?"

"He's sick. Get 'im outa here," he replied, and handed me several small bottles of pills. "Here's his medicine." He stomped off.

The patient sat in the seat behind me, and wouldn't talk to me. He just stared right through me. I looked at the medicine bottles, but the names of the medications meant nothing to me. We all lapsed into silence for the rest of the flight.

At the hospital, the doctor read the medications and muttered. "Oh, my God."

Apparently, I had been cruising at 5,000 feet with a psycho sitting behind me.

* * * * * * * * * *

Sometimes, two medics go if the situation sounds serious enough. Vicki Hall and I were bringing a patient in, and the plane was so crammed that all the seats had to be removed but the pilots. We were on our knees, tending to the stretchered patient when I noticed the pilot ruffling his map. He was new in the area, and apparently wasn't sure where he was. I straightened up and leaned toward him. That's when I noticed how dark it was getting outside. I had flown this route many times and said, "Just keep going straight." At about that time, he saw town lights come into his view on the left. "There's Cordova," he said relieved. And started banking left.

"No. That's the village of Tatitlek. Keep going straight."

"No. That has to be Cordova."

"Trust me. That's Tatitlek."

"Are you sure?" He asked.

I leaned forward, lowered my voice, and cranked up my "this-is-really-important" facial expression. "It's getting dark really fast and we want to make it to the Cordova air strip before it gets too dark to see it. We still have about forty minutes of flying time yet."

He believed me.

I added, "As a matter of fact, I'd like to make it over those peaks up ahead before it gets any darker."

I spent the rest of the time ignoring the patient and directing the pilot back. We could barely see the landing strip as we approached, but we made it. I was a wreck and Vicki was fuming. The pilot got fired.

* * * * * * * * * *

Speaking of Tatitlek: my first medevac flight to pick up a patient there, our float plane taxied up to the boat dock and tied off. I heard light-hearted chatting and giggling and looked up to see my patient (with a splinted broken leg) being brought down the path in a wheelbarrow.

100

Also, a few years later we got a call from the hospital who had received a call from the one phone in the village in Tatitlek. An OB Patient was starting to go into labor and the hospital asked if we could go get her. It was nearly midnight and the only thing worse than the torrential downpour was the screaming wind bending trees and shaking light poles. There were no pilots in town qualified to fly at night or crazy enough to take those little planes out in that weather.

Perhaps we could locate a bigger plane and a pilot qualified for instrument (nighttime) flying. The cops let us into several of their offices to use all the available phones. About half a dozen of us went flipping through pages of phone books covering areas in southcentral Alaska and finally scored with Wilbur's Airlines in Anchorage. Wilbur himself would pilot the plane. He had a bigger plane but could still land on the short runway. I guess he had good brakes.

The weather was less extreme in Tatitlek than it was in Cordova. The village had a short gravel runway but no lights. Wilbur knew the guys from the village would be able to figure something out. He was right. By the time he arrived in Tatitlek at about 1:30 in the morning, the landing strip was fully lit up by using the headlights of 4-wheelers and dozens of tin coffee cans filled with oil and torched off. Everything after that, went well.

* * * * * * * * * *

Everybody in rural Alaska flies, whether they like it or not. There are small planes everywhere.

One nice morning the dispatcher informed me that a small single-engine plane was flying over the Prince William Sound and there was a fire in the cockpit. The pilot had called in a "Mayday," saying he was alone in the plane and was unable to see without sticking his head out of the window. Then, the radio went dead. The dispatcher went on to say that some boats in the area could see the plane. We were assuming that the plane—on wheels, not floats—would try to make it to Cordova, and toned out the department.

It was a clear day and we had all the apparatus out on the apron and stared at the ridge of mountains on Hawkins and Hinchinbrook Islands for first sight of the plane flying in from the sound. It then occurred to me that if the pilot is having trouble seeing or controlling the plane, he might show up anywhere. So I told the guys to drive a couple of apparatus north and space themselves out and prepare to radio us when they first see the aircraft. I sent one pumper to the city airstrip at the lake in case the plane makes it there. Everyone was to remain on their rigs so they can relocate without delay.

I called Valdez Fire Department's Chief Tom McAlister on the regular phone and told him that a boater said the plane was at Knowles Head flying directly east— toward neither Cordova nor Valdez. But if the plane banks right, we've got him. If he banks left, Valdez gets him. As soon as I stepped outside to watch the sky with the rest of the crew, the dispatcher would holler out the window that Tom was on the phone again. Inside, I learned that Tom was asking if we'd heard anything new from the boat. Finally, we abandoned the regular phone and went to the wall-mounted NAWAS phone (National Attack Warning Alert System). It came with a speaker and we chatted on that a bit. The only problem with that was that every town in Alaska was hearing our conversation.

Finally, the fishing boat radioed us saying that the plane was banking left and heading north. I told Tom over the NAWAS, "It's all yours, Tom. Good luck." I chuckled.

"Thanks a heap," he replied.

I learned later that the pilot made a flawless landing, and in fact, the smoke dissipated enough for him to see without sticking his head out of the window. For the rest of that day and for the entire next day, Tom was getting phone calls from across the state, wanting to know how it turned out. We decided if we were to ever use that phone again, we would make one last transmission explaining how it turned out.

EMS CAPTAIN STAN SHAFER IN FLOTATION SUIT, WITH BEAR GUN, PREPPED FOR SAR FLIGHT ABOARD CHISUM'S LEASED HELICOPTER

Photo: Cordova Times

Dive/Rescue Operations

I didn't start diving to perform fire department rescues. I started because a close friend and fellow city employee, Don Endicott, led me to believe there was money in it.

He had started when he was a kid by playing around with a small compressor, pumping air into a metal bucket which he stuck over his head. He found he could walk around under the water providing he didn't bend over. Of course, he couldn't see anything. So he got an old WWII gas mask and found that worked, too. As an adult, he just bought a bunch of gear, a compressor for filling tanks, and began to get jobs as a salvage diver. He never took any classes, but he read a couple of books about diving. When he talked me into doing it, I just bought the stuff but I never read any of the books. Years later, when I took a dive/rescue course, I realized how insane I'd been.

We did some salvage diving: We raised a gillnetter sunk in the flats, used primer cord explosive to shoot the brass propellers off of St. Elias floating cannery. We heard that Japanese like eating sea cucumbers (an animal, not a vegetable) so we gathered a bunch of them, but the sight of them horrified the Japanese we showed them to, because—compared to the ones off the coast of Japan—these were grossly ugly. Having read that razor clams retract into their holes in the sand at low tide, but when covered with water, actually stick their necks out to feed when the tide covered the sand, I figured I could swim along and grab them and harvest a bunch much easier under water. I went alone and anchored my skiff, dove under the water and like a moron, swam with the tide searching vainly for clam necks. I didn't see any, and when I surfaced I could barely see my skiff, it was so far away. I had to swim against the tide for what seemed like hours to get back. I considered dropping my tank and belt to make it easier, but I didn't want to buy more gear, and swore that unless I was being hopelessly swept out to the ocean, I would not do that. You should never dive alone. Especially if you're an idiot.

We intended to dive on the Amaria, so we went in search of it. The Amaria was the flagship of the U.S. Lighthouse Service. The Lighthouse Service was later incorporated into the U.S. Coast Guard. But it was its own entity in 1911 when the Amaria, carrying the newly manufactured lighthouse lens, made in France, to the recently completed St. Elias lighthouse on the southern tip of Kayak Island in the gulf.

The Amaria was cruising from the sound to the gulf along the western side of Hinchinbrook Island when it hit a rock. Taking on water, it swung around and tried to make it to Port Etches on Hinchinbrook, but didn't quite make it. The

103

skipper beached her about half a mile south of the Etches entrance and everyone abandoned ship.

It took another two years for another lens to be made and to get that lighthouse in operation. We went looking but had a lot of trouble because the ground swells coming in from the gulf made it tough diving. So we went to Port Etches and searched for an older ship. Endicott was told by a retired diver about a three-masted trading bark that had sunk in 1854 and was marked as a rock or shoal inside Etches. Virgil Burford, the old diver, stumbled across it while making repairs on a cannery-owned fish trap many years ago. Burford was a hardhat diver, and even though he spotted tin ingots on the ship's deck, he could not get up and walk around on it because he feared his weight and clumsy suit might cause him to break through the deck. Virgil, of course, dived alone, except for the crew that manned the pumps on the boat above. He never got into scuba diving and simply retired his old brass helmet and canvas suit. He was lost at sea June 17, 1969, but he had provided Endicott enough information for Don to research missing ships in the area.

The day we were diving the visibility was so poor nothing could be seen. The weather came up and we tore up Don's anchor and had to tie up to a buoy for the night. The next day the seas worsened, so we went home. We never did go back. However, recently a group of divers from Anchorage found the Amaria and donated the lens to the Valdez museum.

Our next enterprise was to find the Alaska Steamship Company's Olympia, sunk in 1912. The large freighter had just made a stop at Cordova and was enroute to Valdez when it rounded the corner of Bligh Island and hit the reef. Don once said that the Alaska Steamship Company was known for buying old, used-up freighters, sending them to Alaska to work the remaining life out of them. Then they would mysteriously limp up onto some reef where the insurance company would buy them back. I don't know if that's true, but there was an old saying around coastal Alaska, "What's blue and gold and breaks rocks?" Answer: "Alaska Steamship Company."

Anyway, the Olympia sat perched up on the reef for about three years. Before the storms beat her up so badly, sending her sliding down the face of the reef, locals stripped her clean of everything worthwhile. Everything, Endicott said, except her massive propeller, which would bring a tidy sum just for scrap brass. So, off we went.

When we arrived at the Bligh Island buoy, there was a gentle swell and very little wind. A nice day for a dive. Don, Bill Bernard (the 3rd diver), and I suited up for going in. Poised to go into the water, Bill pointed to Bligh Island. A pod of killer whales was just rounding the bend with a trajectory to take them right over the other end of the reef. Without a word, Bill and I took off our tanks and weight belts. Don looked incredulous at our actions and said not to take the orca's nickname to heart. "They're just like big dolphins, they won't hurt you."

"Yeah, but they swallow seals and sea lions whole," Bill said.

"And you look a hell of a lot like a seal or sea lion," I added.

Endicott just shook his head and jumped over the side. We watched his bubbles as he skimmed the edge of the reef. The orcas never did cruise over by him, but they still remained in the area. After he went through one tank of air, we tired of waiting for them to leave, so we left.

The only money I ever made moonlighting at diving was in cutting rope or cable out of propellers of boats, or by brushing sea grass off of hulls so that at low tide, on the grid in the harbor, the boat owners could paint them. Or in kelp diving.

In the early spring, herring spawn all over the sound and often in kelp beds. The Japanese love that stuff (actually, I do too). Take the fresh green kelp, covered thickly with white herring eggs and dip it in melted butter…it's addictive.

When that industry first opened up here, we used grapnel hooks and threw them out in the water and raked in the kelp. We sorted it on deck and cruised over to the tenders or ships that were buying, and then headed back to the kelp beds. However, raking kelp just tore the kelp beds up and stirred up a lot of dirt, which got all over the kelp and much of it had to be tossed away. Someone surmised that if a diver went down with a big wire basket and cut the kelp neatly with a knife—even if this was slower in dislodging the kelp—the diver could do most of the sorting down there, leaving less of that for the deckhands. So it really wasn't any slower, and it saved the kelp beds. Cutting the kelp about 3 inches above the stem certainly promoted faster re-growth certainly than ripping the kelp off by the root.

The first year when some of us dived, and grapnelling was still permissible, there were some real problems. I was in a good bed, cutting and sending the basket up. I was on a hookah (gas-driven air compressor on the boat, sending air to me down the hose). I was cutting away when I heard the engines of other boats approaching the area. I heard my brother-in-law Marty (the boat owner) yelling angrily. Then I saw clouds of dirt sweeping through my area, obscuring my vision. Occasionally, I saw flicks of silver glinting through the muddy water. I surfaced and saw Marty cussing out the fleet of Russian boats that pulled into our spot, pretending they didn't understand Marty, and throwing grapnel hooks.

At night the area where we were (the protected cutoff between the village of Tatitlek, and Goose and Bligh Islands) looked like a city. Boats at anchor as far as you could see. The lights on, music coming from decks, individual clusters of boats tied together, swinging on one anchor line, occupants all gathered predominantly on one deck, eating, drinking, and having a good time. Some boats would have grapnel hooks hanging all over them; others would have rubber diving suits swinging from the yardarms or rigging, drying in the breeze. There was always the buzz of small outboard motors pushing dinghies from one gathering to another.

To the left, you might barely be able to see the kelper on the deck playing folk music on a guitar; from a boat on the right comes the soft jazz melodies from a radio; there would be brief explosions of laughter from somewhere out there in the darkness. Sound skips across the water really well. The lights glowing in the darkening day, the people-sounds being gently carried on the evening breeze, mixed with the always-present creak of the boat on its anchor and the slap of the water against the hull, creates a feeling of timelessness. It is very odd how few settings one sees, pinions one fully in the present. Immediate or distant past almost cannot be called to mind. Planning or speculating about the future is something one could only force oneself to do here. One is totally captured by the immediate. There was no more room inside you for anything else.

The following year, grapnels were prohibited and there was in influx of divers from just about everywhere. Mostly, however, they were abalone divers from California. They had some really neat harness and rig-up ideas that many of us locals copied. But there was a vast number of novices lured here by the promise of quick, big money. They just bought some equipment and jumped in the water. Most of them fared okay. A couple, did not.

A neat little trick the abalone divers used was to clip their hookah hose to their weight belt so that when they surfaced to get back in the boat, they didn't have to try to lift their weight belt over their heads to a deck hand: They just spit out their mouth pieces or removed their masks, unclipped the belts and climbed aboard the boat. The deck hand would pull the hookah hose back up and the belt with it. One young novice diver was found dead on the bottom because his belt hadn't been tight enough. The depth of the water squeezed him smaller, and on the bottom, his belt slipped down to his knees, pulling the mouthpiece out. Apparently he panicked and couldn't find his mouthpiece and drowned.

The second one that died that year was from carbon monoxide poisoning. Most divers did what I did. I lashed the compressor to the top of the boat cabin and had the crew assure that the air intake would be upwind and the exhaust from the engine would be downwind. The diver or crew of this boat set the compressor down low on deck and the gunwales wouldn't allow the exhaust to be carried away by the wind. The exhaust swirled around the compressor and was sucked into the air intake and pumped down the hose and into the diver's mouthpiece.

THE ALGANIK RIVER RECOVERY

One year I was making some pretty good money on weekends and days off disassembling a platform that a pile driver sat on when it was driving sheet pile for a dam in the lake. I was working just under the surface and the visibility was fine. But it was really cold. It was cold, particularly, because I had managed to cut my suit in several places: the boots, knee, shoulder, and of course, my gloves had no finger-tips left in them.

I didn't know anything about hypothermia in those days and was not particularly alarmed when, at the end of every day, I mentally felt a little detached from reality. Also, I was surprised how long it took me to get feeling back in my body when soaking in the tub. It didn't really feel good either. You know how painful it is when your foot falls asleep then wakes up? My whole body felt like that. But the money was good; it meant a new TV and a carpet.

I had taken some vacation time to finish that job and was nearly done when Endicott came up and asked if I would help search for the body of a boy that had drowned.

Two boys in their mid-teens went duck hunting in a borrowed a skiff. About 22 miles out of town, they had launched the skiff in the Alganik River. The Alganik is chilled by the Childs Glacier further north, and as a tributary of the Copper River it is silt-thick and brown.

Some distance downstream from the boat launch, the boys beached the boat and got out to hunt. They turned to see the boat drifting away. One of the boys said, "I'll get it," jumped into the water and started swimming toward the boat, which was now approaching mid-stream. He stopped and turned back as though he'd changed his mind. He had a look of pain and fear on his face. Then he sank.

Don told me that the boy's father and numerous friends had searched for the body by dragging hooks and by laying out a gill net across the river and sweeping for him (hoping he might tangle up in the net). Nothing worked. Don was certain the only way to find him was to dive, so asked me and he got some Coast Guard divers to help us.

It was a bleak day when we all boarded boats (skiffs, air-inflated Zodiacs, and air boats) and headed down the river. The sky was the color of dull aluminum, the river was gray, the surrounding brush was brown. There was a steady wind blowing that reddened everybody's faces, and made noses run. The boy's father and many of his friends were there to help carry lines and diving gear. They dragged a bunch of dried logs into a pile and managed to get a fire going. Mike Noonan had a bottle of blackberry brandy that warmed everyone's belly.

Don said that we would be diving blind, and that we would run a line across the river and the line would have a knot in it every 12 feet. Each diver would hold the knot in his right hand and go as far left as he could, groping the bottom. Then he would come back to center, grasp the knot with his left hand and stretch out

107

to the right as far as he could, feeling the bottom. That way, each diver would cover about 12 feet of the bottom.

There would be a "line tender" on each bank of the river. They would slowly walk the line downstream.

Don did something pretty cool that I've never tried to duplicate, but keep promising myself that I will. In order to tie knots at 12-foot increments—in a very long line—he coiled the line neatly, and holding the coils in one hand, took the bitter end of it, looped it over and through the coils in some fashion, then had a guy take the end and walk slowly away. As each loop was pulled out of his grip, a knot was formed. Pretty cool.

Since the boats had headed back upstream for more supplies, I would swim the quarter-inch "messenger" line across the river. The crew on the other side would use the small line to pull the end of the knotted line across the river. I put my regulator in my mouth, but only intended to clamp it in place with my teeth and not breathe out of it unless I got dunked down for a moment. I opened my lips wide to breathe around it.

I planned for some downstream drag and had started well upstream, hoping to hit the other side near where the crew was waiting. When I had traversed slightly more than halfway, I was surprised how hard it was to pull the line. I looked over my shoulder and saw quite a bow in the messenger. It was creating quite a drag on me. I was breathing like an air piston. I figured I'd better get down to business if I wanted to get across, so I lowered my face into the water and started stroking and kicking almost in a rage. I was breathing from my tank. Once when I looked up to see how I was doing, the cold air hit my regulator and it clicked into the open, free-flow position. The air rushing out on the wet regulator created a block of ice on it. My blowing back into it wouldn't make it reset. By the time I rolled up on the beach on the other side, my tank was very low. So I became a line tender.

We walked the line back upstream a bit and started. It looked like a line of rubber Christmas street lights all submerging at once. In about the same area, estimated to be where the boy went down, Don surfaced, but his arms were held down by the body he held. He nodded and signaled for the Zodiac to move up. Afterwards, we all drank up Noonan's brandy and got a buzz on. We let the boy's father drink first.

COLLAPSE OF THE COPPER RIVER BRIDGE

The hairiest dive Don ever did was at the Copper River Bridge. The bridge, built to accommodate the Copper River Northwestern Railroad, was tall, narrow and riveted steel which had withstood some of the most severe weather anywhere. But it was deemed time to modernize. Beck Construction got the bid to do it.

108

In order to construct new footings for the new bridge, a dam was made to divert the entire river to move under one span so that work could be done under the "dry" one. Unbeknownst to the engineers, this created a tremendous washout under the second span. Three construction workers were moving a truck crane across the bridge spans to reposition it on one of the islands on the other side of the two spans. When it was on the second span over the torrent of water, the entire bridge section and the crane on it toppled over into the water and sank out of sight. One man was seen to surface momentarily but sank again.

Beck wanted the bodies back for the families. Endicott had been contacted by them, and he asked me to help him set up.

He took a large, nylon seine purse line and tied a huge, heavy shackle on the end. He dropped it into the water and let the current push it downstream until it hit something and the line went slack. He assumed it was the bridge span. He took up the slack and tied it off to the span he was working from. He geared up and got into the water. It was insane; water was tumbling and boiling through there carrying large chunks of glacier ice and tree stumps. He looked upstream for a debris-free span and took hold of the line, lowered himself down into the water, and his body stretched out like a flag flapping in the wind. Instead of changing his mind, he hand-under-hand pulled himself under and disappeared.

If he got in trouble under there, an army of back-up divers couldn't help him. There was no telling where he was, since no air bubbles could be seen in that mess. When he surfaced again and got out, he said that he could feel the shackle had come to rest on the bridge span. It was tough under there. He'd wrapped his arms around a steel beam of the bridge and let his legs flap downstream to see if he could feel the crane behind him, but he couldn't.

The next step, he said, was to take another chunk of line down, tie it to the beam, let it stretch back to hit the crane, then hand-walk back down that line until he was on the crane. Then try to claw his way back and forth along the crane and see if he could feel one of the two men still on it. "Then what?" I asked.

"I don't know." He said. He went on to confess that he really didn't want to do it. If Beck insisted, he would. "But I'll charge him a fortune if I do."

Beck thought hard-hat divers would be more stable under that rushing water, so he contacted an Anchorage diving service. They showed up, and from the mid-river sandy island looked at the site and suited up. The two of them stepped into the water just over their heads and climbed back out. They really had no intention of trying to retrieve the bodies, but their standard contract agreement is that they get paid divers' wages if they get in the water.

The bodies were never recovered; the bridge span and the crane were presumably covered by sand and silt in short order and are now a part of the Copper River.

RIP TO "WHAT'S-HIS-NAME"

It was a windy, rainy October night when we were toned out to respond to the city dock because a man had fallen off the dock. The old city dock was where the Coast Guard Cutter Sweetbrier was moored. It was moored on the outside (face) of the dock. The quartermaster's watch shack was at the foot of its gangway. Moored permanently on the inside of the same dock was the St. Elias cannery. The cannery—a "floating" cannery—was actually an old ferry from Washington which had been towed to Cordova and converted to a cannery. While I was enroute, the dispatcher explained that the fall was witnessed by the man's companions who rushed over to the Coast Guard watch shack and called 911. With the cold-water-near-drowning (CWND) physiology, our response would be well under the time-line for a save. About 2 minutes after the page, I was pulling onto the dock.

The dock was so slippery from rain and moss that my car slid sideways as I stopped. The dock was really dark and I could barely make out the small crowd of men there: Coastguardsmen, a couple of civilians, and a police officer. While still enroute, I had decided that this would be our first cold-water-drowning save. I jumped into my Unisuit while onlookers pointed to the spot where the man fell in. He had been stepping from the dock to the boat, but the bulwarks were so slimy, he slipped and fell between the dock and the St. Elias cannery. I was certain that this would be a quick recovery, so even though I didn't zip up my suit completely, I knew I could stand the cold water for a few minutes.

Gusts of drizzly wind buffeted the huddled group of men. A single street light barely illuminated the unusually high swell of the water as it sloshed against the hull of the cannery. I was scurrying so fast, snugging up my lead belt and slinging on my tank, I could only think about the seconds ticking by and the cells of his body organs starving for oxygen. Everything else was a blur. I stood on the edge of the dock, stepped out, and plunged straight down. I was surprised how low the tide was, and how long I fell before I hit the water. I hit the water so hard, I went straight to the bottom. I was stunned for a moment and just lay there.

It wasn't until I caught my breath that I noticed my suit filling up with cold water. It was painfully cold, but I started groping for the guy. I had hoped that since I jumped in right where the witnesses had pointed, I would land right on him. I hadn't. It was completely black down there; I did what amounted to making a "snow angel" with my arms and legs, laying face-down in the mud. He was not within arm or leg reach. I couldn't find him.

After a couple of minutes, I surfaced and yelled to the coasties to get high-intensity lights and shine them down into the water a few feet ahead of where my bubbles surfaced as I swam north along the face of the dock. I dove down again, assuming the glow of lights would direct me. However, the tide was running so

110

hard, that the beams would have had to be directed many more feet ahead of where my bubbles broke the surface. As it was, the area behind me was somewhat illuminated, but black ahead of me. Screw it, I'd move ahead, he had to be close.

It didn't take long for me to stop thinking about him and his oxygen-starving cells, and start worrying about myself. The bottom was strewn with debris that had been tossed in the water for years: old crab pots, rope, cable, fishing nets and such. Not being able to see them, I could only snag up on them and methodically disentangle myself in the blackness. When the valve on top of my tank (behind my head) got snagged up in wire, I had a hard time reaching it. My heart pounded so hard it ached. I tried getting up on my knees (I don't know why), but it felt like I was stretching back the string of a powerful bow. I lay back down on the mud. A feeling of dread swept over me with each unsuccessful attempt to grasp the wires behind my head. I tried twisting my body to lengthen my grasp, but my mask got knocked eschew, and filled up with water.

Twisting around only entangled me more. At one point I almost felt like crying. I lay for a moment like a dog hit by a car, panting and assessing my situation. It was true panic. It took me a few moments of calming myself to realize that if I could not disentangle myself, all I needed to do was unbuckle my tank and slither out of it and head for the surface. Then, I began working methodically.

I finally got loose, and thought about not going any farther. Hell, no one on the surface would know I had chickened out. Then I remembered that if I found him quickly enough, he might be revived, so I kept going. But this time I swung my arms in wide arcs in front of me and moved with cautious agility. Then I was able to feel again how cold I was. About that time I could hear the other divers hit the water, so I headed for the surface and swam for the nearest ladder. With my fins draped over my arm, I climbed up the ladder, exhausting myself with the weight of my tank, lead belt and a rubber suit full of water.

I stripped off my suit, and of course, had jeans and a T-shirt on underneath. One of the medics threw a blanket around me and recommended I get into the rescue truck or the ambulance and out of the rain. I was told that a couple of Coast Guard divers had gone in with Don Endicott and our divers. I knew the guy would be found soon. Endicott had a knack for that. He also had a flashlight. They all did. Rushing without thinking just doesn't pay off.

Anyway, they were searching in teams when a cop came up to me and said that he got a call from Police Chief Bill Bagron who had been watching from the windows of the Reluctant Fisherman lounge and said to cancel the operation for the night and start again at daylight. I refused to do that. We were still looking for a CWND save. Even that aside, I had dived for people the day following a fall, and because of the tidal action, the bodies had disappeared and were never found. I ignored the police officer.

About ten minutes later, Endicott surfaced with his arms wrapped around a limp body. He and another diver rolled the man into the litter suspended at water level—a chore made difficult because of the swells of the water. While the truck company hoisted him, the ambulance crew was preparing for major intervention. But, on the dock, it was obvious that he could not be revived. He had plunged headfirst down and struck his head on the ferry's rub-rail. His neck was snapped and fragments of shells from blue mussels were stabbed into his forehead and eyes.

Bagron didn't speak to me for over a week.

* * * * * * * * * *

You know, people in this business often do foolhardy or reckless things. But they do them impulsively, without thinking, and while in a hurry. This was different. Once in the water, cold and inching slowly, I became two different people. After disentangling myself, I knew I shouldn't be diving alone and without a light. If I had just stayed in one spot and used up my air, no one would know. I didn't owe this guy anything. But even while thinking these things, I kept moving forward automatically, detached from my reasoning self. Yep, irrational feelings of obligation.

More about panic:

Firefighter/medic Mike Gundlach once told me that he had read or heard an explanation about panic. "There is a misconception that people in a burning building panic," He said. "People do not automatically panic in those situations. They quickly consider options for getting out. They experiment and search, probe and think. It is only after they exhaust all avenues of escape that they can think of, that they panic....when there are no options left."

Then Gundlach added, "You've heard the tale that when a person is dying, his entire life flashes before his eyes. I heard a guy surmise that what is really happening is this: Everything that has happened to you in your life, everything you have ever heard or thought, is stored deep in your subconscious mind. A person who is dying, quickly runs through all of these past events searching for a way out of his predicament. He is frantically flipping through his Rolodex of possible solutions."

To forestall panic, expand your options. Firefighters have been taught that if trapped inside a windowless room, because the hallway outside is burning, to kick through the sheetrocked wall to an adjoining room that may have a window in it.

The fire service should be reminding the public that when staying in a hotel, before entering the hotel room, count the doors to the stairway exit so you canf find it in the thick smoke. And, do not set the room key or card on the nightstand. Keep it in your pocket. If you decide to evacuate and step out into the hallway, your door will automatically close behind you and lock. If the hallway is untenable, you would be unable to get back in your room to buy some time. That's what happened to all those hotel guests who died in the MGM Grand Hotel fire. Of course, the best option is to stay only in sprinklered hotels. Planning ahead and taking a moment to think expands your lists of options.

TRAPPED UNDER A CAPSIZED BOAT

I began diving in the early '70s to make a few bucks in my spare time and, in fact, I did make a few. I made several body recoveries when requested by law enforcement, and we all knew those operations were recoveries only; not rescues. None of my attempts at quick underwater rescues ever resulted in a successful resuscitation. Not one. We actually did resuscitate a victim of cold-water drowning once, but he (a Russian capsized in the gulf) was choppered in by the Coast Guard; it was not a rescue from any department divers. But there was one call-out that still haunts the team and me.

After quite a few years of diving alone at night, or in sloppy weather, or through broken ice, I got the money to send a couple of guys to a bona-fide, certified dive/rescue instructors school. Then I bought gear for a team of divers and got them trained and certified. I slowly backed out of making the dives myself. I'm not sure who the dive team leader was (Steve White or Bob Pudwill), when we got an urgent request from the Coast Guard.

A commercial fishing boat had capsized near the entrance of Prince William Sound. The Coast Guard believed the crew had all drowned except a 12-year-old boy who was trapped inside the vessel. A near-by boat—whose skipper knew the family on the capsized boat—radioed the "Mayday" to the Coast Guard and reported that they could hear the boy pounding on the hull. The boy could hear them shout, and they could hear his replies. The hull was bobbing in the waters of Zaikoff Bay on Montague Island.

The Sweetbrier was in the area and running flank speed toward Zaikoff Bay. There was a CG helicopter in Cordova and wanted to take our dive team out there where they would work off of the Sweetbrier to get the boy out. An upside-down boat would have lots of lines and cables and a boom and yard arms dangling in the water that the divers would have to avoid just to get into the boat. Then, getting back out with a frightened boy who was breathing off of a regulator would be a monumental task. I told the team leader that he would be the one calling the shots. If the weather and sea conditions were too dangerous to try it, he could call it off. He nodded and the rescue truck tore out of the station to meet up with the helicopter.

I drove my vehicle up to the base of the ski lift where I might be able to hear some of the Coast Guard radio transmissions on my radio. I was up there for over an hour and never heard a word.

As it turned out, it was too late. Before the Sweetbrier arrived at Zaikoff Bay, the boy stopped responding to the shouts and hull-pounding of the other boat crew. It then drifted onto the rocks. The crew from the other boat managed to climb up on the wreck and cut through the hull, to find the boy dead.

In all the years after that, we never talked about that incident.

CVFD DIVERS WORKING WITH LOCAL POLICE DEPARTMENT

Photo: CVFD

RESCUE OF GAYLE RANNEY AND GARY WILTROUT
AND THE DEATH OF A RESCUER

On Thursday, January 13, 1983, the barometric pressure dropped in the north gulf coast, and created a vacuum into which all pressurized weathers flowed like atmospheric floodwaters seeking the lowest point. From the interior arctic region, the north wind, swollen with building pressure, rolled down the near-continental stretches of frozen Alaska which had lain quiet and lifeless, ice-covered and dead. Like a flash-flood, the wind boiled through the valleys, banked and rolled down the mountain slopes, skimmed through the glacial passes, and spread over the flat lands like a wide cavalry charge, an ominous march across the desolate ice-blue lands the size of several states.

As the front neared the north gulf coast, it moved into the steadily narrowing Copper River Basin, and the winds increased in intensity. The force rose and spilled over the mountain peaks as well, sending plumes of blowing snow off the peaks into the sky like the smoke stacks of racing locomotives. Finally, as the wind was squeezed through the barely 15-mile-wide narrows between Goat Mountain and Pyramid peak, 27 miles east of Cordova, it blew out the other side with such force as to gust up to 85 miles per hour. It screamed at that force down the Copper River Delta and spilled into the Gulf of Alaska.

Here, at the Copper River Delta, the arctic wind intersected with the east-bound clouds, heavy with their snows as they rolled over the mountains and passes and created the silent white explosion over the entire northern gulf.

Gayle Ranney, flying west in her Cessna 180 enroute to Cordova, following the shoreline of the gulf was being slowly swallowed up in the encroaching snowstorm. Leaving the leeward protection of the mountain range that lay north of Katalla, she flew into the full blast of the wind of the Copper River Delta. Her visibility diminished rapidly and the plane pitched widely. She dropped down and skimmed above the waves and islands and marsh. Visibility was intermittent and by sheer luck she spotted a flat scab of land below her and slammed into it, disabling but not destroying the plane.

Shaken, but uninjured—and not entirely sure where she was—she radioed her situation to the airport 15 miles away. It was 1:06 p.m.

In Cordova the large snowflakes dropped lazily and were covering the town with several inches. The atmosphere inside city hall was relaxed and congenial. I was compiling my data of the fire department activities of 1982 to deliver at our annual dinner that evening when, the dispatcher called in to me to tell me about a plane reported down in the delta. She told me that Trooper John Stimson was preparing to go out on the call.

As near as she could tell, the plane was down a couple of miles south of the Copper River bridge. Just in case they were needed, I had her page for the medics and the rescue squad to assemble. The river was frozen and might allow for a ground party if needed.

I went upstairs and found John scurrying around, strapping on his gun, grabbing a chart and throwing on his coat. He told me Chisum Flying Service was cranking up the helicopter to take him out. I asked if there were anything he needed. How about a portable radio with our frequency on it? He agreed, so we walked downstairs and I got one for him. He thanked me, rushed out the front door and was gone.

By approximately 1:20 p.m. John was aboard the Chisum Bell helicopter and pilot Gary Wiltrout lifted off into a snowy but calm sky. They sped eastward through the pass between the Heney Ridge and Ibek Mountain and into the open marsh area south of Scott Glacier. There they caught their first taste of the wind, and at 10 miles out experienced a "white out" and sat the chopper down. Apparently, upon noting a lull in the weather and after a brief discussion a few minutes later, they lifted off again and headed for the delta.

The weather was bad but tolerable until the chopper reached the Copper River 27 miles out. As it pulled out of the lee of Flag Point, it flew right into the muzzle blast of the storm. At that point visibility ceased completely beyond the windshield, which became painted white with a "splat." The two could only see each other. The chopper was flung madly through the air, and when their altitude was about 50 feet, the engine had a total flame-out. Dropping through the air from the height of a 5-story building, Gary Wiltrout radioed that they were crashing. He managed one "Mayday" before impact.

On impact, the fuselage split like melon and most of the survival gear was blown away in a blur as though it were shot from a cannon, disappearing in a screaming white explosion.

The two were strapped to their seats, not quite believing they were still alive, let alone unhurt. However, Gary became aware of a pain in his spine and, at John's instruction, was careful not to move, even though the snow and wind felt like it were sandblasting their faces off.

Their faces were becoming red and swollen. John located one sleeping bag, assisted Gary into it, and burrowed it into the powdery snow in the lee of the chopper wreckage.

John could not locate any other survival gear. His jacket was unlined, the portable radio was gone, he could barely see his own feet in the raging wind and snow, and his clothing flapped wildly like flags. He could not venture beyond

116

the grip of the wreckage and had to keep a tight hold on the parts of the chopper as he frantically thrust and waved his other hand through the snow for any item he might use for survival.

From inside the zipped sleeping bag, Gary—still feeling the impact of the snow and gusts of wind against it—could hear John scavenging blindly and breathing like a freight train.

Since I had dispatched the rescue truck with three firefighters aboard and Medic 8 with four EMTs to Mile 27 at about the same time the chopper left Cordova, they were halfway there when they heard that the chopper had gone down. At 2:36 p.m. Rescue 5 radioed they were at the Copper River bridge along with state public safety vehicles and personnel and airport emergency personnel, too.

The conditions at the bridge, they said, were untenable. Nothing could be seen, and certainly—unless the chopper and Cessna were embarrassingly close—no gun shot would be heard in that roar of wind. Visibility was so poor that in turning around, Rescue 5—following the directions of a firefighter—backed into a snow drift the guide was unable to see, even as he stepped into it himself. AST supervisor, Sgt. Roy Vanderpool, sat staring straight ahead and did not acknowledge hearing a firefighter pounding on his vehicle door wanting instructions. When President Bush was videoed sitting in a classroom after hearing about the twin towers, people seemed appalled that he couldn't move. I understand what overcame him. Other people in this business do, too.

After much effort, shouted directions, and inch-by-inch maneuvering, all the vehicles turned around and started back.

As the caravan blindly groped bumper to bumper back toward the Mile 13 airport, the snow machines and snow machine sleds I'd called for had assembled at the fire station. Rosters were taken, fuel gathered, and they were enroute to the airport fire station which would be the staging area for the operation.

Also, during the previous half-hour the state trooper district headquarters in Glennallen and the Coast Guard rescue center in Kodiak were notified.

Sgt. Roy Vanderpool was in charge of the operation. I offered to handle the ground search if it became plausible to do one. The plan was simple: One snow machine with driver, an EMT, and a rescue sled made up one team. With any break in the weather or increase in visibility, several teams would proceed to the Copper River bridge, drive down onto the frozen river and proceed south in parallel grids to locate the aircraft and occupants. The occupants would be treated and either choppered out by the Coast Guard or troopers, if possible, or taken out on the sleds.

At approximately 4 p.m. a Coast Guard C-130 rescue plane was in the general area, but daylight was lost; the conditions had not improved, so optimism was scarce. The C-130 had, however, made radio contact with Gayle Ranney and confirmed she was still doing well. No contact, however, was made with the Chisum helicopter. The C-130 did pick up the signals from two emergency locator beacons and they seemed to be very close to each other—perhaps no more than one-half mile apart.

But as the hours dragged on there was no slackening in the weather. During that time the Coast Guard cutter Sweetbrier had made it's way around Hawkins and Hinchinbrook Islands out to the gulf, hoping to approach from the ocean side, but the winds would not allow it.

City police in four-wheel drive vehicles patrolled the road to the Copper River bridge all night, stopping and listening for gun shots, signaling their location with their flashing lights, short blasts on their sirens, and calling to Stimson on their radios. Nothing answered and nothing moved but the wind and snow. And, as usual, upon their approach to the river, their headlights revealed snow in absolute screaming horizontal streaks.

Outside the Mile 13 fire station was an assemblage of multi-colored vehicles of different federal, state and local agencies, pickup trucks, snow machines, snow plows and vans, sitting forlornly under the falling snow.

The inside of the station was a cheerless, high-ceilinged metal building cursed with an incessant fan blowing uneven warm air. The army of firefighters, cops, troopers, civilians, and pilots barely spoke to one another. Eyes blurred by smoke and voices hoarse, the rescuers paced and rechecked equipment (a gesture of frustration).

Conversation dropped, team leaders approached, stopped and stared at the silent radio transceivers. Hands wiped clear swaths across steamy windows, and squinting through small snow squalls, a rescuer would blow a sigh and mutter, "Fuck." More pacing. Coffee was poured, not drunk, and left on fire engine fenders and bumpers, on benches and window sills. One can tell the level of frustration by the amount of scattered Styrofoam cups.

The C-130 had finally given up and landed at our airport, and joined the scene of exasperated impotence. The fire station became as voiceless as a cathedral, only red eyes and the fan.

No one wanted to be the first voice to break the silence. Periodically the rescuers would hear the voice of the city police officer radioing into the still night, not expecting by this time to hear an answer.

The same forlornness hovered over the dispatch center in town. Finally, by some invisible expression that signaled a consensus opinion, we radioed to mile 13 to secure for the night. Within an hour the troops had dragged back to town, were checked off the roster, told to reassemble at 5:30 a.m. and left. We had been informed that a Coast Guard H-3 helicopter and a State Trooper Ranger

118

helicopter were expected in the area at first light. The arrival of the trooper helicopter was contingent upon the weather clearing in Valdez.

By 5:30 a.m. the teams had assembled, signed in, geared up and were enroute to mile 13 fire station. The last report from the patrolman was that the weather remained the same throughout the night. The C-130 was already circling the delta area and the H-3 from Kodiak was expected to be on scene at 6 a.m. There was no word yet as to whether the trooper chopper would be able to make it.

The day's routine was quickly established. More of the same...waiting. The only real hope any of us had was the choppers. If the C-130 could spot either of the crashes, a chopper may be able to set down and pick them up.

Aboard the H-3, which was now in the area, was a five-man crew which included the Coast Guard's Doctor Martin Niemeroff, who was a bit of a celebrity in EMS in Alaska. He is the professional who first researched the survivability of cold-water-near-drowning victims. He developed the first set of protocols for rescuers to use when recovering persons who had drowned in cold water. His interests expanded into the area of hypothermia in general, and thus he chose to accompany the chopper on this mission. But the hours dragged on with no opportunity to attempt a rescue. Again, daylight was beginning to wane on the second day. The trooper helicopter landed at mile 13 and the crew joined the crowd inside the fire station. The snow there and in town had slacked off a bit and all eyes watched the eastern horizon which was still black.

While the H-3 flew over a small sandy island at the lower edge of the delta, a clear patch of visibility opened up for a moment and they spotted the Cessna 180. Excited, they spun around and, even in the howling wind, would attempt a landing. Suddenly the helicopter reared backwards, it's tail pointing straight down, and the pilots looking straight up into the sky. The chopper plunged backwards toward the ground. Except for the pilots who were trying to right the aircraft in a flurry of movements, yelling, "UP! UP! UP! UP!" the other crewmembers strapped to their seats flung out their arms grasping anything they could against the weightless feeling of falling through the sky.

Still, to this day, neither of the pilots can remember what maneuver of theirs made the aircraft respond and pull out of the fall, but it did recover from the incident caused by a blown hydraulic hose. It was accompanied back to the airport by the C-130. It was now approximately 3 p.m. and with less than half an hour of daylight left, a quiet rage ran through the crew.

A few minutes later the phone rang in my office. It was my wife telling me to look out my window. I looked and saw the black sky to the east had lifted its hem, slightly producing a small arch of clear sky. I radioed Mile 13 and told them. They replied that they all saw it and the trooper helicopter tore out of there in a last ditch effort before dark.

They found the two aircraft less than one-half mile apart. Gayle Ranney was fine. Gary Wiltrout was stretchered and brought in with a fractured spine (no paralysis), and frostbite of the right leg. John Stimson was found on the windward side of the wreckage, standing, pinned against the wreckage by a snow drift up to his chest, frozen to death.

* * * * * * * * * *

The following day, anyone not aware of the previous events, could have strolled—bundled and warm—across the Copper River, smelled the crisp winter air, squinted at the sun-bright snow wisping gently, and felt that all was right with the world.

And when I think back on the night of the 13th, I feel guilty about being warm and comfortable. I think we have a natural urge to empathetically share someone else's pain. You know, when a child hurts himself, he grabs his injured area and goes howling through the yard searching for someone to recognize his pain. We all feel the need to have someone share our pain—our fear. That's not too much to ask, nor too much to give.

AUTHOR ON THE PHONE WHILE ORGANIZING A SEARCH AND RESCUE RESPONSE

Photo: CVFD

THE FRANK HANSEN OPERATION

After about ten years of search and rescue operations, we (CVFD) became known as "SAR Central." That would sound pretty cool unless you knew what that really meant. It meant that—especially in the summer—we had medics passing in mid-air: A medic would be on a small plane, bringing a patient in from some rural location or fishing boat, and a small plane would be on its way out with a medic to pick up a patient. They would wave at each other as they passed. It meant that department members had to stick around town more religiously, because a contingent of members might be out of town on a SAR. It impacted everyone. Mostly, we didn't mind that much except when something made us cranky. It meant that the employers of the volunteers worked short-handed sometimes all day or for several days in a row. Incidentally, those employers need to be recognized.

We knew that volunteers elsewhere were doing the same things: Unalaska (on the west coast of Alaska) was always up to their asses in SARs. Barrow and the entire North Slope Borough, comprised of eight villages in an area nearly the size of California always had something going on. But in Southcentral Alaska, we were "SAR Central"

It never occurred to me that other towns in our region of Alaska may not have been participating in SARs as much as we, and might be willing to share more of the burden. It never occurred to me until after the Frank Hansen SAR.

On Wednesday, August 20, 1986, a hunting expedition near Port Fidalgo went sour. In the northern part of Port Fidalgo lies Fish Bay. The hunting party had their camp on the north shore of Fish Bay at the base of Billy Goat Mountain and Mount Denson. In this case, the primary hunting area was Mount Denson. This area was nearer to Tatitlek than Cordova. The hunting party consisted of Frank Hansen, from Scottsdale, Arizona, who is the vice-president of Best Western Hotel chain, and two friends, Randy Atwater and Bill Metcalf.

Early that morning, as they milled around near the camp, Frank informed his partners that he intended to head up a creek bed toward a visible waterfall, north of where they were standing to check out the goat situation, and would be back around noon. When he didn't make it back, Bill and Randy searched for him, shouting and firing their guns. When he hadn't returned at day's end, they were certain that he was lost or injured. Even though the terrain down low was not steep, it was thick with trees and underbrush, it is bear country, and the salmon were running upstream. Higher up, above the tree line, the terrain became rocky and steep.

121

The following morning, the two of them hiked the edge of the bay hoping to locate a passing boat they could flag down and use their radio to call for help. However, it wasn't until Friday that they spotted one and signaled it. They radioed the Coast Guard, who in turn called the State Public Safety base in Cordova. So the AST guys flew the area Friday and saw nothing.

On Saturday, city police officer Jim Leuty and his German shepherd, "Phantom", went over there with the trooper and the hunting party members, Atwater and Metzger. Walking the creek with Phantom, Leuty was uneasy. There was bear sign everywhere, and the footprints indicated they were brown bear. The water rushed down the stream, requiring effort for the men and the dog to walk upstream. In a couple of places the water was deep enough that the dog had to swim. Walking against the current became fatiguing. At a point about two miles upstream, the alders were so thick at the waters edge that each side arched toward the center until it looked like a tunnel. The bends in the stream were abrupt and Leuty worried about running into a bear.

The search Saturday and Sunday with the dog did not cover much area, and only one human footprint was discovered a couple of miles up the creek bed. The weather became more drizzly.

Monday and Tuesday, the wind howled and the drizzle became a torrential downpour. Marilyn Mather, Frank's sister, began calling Cordova AST and asking what was being done and why she was not being kept informed. She began inquiring about who the supervisors were of this AST post here. Finally, she flew into town. On the other hand, Frank's wife was being kept somewhat informed. Supposedly, Frank had an $8 million estate and a pending divorce. His divorce was to be final a couple of weeks after his return. In addition to Frank's sister and wife, Ron Evans, the president of Best Western was also being updated—mostly, I understand—by Metzger and Atwater.

Tuesday afternoon Leuty came into my office and began telling me about the SAR operation and complained about AST's reluctance to make a real commitment to the search, their inability to be decisive, and reluctance to ask for assistance. He wanted to know if there were something we could do to push AST into asking us to join in?

Every couple of years we would get a new trooper assigned to Cordova. About half of them were very reluctant to ask for help from us. The other half would routinely call, explain the situation and ask us to "put a package together" for them (organize and conduct a SAR operation).

How could I prod AST into asking for our help? I had not been aware of this situation until Leuty walked in and plopped himself down in the chair. Funny, less than an hour before, I had been on the phone with the state SAR coordinator in Anchorage talking about a formal, written, bona-fide mutual aid agreement with AST. Leuty and I kept talking about potential future SAR operations as though this one here was already concluded. It indeed looked as though AST had no intention of making a full blown search out of it or to call us for any more assistance.

Well, I explained that I would offer any assistance I could if I were asked. Later that evening, Leuty called me at home and told me that he talked with the hunting party and Frank's sister, Marilyn. He said how frustrated they were with the AST and the fact that they just sat on their asses and refused to do anything. He wanted to know if I could offer them anything. He asked if I would call them at the Reluctant Fisherman Hotel.

So I called and talked with Bill for a while. He asked what, in my opinion, could be done. I suggested a couple of things: First, find someone familiar with the area who could lead a search party there. Second, try to find an infrared viewer (thermal imager) and spot Frank from the air. Third, get more search dogs, the SEADOGS out of Juneau, to help. He asked if I could do any of these things, and I said I would try within the limits of my authority. So I hung up and called the AST post and asked if they would be interested in an infrared viewer to look for warm objects from the air and he said sure.

So I called Elmer Hurd at his residence in Anchorage. He is the head of forest firefighting for the Department of Natural Resources. I asked him if we could borrow the "Probe-Eye" and explained what for. He told me of another item they used called "FLIR" (Forward Looking Infrared something-or-other) which they had attached to their T-28 (a fast, little single engine airplane). It picks up warm items and indicates them on a screen like radar, and you use Loran navigational coordinates to transpose the location on a chart. He said to the best of his knowledge the item was not in use now. I explained the situation and he said that he would make it available to the troopers. I told him what the weather was like in this area and he was skeptical whether the T-28 could be used. So, I gave Elmer the trooper's phone number and asked him call and offer the hand-held "Probe-eye."

Then I called a local trapper, Larry Kritchen, to see if he was familiar with the Fidalgo area and he said no. But he suggested Kenny Vlasoff or Carroll Kompkoff. I called Kompkoff and he said he hadn't lived in the village (Tatitlek) for 20 years, plus he is 60 years old. He gave several names of guys who live in the village and hunt that area. But those calls would have to wait until morning. My daughter, Cindy, was going back to college and would be on the ferry to Valdez that night. The ferry was to leave just before midnight. I typed a little note to the captain of the Bartlett asking him to keep an eye on the Port Fidalgo area for any campfires or distress signals. I asked Cindy to give it to the ship's purser to give to the captain. I then called the trooper's house and woke him up to give him the phone numbers to call the SEADOGS.

123

It was now after midnight. I tried to go to sleep, but I repeatedly looked at the clock on my night stand, and calculated where the ferry might be at that moment. I visualized them scanning the blackness that was landfall and spotting a signal fire. I imagined my phone ringing and the dispatcher telling me the fire had been spotted. The phone never rang.

I'd spent a miserable night or two in my life and could not fathom how miserable Hansen must be if he were still alive, possibly with a broken leg, and hunkered down in the cold rain. I felt so guilty, being comfortable in my bed that I had to get up. I felt that I should be doing something, but not knowing what, I just paced and drank coffee the rest of the night.

The next morning, Wednesday (August 27), I was talking about the situation with Groff. I showed him the approximate location on the map and he went to the forest service and got some more detailed maps and a series of aerial photos. While he was doing that, I called the village (Tatitlek) and could not contact any of the suggested names because they were all out fishing.

I then got a list of boats possibly available to use as a base camp in Fish Bay if a full-blown search were to be initiated. Then Atwater and Marilyn came down to the office about the time Groff got back with the aerials. We talked for awhile. Atwater told Groff and me the search had been conducted up to this point and where he thought Frank might be found. Out of Marilyn's earshot we mentioned the poor prognosis of Frank surviving hypothermia in this driving rain. They asked what else I might do without being requested by the trooper. I told him I'd think about it.

After they left, Groff and I called lots of boat owners and were able to get tentative commitments from several people who said it would be a terrible inconvenience (commercial fishermen make their annual income during these short summer months that feeds their families for the rest of the year—I think "inconvenience" is too soft a word) but that they would do it in a pinch. Finally, after a call to their headquarters in Anchorage, the Alaska Department of Fish & Game said we could use their vessel M/V Montague for a couple of days. Local fisherman Ross Mullins offered a vessel. I got Valdez FD to offer 6 guys, and the U.S. Forest Service to offer 4 guys. With 2 police officers and a couple of firemen, 14 persons were lined up, not including the hunting party or AST.

Valdez Fire Chief McAlister checked into the cost of leasing tour boats from Valdez for base camps and they were expensive—$1000 per night for the 12- to 15-person boat and $450 per night for the 6-person boat. He also checked on a helicopter service out of Valdez for $450 per hour which was $250 cheaper than the Anchorage charter that the hunting party had located. I gave that information to the hunting party at noon and also to AST, telling them that I had checked these things in case they (AST) decided to move in that direction. They told me that authorization for the SEADOGS and other expenditures was being handled in Glennallen (trooper headquarters for this region) and that everything was on hold pending decisions from "up there." Having heard that, I wrote down the phone number of the Commissioner of State Public Safety, Robert Sunberg (appointed directly by the governor), and his assistant commissioner Jim Vaden. I also wrote down the number of the governor and gave it to Atwater and explained to him that it might be advantageous to make some calls.

By afternoon, half the state was calling wanting to know when this search was going to get underway. I told them we were just waiting for word from AST to go. The "Probe Eye" had been sent; the SEADOGS had been called for, but were not contacted early enough to make their 10 a.m. plane out of Juneau (even though AST was told by Groff at 9 a.m. to hurry up); and boats, helicopters, and planes had been arranged for. I was told that Ron Evans, president of Best Western was due in town at 1 p.m. Also, when I found that the SEADOGS missed the opportunity to make the commercial flight, I called the trooper and told him that the Coast Guard on Annette Island would be more than happy to fly a C-130 to Juneau, pick up the dogs, and fly them to Cordova. I explained that they are required to fly a minimum number of hours anyway, and if they don't get a mission they literally go up and fly around in circles. He said that was certainly "worth thinking about." I also informed the hunting party of the SEADOGS problem, but predicted that the trooper would never call for a C-130.

By mid-afternoon I needed a nap, but by the time I left the office at 3:30 p.m. I had gotten word that the local AST office was receiving phone calls from state senators and a congressman. I thought those phone numbers I gave to the hunting party had something to do with it, but as it turns out, the owners of the Reluctant Fisherman, Dick Borer and his wife Margy Johnson called the governor, and U.S. Senator Frank Murkowski and Congressman Don Young.

The State Commissioner of Transportation, a retired Coast Guard officer who had been in charge of search and rescues in Alaska, called AST and informed them that the Coast Guard helicopter, which had been out of service for routine maintenance, would be flying tomorrow and would be in the Fish Bay area to help.

Before I knew that these other arrangements had been settled, I called Leuty from my house that evening and told him that we fire department members will not act as officials of the fire department—and consequently of the City of Cordova—we will just be "residents" who are going to help some people find their buddy. In a non-official manner, that is what we would do in the morning. Ron Evans said that he would pay any expenses, so we didn't need AST for that. This would be a private operation.

Of course, you always worried in the back of your mind about not being insured if one of your people gets hurt or killed. But we would risk it. After I talked to Leuty, I got a call from the Coast Guard Rescue Coordination Center in Juneau. They told me that the SEADOGS would be arriving at the Cordova airport in a couple of hours. JoAnne Havens (firefighter/medic, and wife of Deputy Chief Roger Havens) made arrangements for them to be picked up and arranged for rooms at the hotel for them. Then the trooper called and wanted me to arrange for someone to help the SEADOGS and handlers up the mountain. I got the forest service to donate a guy to do that.

Thursday morning, at 7 a.m. I called the Reluctant to have the dog handlers come to the fire station to look at the aerial photos before they left. They said they had to meet with AST at 8:00, but would be down after that.

At about 9 a.m. Bruce Bowler, head of SEADOGS, and Jeff Newkirk, a Juneau firefighter, showed up. Metcalf, Marilyn, and Atwater came too. Everyone went over the charts and photos. Groff and I had already drawn sectors to be searched with the dogs and a beach area to be searched with a group of volunteers. Bruce asked if I would organize and run the beach search and coordinate the overall operation. I explained that I hadn't been asked by AST for any assistance yet, and was only making suggestions to them and alerting people in case we were needed. Bruce asked me to do it, and since he operates through AST, that was official enough.

I had the crew toned out to assemble and called the forest service. I selected Hoock, Zeine, Cheshier, and Kirko to go. Kirko was to be the coordinator on-site. I called Valdez and told them that in about an hour a plane would pick them up at Robe Lake. Because of the legalities of transportation for the forest service (they can ride only on "carded" planes or helos), I cancelled them and got three Coasties to go. I sent Belgarde off to find the Montague because the trooper called and said the Montague left and he could not contact them. They were fueling up and said they would be ready at 1 p.m. I sent JoAnne and Marilyn shopping for several hundred dollars worth of groceries and dog food, to be delivered to the boat. I filled up an ice chest with ice from the Reluctant and delivered it to the boat because their refrigerator was out. We assembled all sorts of gear; a shotgun, radios, spare batteries and chargers, mosquito nets, gloves, space blankets, surveyors tape, binoculars, charts and maps, rosters, and aerial photos and packed them up to go.

Vicki drove everyone to Mavis Island, on Eyak Lake, to board the float planes. While the search parties were boarding, Vicki shook her head and said, "Boy, the troopers are gonna bite the big one on this."

At first, the crew loaded up into a Twin Beach aircraft and started taxiing across the water, but it was sputtering so badly it couldn't get up speed.

Nikki Cheshier asked the pilot, "Is this thing going to make it?"

"Well, I'm trying." He said .

Nikki said, "Turn this thing around and take us back" Which he did.

The crew waited and then flew out on a Beaver. The Twin Beach, after the pilot drained water out of the carburetor, flew to Valdez and transported the Valdez crew to the site. (Weeks afterwards, Chief McAlister remarked what a piece of shit that plane was. I told him my crew refused to ride in it.)

The Montague left at 11 in the morning and got to the site at 4:30 p.m. We used them as a communications link with the search parties.

Also going from Cordova were Leuty with Phantom and Brady with his dog Angie (that made four dogs total) and Bill, Randy and Marilyn. Scheduled to come in at 1 p.m. were two guides from Yakutat. I was at my desk in my office when the guides came in. The main guide used the nickname "Bear," and without looking up, I directed someone to issue him a portable radio and designated them as "Bear One and Bear Two." The only reason I detailed this meeting was that a few years later, Bear called me and asked if I would be a character witness for him for his trial. I didn't even asked what charges he faced, I just explained to him that I didn't even know what he looked like, let alone could I testify about his character. I've never heard from him since.

Anyway, two more SEADOGS were scheduled to come in on Friday. By 6 or 7 that evening, the beach searchers were done and the dog teams were headed down the mountain for a debriefing and to plan for the next days search procedure. One human footprint was found at about the 2,000 foot level, and another four or five were found a little higher up, heading north. Incidentally, it was a beautiful day and the next day promised to be the same. Bear 1 and 2 decided to camp up on the mountain rather than sleep on the boat.

The Coasties and Frank's friends flew back to Cordova for the night and the Valdez team flew back to Valdez. That left about nine people on the boat and two on the mountain. At about 9 p.m. Brady radioed back that he wanted his climbing gear, two more climbers, and enough gear for four total. I had Plumb and Gundlach gather the gear, and arranged that they fly out in the morning on the first trip. By this time two more dogs arrived and they would fly out with the climbers.

The nine sleeping on the boat had a tough time. Mark Hoock snored so loudly that Zeine, Kirko, and Cheshier could hardly sleep. When Kirko got up to take a leak, Brady's dog, Angie was wagging her tail and wanted to play even though it was the wee hours. But Phantom, who was sleeping at the top of the stairs was livid because he couldn't sleep with all the snoring. He lunged at Kirko, bared his teeth and growled menacingly. So Kirko went back below and, using his flashlight, found a forward hatchcover, and went out on deck that way. Hoock, when he woke up with a full bladder had to take the same route out.

The next morning, Friday, at 7:30 a.m. more groceries were purchased and taken out with the dogs and climbers. On site, the crew was late getting started because the chopper in Valdez had some needed maintenance and the charter guy needed authorization from the Cordova trooper to fly. I called the boat and the trooper said that Ron Evans was paying for the charter and that Evans was enroute to the site, but to go ahead and inform the charter guy to send the chopper. It was needed to shuttle searchers up the mountain.

Just about the time the chopper got there, Bear 1 and 2 crested a ridge and saw Franks body on a ledge about 1500 feet down a vertical drop-off. It appeared as though he had landed on his head. A bear was eating him. They shot the bear, which toppled off the ledge.

Brady and a Valdez firefight dropped a line down, used an ice axe as an anchor, and belayed down to the body. Brady and the firefighter wrapped the body up and, with a line and harness, had to set up a pendulum swing for a better position to enable the chopper to lift the body out.

Later that day, Hoock told me that on the boat, Marilyn was enraged at the delay in the whole operation and the fact that so many people were working so hard and that two ASTs were sitting in lawn chairs on the beach. She was voicing her frustration to her attorney who was also on the boat.

The body and some of the searchers were back shortly after lunch time. About six or seven were coming in on the Montague and due in at about 8 p.m. At about 4 p.m. I went to the airport and said goodbye to Bruce Bowler and the other three dog handlers. Later, I called Ron Evans and gave him my condolences. He thanked me for my assistance. I felt too uneasy to call Marilyn. Since I had a band gig that evening, Vicki offered to meet the crew coming in and supervised the stowing of all the gear. She did this after just completing a 5-hour EMS medevac herself—a man at the Tsiu River was run over by a forklift from his foot to his chest and back again. She had to fly out to get him and bring him back.

* * * * * * * * * *

McAlister and I chatted about this event for quite some time on the phone. I mentioned that I had been contemplating an official mutual aid agreement with the troopers. That might make them—especially new ones just transferred to Cordova—more prone to call us for help in a more timely fashion. I explained to him that I had done all the preparations for this SAR package prior to being asked. So, when I was officially asked for help (by Bruce Bowler at 9 o'clock Thursday morning August, 28th) I had the first of the searchers enroute before noon. Boats, aircraft, and provisions were ready immediately. Less than 24 hours after our team arrived on site, Frank was found.

Anyway, McAlister had a better idea. Why don't we draw up a mutual aid agreement between each of the communities of the north gulf coast—Yakutat, Cordova, Valdez, Whittier, and Seward—so that we could draw resources from more towns and spread out the responsibility a bit. This agreement would then be offered to State Department of Public Safety. It was such an excellent idea that McAlister had a draft written up by the next day. Each city signed it and so did the State Commissioner of Public Safety. The geographic area covered by this agreement is about 50,000 square miles.

THE FLORIAN RESCUE

HASTY TEAM/SUPPORT TEAM CONCEPT

I can't remember when we started using a technique called "Hasty Team/ Support Team" that we borrowed from avalanche rescue groups. But it was such a great routine, that we used it repeatedly. When the circumstances of the event lends itself to the Hasty Team/Support Team approach (you know where the injured person is), it gives you the best of both worlds: Quick access to the injured party and broad support in the field for the operation.

Here's what I've seen in the past: Precious time is spent while every conceivable piece of equipment is gathered by the army of rescuers before they leave the station to get out to where someone may be hanging on by a thread. Naturally, you don't want your people stranded in a remote area without the equipment and provisions to sustain them.

So, here is what we began doing. Assemble a Hasty Team of 2 or 3 people who carry the minimum amount of gear to treat the patient's injuries and to hunker down for a short period of time. The Hasty Team can be out the door in 20 minutes to drive, climb, or fly to the patient. They don't have enough gear or manpower to get him down off the mountain, for example, but can keep him from dying.

While they are enroute, you organize the Support Team. This is the safari of people who lug the stokes, climbing gear, food, and all the crap to get him or her out or to bivouac overnight, if necessary.

Of course, this Hasty Team/Support Team is not entirely risk-free, especially if you are in too much of a hurry. One fall evening, I dispatched our Hasty Team a little too quickly. Two visiting Scandinavian climbers (Ysbrand and Florian) had walked up to the Heney Ridge behind town, when Florian fell and injured his left leg. Ysbrand came back to town seeking help. At 8:05 p.m. the assembled Hasty Team—Bob Plumb, Dana Smyke, and Ysbrand, took the needed equipment—Medic pack, SKED (a roll of thick plastic cut and designed to wrap around a patient for carrying, belaying, or helicopter hoisting), 300' of static line, two portable radios with spare batteries, 2 figure 8's, harnesses, carabineers, space blankets, matches and lighter, head lamps, flares, and their packs of personal gear—and headed for the Coast Guard helo station at Mile 13. Ysbrand would direct the chopper to the accident site, then remain in the helicopter. Since it was early fall, I figured there would be ample time for the Support Team to rendezvous with them if belaying the patient down the mountain were necessary or if bivouacking overnight were the only options.

At the same time, Bob Pudwill and half a dozen other guys geared up for the climb up the mountain. When we looked at the small group of guys and the mountain of equipment we wanted up there we radioed the chopper to see if they could haul the gear up to a base camp for us. They said they would try, but there wasn't much light left.

They came screaming over town toward the hospital's heliport as we drove quickly down to meet them. On the pad, they never shut down, and the chopper's "swimmer" motioned us frantically to get over there and load up quickly. The crew just started throwing gear in as the swimmer stacked it: 6 more harnesses, 8 pulleys, several figure 8's, (in case belaying were necessary the next day), 4 heat packs, over 2 gallons of water, dried food packs, stove with 2 extra fuel tanks, hoods, mittens, 4 mustang suits (containing some strobe lights and ELTs), a tent, 2 bundles of blankets, 2 plastic tarps, webbing, carabineers, 2 sleeping bags, and chemlights. The inventory ranged from being inadequate to completely useless. But somewhere between the extremes, they would have what they needed.

The chopper launched like a rocket and headed toward the mountain, which was quickly fading black. We drove back to the fire station and all stood in the parking lot and watched. The chopper found a snowy bowl below the accident site and dropped off the gear.

By now it was totally dark, so the chopper ascended the face of the mountain and turned on its "midnight sun" light and saw Dana and Bob slowly making their way to the patient. Hasty was very grateful for the light. With the help of the light, Bob spotted a ledge below them that was big enough for all three of them to wait out the night. With some effort, Bob and Dana were able to get Florian, wrapped up in a flight suit, down to it, and make themselves as comfortable as possible. The chopper, as it rose higher to go up over the mountain and head for the air base, radioed Dana and told him that it was snowing higher up. It was raining on Hasty. It was just a light mist on Pudwill and the Support Team.

It was 11:15 p.m. when Plumb radioed a patient report consisting of stable vitals, a deep laceration on his left leg (showing no indications of fracture, but would be splinted anyway), lots of scrapes, and pain in his right hand. They would be bundling up for the night.

When Support made it to the gear drop-off site, they set up a comfortable base camp and chatted a bit with Hasty over the radios. Hasty was miserable in the rain. The rescue blankets that they had, made of plastic and paper, dissolved in the rain. The piece-of-shit radio they had was on its second (and last) battery and was losing power with every transmission. They had several large Hefty garbage bags that they wrapped themselves in to keep dry. They planned that at first light, Support could try to make their way up to help in a technical descent down to a spot where the chopper could pick up the patient. Hasty radioed us and said they were turning off the radio for the night to save what was left of the battery.

At 10 minutes after midnight, I headed for home.

At about 5 the next morning they were all on the move. With daylight, Hasty saw that they had nestled in near the edge of a sheer cliff that plummeted straight down several hundred feet.

Dana and Plumb were on the radio saying they were starting down. Support could not reach them from where they were, although they'd made their way up to within sight of one another. Hasty rigged up to start belaying Florian down. While Hasty did that, Support was rigging for further descents. The weather was clearing and it was almost totally light as the two teams joined and made their way toward the bowl carrying the patient. The chopper was right on time, picked up the patient and flew him to the hospital.

The chopper began a series of shuttles back and forth bringing in crew and gear.

So, the plan was good, but Hasty and the patient were still miserable and hungry the entire night, because we couldn't predict that the Support Team would be unable to rendezvous with them and set up a comfortable camp.

So, Hasty Team members must gear up as though Support Team will be delayed. No exceptions. In all, the Hasty Team/Support Team concept can save lives by getting critical help to a victim without delay. But be a little prudent. Too Hasty makes wasty. Sorry, I couldn't help myself.

LOADING GEAR ONTO A COAST GUARD H-3 HELICOPTER
Photo: Cordova Times

132

BLOWN OFF THE BRIDGE

It was about mid-day one winter, when the sky was clear, the sun was shining, and the cold air was windless when we got toned out that a woman fell off the Mile 27 bridge. On the way out there, I was convinced that there would be nothing we could do, because no one survives a plunge into the Copper River. The temperature creates instant and fatal hypothermia, the silt packs into clothing like cement and drags the body straight to the bottom. And finding the body later would be nearly impossible because the speed of the current is usually ripping under the bridge. And even though I had 27 miles to drive—over half of that on an unpaved road—I still drove like a bastard.

As per protocols, I radioed my locations periodically, and when I was about 5 miles away, my son Jason, Lieutenant of Engine Company 2, radioed to remind me that the rescue truck was only a couple of miles behind me with the divers and diving gear. A short time later he radioed that to me again. I was getting a little pissed at his way of telling me not to suit up and jump in without the rest of the dive team. Even though I usually did that, I'm not insane enough to take unnecessary risks. Without a tether line, I would end up in the Gulf of Alaska in a frighteningly short period of time. Plus, I know that the body would have come to rest quite a ways downstream and shortly be buried in the sand on the bottom. Nevertheless, I said, "10-4," both times and kept driving.

When I approached the bridge, I heard the trooper radio that there were hurricane-force winds on the bridge. I could see the bridge up ahead, and where I was there was no wind…zip…zero. At the bridge, though, I felt my car rock in the wind, but still, I felt their report was a bit exaggerated.

I stopped on the bridge; I couldn't open the driver-side door against the wind so I got out on the passenger side, and the sound of the wind was almost deafening. As soon as I was next to the hood of my car, the wind caught my helmet and, with the chin strap under my chin, it snapped my entire upper body back like a lower case "r." I had to drop to my knees, in the lee of my car, to right myself. I made it back into my car and took my helmet off. I reached into my dive gear bag and pulled my rubber dive hood onto my head. Back outside, I could not look upriver, into the wind. Either dust or ice crystals scoured my face and eyes and I could only look down. So, I got my diving mask out of the bag and put that on. Of course, my turnouts kept me warm, but I looked like an idiot. I crawled to the railing, grabbed hold, and hand-over-hand made my way to the section of the bridge over the island of sand that the second support was buried in. I crawled over it and made my way down. The river was frozen solid and the wind had blown it to a sheen of ice.

133

Even though, while I was getting my hood, I'd radioed to the other responders what the conditions were like, when the rescue truck pulled up behind my car, 140-pound Robbie Mattson stepped out of the back and shot straight across the bridge, slamming into the railing on the other side. As he clung to that, others in the truck threw him a line for him to pull himself back to the rig. They then secured the line to the near railing and used that while going through the laborious job of unloading the stokes and all the needed gear.

The divers were protected by their clothing, and some of the others put on Mustang suits. Those left in turnouts discovered that helmets were a hindrance and either wore nomex hoods or nothing at all. They made their way to where I was. Before we arrived on scene, the troopers had already been there for some time. They were wearing lightweight jackets, cotton pants, and no hats. The conditions were brutal for them, yet they never budged from their positions.

I think it was Plumb, Smyke, and Mattson who elected to go after the woman. Apparently, the woman and her boyfriend were in his car having an argument when she got out and was blown over the edge and she hit the ice below. She was not moving when we got down there.

We tied a line to the foot end of the stokes and a line to the head end of the stokes. Smyke fastened the end of the line to himself, and he and the other two lay prone on the ice, and stabbing the pick end of the fire axes into the ice, they slid their way across the ice. They began well upstream (and upwind) of the woman and slowly picked their way across to the bridge abutment that the woman was behind. When they were next to her, they used the line and pulled the stokes over to them. We held onto the second line to be able to pull it back.

The woman was frozen face-down. Her warm body had hit the ice and warmed it for a few minutes, and then the ice re-froze around her. They yanked repeatedly to free her, all the while holding onto their axe handles to keep from getting blown away. Finally, she was ripped completely out of her clothes and they wrestled her dead body into the stokes and strapped her in. The crew on the beach pulled on the line and the stokes zoomed effortlessly across the ice to us. Smyke and his crew made it back quickly enough to help carry the stokes up to the bridge. As expected, it was a chore getting the stokes to, and then into, the rescue truck. The rescue truck backed off the bridge and into a calmer area where they placed the woman's body in the ambulance.

Samantha Lobe, 17, our newest and youngest member was on the run, and she rode in the back of the rescue truck with the frozen nude body of the victim. After the body was taken from the rescue truck and into the ambulance, I put my hand on Samantha's shoulder and asked, "You doin' okay?" She said she was. Back in town, my son Jason told me that Samantha (his girlfriend) revealed to him that if one more person came up to her and asked if she was alright, she was

134

going to get really pissed. I spread the word to the rest of the crew. I guess it's natural to feel fatherly to someone like that, but what is not so apparent is that assuming a 17-year-old girl isn't as tough as everyone else is insulting. Here's a better idea: Assure that a new member is watched by their immediate supervisor for any unsettling behavior, or post-incident stress. That's all that is needed.

Oh, and by the way—I know I have mentioned this elsewhere—but these State Public Safety guys get very few accolades for taking the beatings that they often do. Letting their boss know wouldn't kill ya.

I should mention that Samantha is now my daughter-in-law, and is a paramedic/dispatcher for the Anchorage Fire Department.

TWO RESPONDERS TO THE BRIDGE CALL—AUTHOR'S SON, JASON, AND SOON-TO-BE DAUGHTER-IN-LAW SAMANTHA LOBE WHETSELL. HEY, IT'S *MY* BOOK!

Photo: CVFD

135

FALL FROM MOUNT EYAK

The day before Father's Day, we were notified that a young woman named Peggy Hall, age 21, had fallen from the peak of Eyak mountain and killed herself. Five young crewmembers that went up there were working a summer job on a fish processing ship that was in town, and on such a nice day off, decided to climb up Eyak to look at the sights. There were five of them making the climb and even though it was mid-June, the snow was deep—yet hard—near the top of the 2,300 foot peak.

Wearing cannery boots—unlined, thinned-walled boots with minimal treads—they plodded first up the grass and rocks then through the hard-packed snow to the top. Peggy and one other student were on the very top, while the other three were on a ledge right below them. Peggy slipped and fell headfirst right past the lower three sightseers. Taro Egawa impulsively reached out to catch her, and he was yanked off his footing and plummeted down, too. The remaining three watched in horror as Peggy and Taro tumbled and cart-wheeled together down the nearly straight-down west slope of the mountain. They collided repeatedly. One of the witnesses later said that at one point Taro stopped, but that Peggy's body went noticeably flaccid and lost all muscle tone. Even untrained, they could tell that her neck snapped. She then disappeared into the mist below.

No one knows how Taro managed to arrest his fall, but he did. With his friends' help, they all managed to descend the mountain to get help. Taro suffered broken ribs as did one of the others (from a fall during their descent) and both were taken to the hospital. The other two summoned the fire department. It was now 7:20 p.m. They described the fog rolling in as they descended. They were absolutely certain that Jill died in the fall and that she would be difficult to find.

The team I assembled consisted of Plumb, Gundlach, Pudwill, police officer Fred Brady with his dog Angie and AST's Greg Hamm and his dog Mink, and me.

After the briefing of the conditions—an evening drop in temperature, fog now enshrouding the mountain top, and the snow hardening to an ice-solid face—we geared up. Here is a neat technique I picked up some years before. I needed to be a wearing raincoat and pants, but since the slope is so steep and so icy, if I were to take a tumble, I would slide so fast I would never be able to arrest my fall. So I put on very lightweight nylon rain clothes over my wool long underwear. On top of that, I put on jeans and a Carhartt jacket. My outerwear could be totally soaked and I would never feel it. And wearing regular outer clothes gave me access to pockets. I didn't have any of that expensive mountain-climbing clothing. Since I knew I would be trudging creeks and marsh as well as through hard snow, I put plastic sacks over my socks, then put my climbing boots on. I'd been doing that for years and my feet have always remained warm and dry as I've waded across creeks made from melted snow.

We all had harnesses, ice axes, climbing rope, and the SKED. Brady had his dog and Hamm had his dog. We went up the south side in record time. And the only snow—which slowed us a bit—was just on the top few hundred feet. But once at the top, we looked over the cornice at the ominous west side. The fog was thick and eerie looking.

We would have to rappel over the cornice and rope up together to traverse the face of the mountain. No one seemed to mind this except the dogs. After Plumb reiterated how to use the ice axe to arrest a fall and had all of us practice jamming the axe into the snow and slam our chests on top of the axe and lift ourselves up onto our toes, we geared up to go over the cornice.

We drove the pointed handle of an ice axe into the snow, up to the hilt, to act as an anchor for the rappelling rope and Gundlach went over edge, got good footing below, and went off belay. Then the dogs were harnessed up, and they were okay with that until we dangled them over the cornice one-at-a-time, at which time they went ape shit. They were very happy when Gudlach wrapped his arms around them below.

Down on the west face, we all clipped in together with our ropes, but since the dog handlers weren't experienced climbers, they did not go first. Plumb led the way, followed by Gundlach, then me. Fred and his dog were behind me, and Greg Hamm was behind him. Pudwill was last. The dogs handled the steep slope as though it were flat terrain.

The fog was so thick I couldn't see the first or last guy. I could only see Gundlach in front and Brady behind. We were leaning way into the slope and using the handles of the ice axes like canes as we walked (except for the one axe which was left on the cornice). We were looking for slide marks, which we expected we would cross over. Then Hamm yelled for us to stop. His dog alerted on something that four of us had just walked over. There were no slide marks there, because the snow was so hard. But the dog was scratching and digging in it. Greg bent over as we all stood motionless where we were. He stood up holding his hand up, but we couldn't see what he held. It was a long, black hair. The dog was barking and looking down the mountain into the mist. Greg gave the dog a command and watched as the dog made his way down and out of sight.

We all converged to Greg's spot and started down. It didn't take long to descend to where Peggy's body was. It was a bit tricky getting her body into the SKED because the plastic was so slippery on the snow and ice, but we eventually got her cinched up inside like a papoose. Then it was decision time. It was

137

a pretty short distance to the top, where a trek down the south face would have been comparatively easy. But it was up and very steep. It would have required a technical ascent using lots of anchoring, pulley set-ups, and we weren't quite sure how to get back up over a cornice, which of course is an overhang. Down, on the other hand would be a much longer distance and pretty tough when we got into the tree line. We looked up, tried to make a few yards up and that was excruciatingly difficult. The unanimous decision was to go down.

Descending was minimally technical, consisting only of using ice axes as anchors for belaying ourselves and the victim down. But eventually, the terrain began to level out and the snow disappeared and we were trudging through soggy muskeg, and fast-moving, chilling streams. We got to the tree line about midnight. In Southcentral Alaska, even in June, midnight is nighttime. And under the canopy of the trees, it's flashlight time. So we had our head lanterns on and we made our way through the thick brush and over what seemed like millions of fallen trees, lifting the SKED oftentimes chest high, or dragging it under bowed up roots. Occasionally, we came across a much welcomed little open meadow. It was in one of the meadows when both dogs froze and stared off into the darkness to our right. The both growled and the hair went up on their backs. Brady's dog started whining and looked up at Fred. "Mr. Brown's in the area," Fred said, referring to a brown bear. Greg was the only one of us that was armed.

"You got your gun, Greg?" I asked.

"Yeah. It's in my pack," he said, spinning it off his back and onto the ground.

"In your pack?" someone asked, incredulously.

I had already spun around and found the tree I would climb and I noted others were too. But apparently, Mr. Brown wasn't interested in such a pitiful-acting bunch of buffoons and left. The dogs relaxed and we moved on.

The team complained that I never let them stop and rest. Whenever one would ask about stopping, I would get testy and blurt out, "No. I want to go home. I'm tired and I want to get out of here." Any time I have to do a marathon of exertion, I don't stop, because I can't make myself get started again. So I found if I get a momentum going, just keep going.

It was still pitch black out when we stopped at a deep precipice and had to belay down. Brady would be our anchor while I would be a few feet in front of him and tend the line while the others took the body down. The only thing we could see was what was illuminated in our head lamps.

Fred dug in behind me and the others went over the edge with the SKED. Then I heard Fred say, "Ooooof." The next thing I saw was his size 12 boots, one on each side of my waist. He slammed up against me and I fell backwards on top of him and lay on top of his body like a sled and rode him over the edge. We all

138

went flying down into the darkness. It seemed like we flew through the air for minutes, having no idea how far we would go or what lay at the bottom. We all hit with multiple "thumps" and sounds expelled from our lungs. Dead silence in the black. I asked, "Is everyone okay?" No one answered, and I got scared. "Is everyone okay?" I repeated. Then I heard a chorus of nervous snickering. I still had my flashlight and scanned the area. Fred, to my right side, was moving, and the others, to my left were in a pile, and were okay. I don't think we fell more than 12 to 15 feet, but were very lucky. I was now about 1:30 a.m. and we had no idea how much farther we had to go to reach Orca Road.

The farther down we climbed, the thicker the trees and brush got. Using only what I describe as controlled rage, we plowed our way through like bulls, cursing and growling. We were all drenched in sweat and our muscles were about spent, when the terrain flattened out and we heard the river running ahead of us. We reached the river at about 2:30 a.m. and it was now light enough to see. We carried Peggy more easily now, walking the banks of the river. Fifteen minutes later a group of firefighters met us and took over the carrying. We all reached the road at about 3 o'clock—seven and a half hours after we began, but it seemed like days.

Several weeks later, when the weather was good and more snow had melted, Pudwill went up and drove some pitons into some boulders at the top and stretched a cable between them for sightseers to hold on to. This was one of three incidents occurring at the same spot. Two ended in death and one in serious fractures that required a helicopter evac.

Later that summer Taro returned to town, his ribs on the mend, and he went back to the hospital complaining about an infection in his buttock. The doctor looked at the festering, swollen lump and started digging around. He extracted a tooth which had broken off from Peggy's mouth and gotten embedded in his butt when they were colliding into one another during the fall.

The peak of Mount Eyak has a breath-taking panoramic view, but even though it isn't a particularly tall mountain, hikers need to be wearing proper boots and stay on the south face, regardless of the time of year.

* * * * * * * * * *

Of course, those safeguards are not effective against stupidity, like when 20-year-old Brad went up one August evening and there was virtually no snow on the peak. He was from Palmer and was in town because he was crewing on the fishing vessel Pamela Lynn. At about 100 to 150 feet below the summit, he felt compelled to roll a boulder down to see if he could start a rock slide. Unfortunately for him, the boulder only rolled a couple of feet (enough to make the guy fall forward), it turned and rolled sideways over his leg and snapped his femur, then it rolled farther down the mountain. It was 7:30 p.m.

Brad's companions, Bill and Bob, used their clothes to cover him to keep him from getting cold since he was only wearing shorts—it had been a sunny, warm day. They also built a rock barricade to protect him from the wind before one of them went for help.

The fire department was toned out at 8:20 p.m. Nikki Cheshier was on duty as a medic that afternoon, and Mark Kirko as firefighter. They made up the Hasty Team. They dressed and geared up for the occasion and grabbed minimal medical gear. Kirko had been rock climbing at our drill site near the ocean dock when he got toned out. At the station he started to take a bag of climbing rope and some high-angle gear but was told by the officer on duty that he would not need that. A Coast Guard chopper was waiting for them at the base at Mile 13, so they headed out. No need to assemble a Support Team to go up. Up near the top, they spotted the friend of the patient and the chopper lowered Nikki and Mark and a stokes litter. The chopper pilot told them he had only taken on enough fuel for the first trip up. While they were tending the patient, the chopper would go back to the airport for fuel. The patient was lying on a narrow rock ledge several yards below them.

Nikki and Mark inched their way to him, dragging the medical gear and stokes over the layers or fractured shale rocks to the ledge the patient was on. He was lying there taking up most of the space. There wasn't much ledge left. It was so precarious up there on the ledge that it took an exorbitant length of time to expose the injury and apply a traction splint. They took a full set of vitals, of course, and relayed the information to dispatch. They were so busy doing that while balancing on the ledge and trying to hang on, they didn't notice that the afternoon fog was enveloping the mountain. It was getting thicker and thicker. A feeling of dread came over Mark. "The chopper can't drop a basket through fog like this." They knew that in a few more minutes, they wouldn't be able to see well enough to climb back up, there was no Support Team coming with equipment to manually lift them or the patient back to the peak; they were starting to get soaked and cold, and so was the patient. In addition, they were getting muscle cramps from just trying to remain perched on the few inches of ledge they each had. They couldn't even tie off together because they had no rope. They warmed themselves somewhat by ripping off long lines of profane statements about the ever-worsening circumstances and just piss-poor preparations, none of which they were responsible for.

The patient was looking shocky, and Mark and Nikki were silently staring like ghosts into the fog by the time they heard the chopper. The chopper couldn't see them and Mark could only tell them over the radio which way to maneuver. Eventually, the chopper was right overhead and they could see one another. The rotor wash (about 120-knot wind) was dislodging broken rocks all around them, and Mark worried about getting hit. Small pebbles were flying all around them. Finally, the hook hit near them and he locked it in on the basket.

It was a great feeling for Nikki and Mark to have the patient hoisted up and out of the way and the rotor wash and deafening sound disappear as the chopper headed for the hospital. While it was still light enough, Nikki and Mark made their way back to the patient's friend.

It was 10:40 p.m. by the time the chopper got back and hoisted the three of them back aboard.

That was a two-and-a-half-hour operation, thanks to the availability of the helicopter. It not only saved hours, it saved a hell of a lot of brutal physical exertion. But had the chopper been grounded by fog or a mechanical problem, it would have gone bad fast.

Here was the same potential as with Florian on Eccles. Start your backup, or your Support early (while it's still daylight). Take extra gear. What's the harm?

Also, I know that other agencies wait until daylight to begin rescue operations.

Some patient won't hang on that long, especially if they are alone and don't know that help is coming.

AVALANCHE AT THE TUNNEL PROJECT

Spencer Tracey played the lead in the movie *Captains Courageous*, which I thought was a cool movie character when I saw it as a kid. Rugged, wind-blown, "thrashing hair and whale-blue eyes", with a glint of humor always. Oblivious to the rain, seas, and tough life, that was old Spencer. Maybe that's why I thought Gary Stone was cool. Wool shirt, black jeans, tousled hair, broke as hell, but never whipped by it. A logger from Idaho, with a pod of kids just like him. All were twinkle-eyed, soft-spoken, civil, and solid.

When logging went to hell in Cordova because of the uproar against it by local environmentalists, Gary picked up work with a contractor who spoke of Gary the way everyone did. The first on the job in the morning, he independently took care of the equipment and the job. You know, it's like the dependability of a close relative that you'd never think to question. Of course, the contractor wasn't exactly rolling in dough, either. So Gary was barely getting by. In fact, finances were so bad that when his son, Micah, went to the state finals with the high-school basketball team, Gary wasn't sure how he could go watch...all the way up to Fairbanks. He really wanted to go, too. Anyway, Gary's boss had to have a rig picked up in Valdez and brought over on the ferry, so he got Gary a round-trip ticket to Valdez. He really only did that to get Gary onto the road system, because Gary told him that somehow he was going to go to Fairbanks to watch Micah play his last games. So he hitched a ride the 400 miles up from Valdez, and rode most of the way back to Valdez with my wife and me.

I never gave it much thought when he never bought food at any of our stops coming back. I didn't know he was broke. On the way back he confided in us that he might have to go out of state to find steady work, and that Micah was going to remain to finish school. He wasn't sure he could come back. But, he added, Cordova was like a family in itself, and that Micah always had people like us he could talk with. We were flattered.

Micah's life was not without its own share of disappointments and hard times. He, too, was never victimized by it...he was a working man. Hell, just before Micah graduated, he came by my house and handed Lou and me an envelope with the graduation announcement. I, like always, felt obliged to make a crack (not knowing how really broke he was) and quipped, "What're you doing hand-delivering announcements, saving thirty-two cents?" I didn't know that was exactly what he was doing. That evening, we visited for an hour or so as he explained his plans for the future. Gary had been out of state for awhile then and Micah was just trying to get by, going to school and supporting himself. Tough game for an 18-year-old.

142

Anyway, Micah graduated, and Gary came back when a construction project opened here. The electric company was starting their second hydroelectric project out past the end of Power Creek Road. Power Creed Road is a narrow, winding, dirt road that heads south skirting the edge of the Eyak Lake for about 6 miles, then follows the river that feeds it, toward the mountains. Where the road ends, a trail begins into the apex of steep mountains that loom about 2,500 feet above the river. Whitewater Engineering (the construction company) punched a road farther into that area near where the trail was and began the tunneling project. The crew was a grungy bunch of tough, hard-drinking workers—some locals, and some here from other projects with Whitewater. On March 15, 1999, Gary got Micah and some of his young, recent-graduate friends work there as well. He was a great role model for them. Micah enjoyed cutting timber with his dad. They weren't outdoorsmen like the hippies who think they can accomplish world peace by canning vegetables and playing the tambourine. They loved the brisk, and even drizzled mornings, the scream of the chain saw, and the smell of the blue exhaust. Clearing the area for the tunneling was the most fun for both of them.

Problem: A lot of the locals who are hikers and skiers were alarmed about the location of the project. Of course it was a perfect location for the tunnel and turbine because all the rain and melting snow from those mountains converged at the apex into a fast-moving river that tumbles, then winds its way into Eyak lake. All this water running down the mountain slopes would be directed into the 1,000-foot-long tunnel, gaining enough force to spin the turbines. Some of the locals realized the area was an avalanche catch-all. Historically, the valley fills up with debris in the winter and the snow on the valley floor remains 30 to 40 feet deep well into summer. Experts call an area like this a "terrain trap"—a steep-sided box canyon that terminates in a pit, with only one way out.

Bob Plumb was very familiar with the area and had grave concerns about the avalanche hazard. Mike O'Leary, even more of an avalanche expert protested to a number of people about the potential. Before the project began, Whitewater representative Dick Potter came to a fire department business meeting in November and showed us the layout of the project and asked about our rescue capabilities. He was a very likeable guy, and brought his wife to the meeting as well. The congenial meeting created rapport between Whitewater and the CVFD. We discussed some of the hazards out there and explained our view of the avalanche potential.

That evening, after he left, and numerous times after that, we discussed our planned response to an avalanche there and had Plumb purchase more conduit for use as probe poles. The Hasty Team/Support Team approach we had been using originated with avalanche rescue groups. Rescuers are summoned, and while the largest body of rescuers are being equipped with probes, beacons, shovels, maybe search dogs, and are briefed in running probe lines of very fine spacing, the

Hasty Team (a smaller group) is at the most likely site of burial and are randomly probing, hoping that by chance, they might just locate the victim.

Anyway, the concerns about avalanches prompted an official evaluation which read:

"The project site from approximately 300 feet west of the powerhouse to the exit portal is at risk from avalanches off the mountain south of the job sites. I have not obtained the topo maps yet, but would estimate that this mountain has a vertical rise in excess of 2,000 feet. This vertical would allow sufficient momentum to be gained in an avalanche event to create very large and destructive force which could affect your personnel and equipment during the construction phase, and is likely to affect the finished facility in the longer term. Any large avalanche event is likely to inundate the entire valley floor more or less simultaneously. Frequency of major avalanche events is highest in the area next to the bridge and portal. Avalanche frequency would decrease gradually as you move down valley from the portal staging area."

That report, which was not widely distributed, was written March 8th; less than five weeks prior to the incident.

Thursday, April 15, followed a week of bad weather conditions that grew worse with each day. The night before, the area was buffeted by a rainstorm, and the tops of the mountains were loaded. Gary said to Micah, "I really don't have a good feeling about this." Around 9 a.m. Gary sent Micah and some of the younger workers home. Gary remained with some of the older workers. Around noon, Gary had been working with the excavator next to the bridge when the mountain let loose. The other workers, who miraculously were not in that immediate area, turned to see the huge wall of snow careen out of the cloud bank and slam onto the valley floor. They slogged though the 20-foot-deep snow to the excavator to look for Gary. They were able to see the top corner of the excavator and dug down to the door of the cab, but saw that he was not in it. Using a cell phone, they called in the alarm.

At 12:10 p.m. we received the page. While we were assembling at the fire station, Deputy Chief Bob Plumb called Points North Heli-Adventures company for their assistance and the Eyak Ski Patrol to respond and assess conditions and establish probe lines. I designated Plumb to be operations chief and to set up a command post at the site, while I established the EOC at the station. The CVFD would be the Hasty Team and the ski patrol and Points North would supplement them as soon as they were outfitted and ready. The Hasty Team, comprised of 4 vehicles and 16 people were enroute at 12:19 with an estimated response time of 30-40 minutes.

While they were enroute, we received another call that perhaps 3 other people were buried. However, we soon ascertained that all other persons were accounted for. At 12:30, Points North checked in at the station with 10 people; we listed their names and the needed personal information about them and they were enroute at 12:35. We contacted the local state public safety officer Sgt. Jeff Babcock who, in turn contacted the Anchorage post and put them on standby. About the time that Plumb had set up the command post and radioed back that Truck Company 5 was doing random probes around the excavator and that fire department medic, Dixie Lambert, had her search dog, Gypsy, working the area, I got a call from AST Sgt. Paul Burke in Anchorage, offering the assistance of another dog, an avalanche expert, and himself.

Since neither Gypsy nor the random probes were netting any results yet, I said yes. I was really disappointed that Gypsy was not getting any hits. I kept thinking about the time that she had detected the location of a logger at Two Moon Bay buried under 30 feet of mud. Of course, Dixie reminded me later, that it took many days for the scent to rise up through 30 feet of mud. And this snow had the consistency of cement. Dogs save avalanche victims who are buried under powdered snow when a detectable scent is at the surface in just a few minutes.

By 1 p.m., five members of the ski patrol had checked in and were on their way to the site. Mike O'Leary, senior member of the patrol, was very apprehensive about having people working that site. He is a local expert in avalanche conditions and paced and muttered during the few minutes he was at the station. "Too dangerous," he kept saying.

Before they arrived on scene, Plumb, himself very cognizant of the dangers, had been doing all he could to assure the safety of the responders. Armed with compressed-air horns for signalling, and binoculars for scanning the mountain slopes, four of our members watched for subsequent avalanches. However, the low ceiling of clouds obscured the view of the mountains above 400 feet. It was still snowing. Mattson was tasked with watching the probers to track where each person was. The creek was dammed up by the snow, and the water behind it was rising. A probe line was set up on the upstream side of the excavator. A rope was used to keep the line straight and the probers worked from the excavator towards the rising water.

A small slide was spotted coming down, air horns sounded, and the probers scattered. They returned, this time accompanied by Points North, and Plumb pulled everyone back out of the slide area who was not wearing avalanche beacons. Another large slide was spotted and the searchers retreated from the area. One CVFD member was stuck in the snow but was pulled free, and Gypsy got bogged down and had to be lifted out. She injured her hind legs permanently. This slide piled an additional 15 feet of snow on the previously searched area. The toe of the new slide was now 20 feet from the excavator. Plumb moved the staging area further back down the road.

145

In the meantime, at the EOC I set up the tracking board with Post-its of names of all the responders and their locations and times; had radio announcements requesting volunteers for the next shift; set up a card table at the entryway to city hall, manned by a Coast Guardsman, for listing those who came in to volunteer; and sent people to the hardware stores to purchase all available snow shovels and transport them to the site.

The ski patrol arrived at the site and strongly recommended that all personnel be evacuated from the area. Plumb began pulling everyone back. Points North requested they be allowed to search in the area of the logs (which had been part of the bridge, and had been ripped up and moved by the avalanche). Plumb gave them an okay for 30 more minutes. Another slide came down and the searchers evacuated again. This time, the slide stopped above the area, and the search resumed.

The water, which was dammed up, was seeping through the snow and rising in the area of the excavator. By 1:45 p.m. the water was inside the cab of the excavator and was rising at about the rate of 4 feet every 30 minutes.

Plumb was under a lot of pressure about now. The ski patrol was pressing him to remove everyone from the area; Points North wanted more time in the area; now the Whitewater crew wanted to move a crane into the area to start pulling the logs out, hoping that Gary might be in a void under them—never mind that the water was now 10-12 feet deep there. Plumb gave the okay to Points North and provided them with beacons for the operation but by 2 o'clock Points North was convinced it was useless and everyone was pulled out of the area.

Before Whitewater could assemble the crane and equipment, Plumb pulled the pin: Nobody goes in there. Whitewater personnel were livid. One of their own was in there. Plumb consulted again with O'Leary and the city manager (who had responded with the Public Works crew delivering a pickup truck full of shovels), and stayed firm in his resolve, enduring the remarks of the Whitewater crew. It had been over 2 hours since the initial slide. Here are the odds of living through an avalanche: A victim buried in 6 feet of snow, or buried for 30 minutes has a survivability of about 10 percent. If buried deeper than 6 feet and for 30 minutes survivability is less than 10 percent. The snow here, weighing tons, was 20 feet deep, and entrapment was over 2 hours—not to mention the water. It was done.

While this was going on, Micah came into the office and I led him over to the administration offices. Learning that the mayor was not around, I put him in the mayor's office and told him he could feel free to use the phone, have people in to keep him company, or whatever. I then called my wife and told her what happened, and suggested that she leave work and go sit with Micah. Lou came over to my office after seeing Micah and said that Michelle O'Leary was with him and his friends. I learned that Micah had worked as a deck hand for the O'Learys during seining season. Later, Michelle came over and asked me to tell Micah and his friends that it's too dangerous for them to go out to the site to look for Gary. I did that. They were not convinced. I informed them that the troopers ordered the road closed. Since it was a search and rescue operation, they had the authority to do that.

In search and rescue operations, the responsible agencies—between local responders and State Public Safety—are often loose and informal. This was the case when Sgt Paul Burke arrived in town about 4 p.m.

Paul Burke was a cool guy. Before his arrival, Plumb and I discussed the jurisdictional question in more depth and with greater intensity than usual. Okay, so this day we were like others we used to view with disdain: those "bureaucrats" who seemed more focused on ridding themselves of responsibilities than on the actual tragedy itself. We took the bull by the horns in this incident the way we usually do. In fact, the only SARs I can think of that we did not take the initiative was when State Public Safety was already working an operation outside the city limits. Here, however, we were the ones responsible. Even being in a remote area, it was within the newly annexed area and accessible by road. It was ours, and we took it on.

However, years ago AST was saddled with the responsibility of SARs in Alaska because there is so much of Alaska that is not within any other jurisdictional area but the troopers. And it wasn't all that long ago that many, many local jurisdictions were incapable of handling such operations. The idea is that AST must see that it is done, if no one else is doing it, even if it's inside a local jurisdiction. We have taken advantage of this before. Any time we predict that a SAR (in our jurisdiction) could cost significant money, we contact AST and tell them of the situation. We offer to run the operation for them if they want. If local AST is available, we will run it together. On occasions when they are not available, we would get the "mission number" from the area headquarters in Glennallen, and attach a list of expenses for reimbursement.

This discussion did not have the noble topic of finances. Neither Bob nor I wanted to be viewed as timid, over-cautious, and even cowardly in refusing to go back in. We knew the pressure to do that would be great. If we were to turn this operation over to AST, Burke could take that heat. It's a great relief to be able to say to people, "I agree, but they have tied my hands on this. There's nothing I can do." I have always hated people like that, especially when it was so obvious that they were bullshitting me. But now I was plotting to be one. And to do it at the sacrifice of a trooper I hadn't even met yet…planning to let him be the object of people's anger.

Burke was led into the apparatus room accompanied by a K-9 and handler and an avalanche expert. Our crew was waiting in the training room, eating the dinner that one of our guys prepared for them. Burke started right off with, "Well, what do you need from me? I'm offering to help you in any way I can, from providing advice, providing logistics, to taking over the whole operation if you want."

I said that even though the site was within the city limits, and because of our inexperience in avalanche rescue, "If you don't mind taking it over, I'd just as soon you do that." And then I just blurted out the underlying reason which came as no surprise to him, "Everybody agrees that it's too dangerous, and the potential for more avalanches is great. It's easier for us to blame you if we don't go back in." There, I said it....right out in the open.

He didn't flinch. "No problem," he said. Yep, Burke was cool guy.

It took him about 20 seconds to plan his next few hours of work. He told the team leaders in the training room, to fill out Unit Logs (document what each team did and the area they covered), then they could take some time off, and be back at 7 p.m. He spoke briefly with three Whitewater guys who had been summoned in. They spoke hesitantly to Burke and barely acknowledged Plumb or me—still pissed at the cowardly firefighters. Burke then took the avalanche expert, Doug Fesler, and Plumb in tow for a Points North helicopter overflight, but I interrupted his walk through the building and directed him to the mayor's office to assure Micah that everything was being done that could be. I didn't expect that level of candor, as he put his hand on Micah's shoulder and flatly said, "I'm sure we're looking at a fatality, here. Can you accept that?" Micah stoically indicated yes. These guys are solid.

Upon completion of the over flight, Burke affirmed that pulling out of that area was not only smart, but had been delayed too long. The cloud cover had lifted by the time the helo flew over the mountain peaks and the sight made Plumb shudder. The upper 1,000 - 1,200 feet of snow had not released and was dangling like the sword of Damocles. Exonerated. And looking at the water level, the avalanche expert remarked, "You don't need probers, you need divers." Upon their return, Burke put the operation on "hold." We informed everyone that they need not return at 7 p.m.

There was a fair amount of drinking going on the next couple of evenings and rumblings of townspeople planning on going out there to search without permission. About two days later, another over- flight revealed that the water had melted much of the snow and had seeped away; Gary's body could be seen on the ground between the excavator tracks and the logs of the bridge. The bridge had been ripped off of its foundation, slammed against the excavator, and the entire thing shoved about twenty feet. One of the hydraulic-operated rams had been snapped in two. The force must have been tremendous. It appeared from a distance that Gary's body may have been wedged under the track of the excavator. To remove it, Burke called the Anchorage Fire Department for the use of a couple of their guys and pneumatic lift bags. He was surprised to learn that we had such equipment in Cordova. But he hadn't actually scheduled them to fly down yet because he wasn't sure when it might be safe to enter the area. Everyone was hoping the upper part of the mountain would let loose and be done with it.

Burke flew in and out of town so often during this time that I was never sure when he was here. Plumb, however, had him on the phone often. At least the media quit calling. Burke made it clear to them that he was the one to contact regarding this incident. However, "persons interested" in the event were appearing almost hourly. There was a network of connections whose representatives needed to talk to us: Cordova Electric Cooperative (who was having the job done) and their insurance agent or the agent's attorney; Whitewater Engineering (of course) and their insurer's attorney; State Public Safety (responsible for the recovery operation); State OSHA and someone from the States Attorney General's office; and people whose connection to this was never clear to me. Burke, who always seemed mellow and in good humor, could talk to all of them simultaneously. I never saw him "lose it" when having a conversation interrupted.

Another day dragged by. Word spreads fast around here. People knew the body was visible and were getting more disgruntled by the lack of action. With the vision of a bear finding him, Plumb and I even discussed dashing in there, retrieving the body, and getting out before anyone even knew. About the same time we were discussing it, a local state public safety officer and another man did just that. Gary had not been pinned under anything.

Exoneration: that's what this divulgence of detail is all about. We could not save Spencer Tracy. But it was not ineptitude nor cowardice….a bit clumsy, perhaps (our rescuers are drawn from the community—none of them look heroic, there's not a green beret among them) but they were there under the looming snow…nerves jangling. The approach was systematic and all the bases were covered. We just couldn't save Spencer Tracy. I guess, ultimately, I hope Micah doesn't think we could.

AVALANCHE IN A RESIDENTIAL NEIGHBORHOOD

January, the following year, it had been snowing so hard that all I had been doing was shoveling hours a day, day after day. And with the recent storm, gale-force winds and heavy wet snow, shoveling was excruciating. I had just laddered my house to shovel my roof when my pager went off and dispatch reported that a "big" avalanche had just occurred at Mile 5.5, Copper River Highway (a spot known for previous avalanches, a fact ignored by a number of residents who lived there).

Just as I reached the bottom of the ladder I heard Deputy Chief Plumb was enroute. A few minutes later, when I heard him report massive destruction there, I asked the dispatcher to contact the city manager to activate the Emergency Operations Center (EOC).

Mile 5.5 is actually a dirt road which is a by-pass or detour off the highway. The loop road runs parallel to the highway but closer to the lake. The area was fully developed with structures (homes and warehouses) between the loop road and the banks of the lake and also between the loop road and the highway. It is about a mile long before it rejoins the highway.

I pulled in to the loop road next to Plumb's vehicle. We could not drive any farther. Large trees were across the road and snow was piled high beyond that. I told Bob to set up the staging area at the intersection and I would go do a size-up. I climbed up onto the first fallen tree and started the trek across the snow. I turned to see the city manager joining me. Apparently, he'd heard the tone-out on the radio in his office and never got the message from the dispatcher about the EOC. Anyway, the city planner—our Plans Chief—could get things started.

Walking across the snow was surprisingly easy. It was packed so hard, it was like concrete. As I headed east, I made a quick sweep though buildings damaged but not destroyed, and reported each building as apparently vacant. About halfway down the road I approached a group of people standing in the middle of the worst section. One was the recently retired police chief, Kevin Clayton. He had been making a tally of residents. As I talked to him, I got completely disoriented. "Where's your house?" I asked.

"There," he said, pointing to a single-story green structure. It just didn't register. His house was a 2-story home. It had been pushed forward, collapsing the bottom floor so that his second story now rested on the ground. Farther east had been a very large 2-story home that was now barely discernable. Between them had been Jerry LeMaster's (and Martha Quales') home, but there was nothing there but debris. I just couldn't get my bearings. Jerry was a 14-year veteran of the CVFD.

As Kevin recited the names of residents he had not seen, I relayed those to the dispatcher for police officers to try to locate, or for the EOC to locate by phone. The rush was on to account for everyone. One resident reported in an odd way. Kirk Gunnerson had been in the 2-story house east of LeMaster's, upstairs on his cordless phone when he felt the house shudder from a gust of wind, then explode. Stunned to find himself standing in the shambles, he stepped out through the tear in his roof and onto the snow. He stood there in his underwear, dialed the store where he worked, and said he would not be in to work today.

Kevin knew LeMaster's property layout well and pointed to where Jerry's driveway was. Firefighters were arriving with our makeshift probe poles (conduit). I told them to probe the driveway to see if his vehicles were there (hoping that one or two of them were gone, indicating they weren't home). Dana Smyke had armed himself with the compressed-air horn positioned himself away from the crew, and stared at the mountain slope; its top two-thirds was obscured by the cloud cover. He never moved or took his eyes off the slope. If a subsequent release happened, there would be only seconds to run out of the way. Bob Plumb was doing an excellent job summoning loaders and backhoes and had already summoned the ski patrol and Points North Heli-Adventures company to respond.

One by one, dispatch and our staging area were radioing the names of persons now accounted for. We were down to maybe 4 or 5 when the probers shouted that they'd felt vehicles under the snow. Jerry and Martha had been home.

I wondered if Jerry had his pager on his belt when the avalanche hit. I asked everyone to be quiet, all firefighters to turn off their pagers and portable radios, then I asked the dispatcher to page the fire department. We all knelt and stared at the boulders of snow and ripped lumber, listening. Nothing. By now, lots of people were there—department members, residents, skiers, equipment operators, troopers, police, Coast Guard, just plain neighbors, and residents. They stared at me for directions. "Dig," like I know everything.

I noticed the city planner there with a shovel. So much for the EOC. I radioed Plumb and asked who was running "resources." He said the apparatus room had been opened up and our people were on it. Just like any search and rescue, only this was not out in a remote area. The process is the same, and we'd been doing it for years. Vicki Hall, Dick Groff and a couple of aids were handling all "check-ins" and dispatching resources to our staging area. Melanie O'Brien and Penny Oswalt were checking them in at the staging area, and Plumb would ask if they should be sent in to the hot zone. Plumb also sent the few avalanche beacons we owned to the hot zone for our people.

I was glad to see Mike O'Leary show up. He said that often the 80-mph winds generated by a slide will blow people right out of their buildings and they would be found some distance downwind. He recommended the ski patrol probers run probe lines from the road to the beach. He had his wife Michelle run that operation. I began to call that "Zone 3."

We needed to open the road so that heavy equipment could get in. By now, everyone had been accounted for but Jerry and Martha. I had loaders and backhoes start from each end of the road and begin clearing snow, trees, and other debris. It took quite awhile for the heavy equipment to get near the area where we were digging. When they were about 50 yards from each other, department safety officer John Thompson said that since he was very familiar with excavation operation, he would like to coordinate the use of the loaders and backhoes. I told him to assign spotters for both of them. The spotters were to assure that the loader buckets would scoop slowly, by inches, allowing the spotters to look for victims as the buckets were filled, and then to search through the cut banks as soon as the loaders backed up. I didn't want victims traumatized by the heavy equipment. That operation became "Zone 2."

Zone 1 was the hand-digging and debris removal, and was the most manpower intensive. Oftentimes, a loader was asked to stop clearing the roadway and tow on a cable strap or chain to help pull a heavy piece of debris from Zone 1. About every 15 minutes, I would have all equipment shut down, all pagers and radios turned off, all searchers stoop down, and order another pager tone-out. Or whenever a piece of debris was removed, revealing a void beneath it, I would have the digger lie down and stick his head as far into the void as possible, and I would call for another page. Nothing.

City Public Works used additional loaders and men on chain saws to remove trees that had not been completely uprooted and were hanging over Zone 1 at forty-five degree angles. State Department of Transportation and the city were using loaders to open up the highway which was buried under what looked like about 30 feet of snow.

Fire Department EOC had trucks deliver a constant supply of new shovels as diggers were breaking theirs regularly. Supplies of bottled water, coffee, donuts, and sandwiches were arriving and our on-site medics set up the rehab sector. Belgarde, our assistant safety officer, strolled through each zone, but primarily Zone 1, and monitored all activities. Crews working with chain saws, while standing on unlevel and slippery terrain and in close proximity to digging partners had to be cautioned regularly.

The lookout with the air horn never budged nor was distracted from gazing at the slope. At about 11:30 a.m. someone shouted, "Over here." He had spotted the dark blue, lifeless hand and arm of Martha Quales. I felt for a pulse. Not surprising, it was cold and pulseless. It took another fifteen or twenty minutes of work to be able to free her. EMS Captain George Keeney was ready with his equipment and crew. When she was finally extricated, George followed the full protocols (treatment for hypothermia, attaching the heart monitor/defibrillator, intravenous line, etc.) even though the obvious extensive trauma looked fatal.

As the ambulance left with Martha, the routine continued. Shortly after, I was tapped on the shoulder, and I turned to see a face familiar from nightly TV. I was surprised that Channel 2 News from Anchorage had their reporter on scene hours before the rescue dogs that I had called for arrived. Dan Fagan and his cameraman were quiet, respectful, and unobtrusive. In a matter of minutes they became almost invisible. I kept radioing dispatch for an update on the arrival of the dogs from Anchorage. The state troopers in Anchorage were in a scramble, and eventually located dogs that would be available from different search and rescue groups. Then transportation and logistical support had to be arranged. I do not fault them for the terrible time lapse, but it was frustrating. Dixie's dog, Gypsy, was getting too old for this, and had been injured in the last avalanche search by being caught in a subsequent slide, and could not help us now. Having dogs immediately available would have sped things up and saved tremendous physical effort of nearly a hundred people.

Time dragged on and searchers were getting that glassy stare that people get when their brains go numb. They were becoming robots. Me too. I was acting....acting like I had plan, a system, like I knew what I was doing. In my entire career, I had never been so stumped by an operation. I was faking it completely, but I kept the demeanor.

Safety Officer Belgarde walked up. "We've got less than two hours of daylight left. You better come up with a plan for tonight. You're going to have to scale this down some."

"Okay. Here's what we'll do: The search dogs'll be here soon, so I'll have Truck 5 set up the lights, and the firefighters—ya know they won't leave without finding Jerry—will divide up into the search team and relief team. And we'll work with the dogs. The search team will consist of no more people than we can supply with beacons, since we won't be able to see the mountain. And the relief team will have their own shovels and probes and be way out of the hot zone. They'll be the backup for the search team. As soon as it gets dark, we'll move everybody else out of the hot zone, maybe send them home until daylight."

Digging and paging continued. Jill Fredston, avalanche expert who just flew in from Anchorage, walked up. She and her partner, Doug Fesler, had been down here after last years' fatal avalanche. She just shook her head as she looked around. Jill authored *Snowstruck*, (Harcourt Books, 2005), containing this incident.

The depressing tedium of the routine was broken once when my portable radio squawked "Medics are needed at the Jack Kimmick residence." (200 feet from me) "He's passing out."

"I got it," I announced, as I took off running toward that house. I was joined by another medic. But before we got there, the person on the radio cancelled the call as a mistake. That was the best I'd felt all day. I even mentioned that at the critical incident stress debriefing (CISD) session several days later, that I really needed to run. What made that slightly ironic is that we don't allow running during emergency calls because of the potential for injuries and it doesn't look professional. In addition, running creates an air of out-of-control urgency for everyone in the area and makes people do stupid things. Incidentally, so does yelling, so we don't allow that either.

The loop road was finally fully opened and the heavy equipment slowly sliced into the bank of snow and debris, methodically making way toward Jerry's house site.

At about 3:30 p.m. no one spoke as the army of diggers robotically tackled the task, like ants moving a mountain. Plumb left his area, which was less frantic now, and came up to the hot zone and chatted with Jill Fredston.

Time for another page, chain saws and diesel engines shut down, pagers and radios turned off and without expectations, paged again. No tone was heard, but in the lull—and the scene videoed and played over and over again on TV—firefighter Kyle Marshal shouted, "He's over here, I just heard him…He's right under me!"

There was a massive convergence to the spot, and I finally got to do something chiefly. "I want six people digging on that mound, and I want a debris line over here for moving shit." I radioed for Medic 7 and crew to back up to the hot zone and prepare for a patient.

As people started digging frantically, and others heaving on building debris and yarding it down the debris line, Plumb shouted for everyone's attention. He explained that Jill Fredston wanted to advise us. She cautioned the team to be careful about moving about in the area since we never know how stable the void is that Jerry is in. A wrong shift in weight could collapse the void in on top of him. It's a good thing she was there. I directed the truck company to fetch the lifting bags and cribbing and set them on a tool tarp nearby.

After about half an hour of digging, they spotted a small hole in the snow where the sound of his voice became much clearer. Shining a flashlight down the hole, the top of Jerry's head was visible. Mark Kirko took his helmet off, lay on his belly, sticking his head into the hole a bit, talked to Jerry. Mark would then pull back and report that Jerry was conscious and lucid but was in a lot of pain and very cold. Joanie Behrends traded places with Mark and did a physical assessment based on what Jerry was saying. EMS Captain George Keeney wrapped

154

oxygen tubing around a heat pack and gave the oxygen mask to Joanie who slid it down the hole. She could not reach his face, but rested the mask near the top of his head so he might be able to inhale some of the gas as it flowed around his head. She was able to get some heat packs to touch the top of his head also.

After a few more minutes, she reported that Jerry said he was wedged tightly under a piece of wood and could not move anything. The diggers by now had revealed a wooden door which lay on top of him. It was held down by a large water heater which we were yarding out of there. Also, holding up one edge of the door was a furnace.

Joanie jerked up and announced, "Get me the suction, I smelled vomit." She stuck her head back into the hole: "Jerry. Jerry. He's unconscious." We reefed on the water heater and Keeney revved up the chain saw, jammed the spinning chain into the door and ripped it lengthwise. Everyone winched, picturing Keeney inadvertently severing one of Jerry's limbs. The door was flung away and Joanie said, "He's not breathing." All concerns for spinal stabilization were disregarded and Jerry was reefed up out of the hole and onto a backboard. The bag-valve-mask was used to provide ventilation, and a pulse check revealed he was in full arrest. Compressions were started, but his head turned the ominous purple of complete cyanosis. A chain of diggers and firefighters stretched out their hands to wench us stretcher-bearers out of the deep hole we had dug ourselves into, and a moment later Jerry was in the ambulance enroute to the hospital. It was six hours after the avalanche. We figured he was dead.

We just stood there. I saw Kirko had tears streaming down his face and that got me started. Christ, I hate that. We hadn't moved yet when I heard on my radio, Medic 7 reporting to the hospital that the ETA was about 5 minutes; that they were still providing ventilations but they had restored the patient's heartbeat. "Let's pick this shit up and get back in service," I directed.

By the time Jerry was wheeled into the hospital, he was talking again. Demobilizing an event that utilized about 100 people and lots of equipment is time consuming. While the crew was doing that, I met a convoy of vehicles coming in from the airport carrying visiting high-school basketball teams here for the Elks Tip-Off tournament. I handed out portable radios to the convoy and had PD on the other side of the slide area, and front-end loaders standing by before I let them go through the slide area.

Jerry was stable enough to be flown to Anchorage for surgery on his right arm, which had been crushed and without circulation for a long time. Odds were that it would have to be removed.

My son Jason was at the hospital in Anchorage when they took Jerry in. "He was fine, except he complained that it took you guys too long to get him," Jason said on the phone.

The surgeons did some miraculous work, split his arm open lengthwise and replaced some blood vessels, restored circulation, and his arm, later, became fully functional. Jerry never talked about the event. Lots of phone calls came in to the department from newspapers, TV reporters, magazine writers. One person called repeatedly, adamantly wanting to interview him about a near-death experience. Plumb just wrote down names and contact numbers and gave them to Jerry. Jerry never returned any of the calls.

One TV station sent a crew down and interviewed me about the event. When the reporter asked me to describe what it was like when we found him. I said no. They never aired the interview. However, when "Real TV" sent a crew and interviewed Bob, George, and me, they did a good job. The trouble with that was that "Real TV" is not one of the programs we got in Cordova, but friends from out of town recorded it for us and sent it down.

Jerry finally agreed to be interviewed. A crew from a London "Discovery"-type program came over and picked up some more footage and put this event in the middle of a one-hour documentary about avalanches. Jerry was in that. It was aired in Europe, but not here.

After 14 years in the department, I knew Jerry pretty well. He's always had a taste for the grape, but he really started hitting the sauce after he healed from the surgery. Vicki seemed to be pretty concerned about Plumb, too. "It really affected him," she said. And even though it had—maybe he felt somewhat responsible since avalanches are one of his key interests, and perhaps he felt he should have complained more about dwellings being in that area. But I think the person most affected by it was Vicki. She seemed nearly obsessed with how poorly others were dealing with this event. However, she was the one who doggedly and steadfastly, was at the site every day for the next three months, digging for items that could be salvaged. I was there, too, and so was Groff.

While others came by from time to time to help, the three of us never failed to be there: Vicki, because she could not tear herself away from it; Groff and me because Vicki could not tear herself away from it. Kirko could not drive to the area for several weeks. When he finally was driven through that spot, he kept his eyes closed. Joanie Behrends, quiet, calm and solid as always—reliable as gravity—tended her routine of drying all of Jerry's personal papers. In the basement of the hospital, we had constructed racks of chicken wire. Everything we scraped out of the snow for the following months, we took to the hospital basement. Joanie spread papers out on those racks for drying, then stacking them and placing new papers on the wire racks…day after day after day. She saw to it that the Ladies of the Moose Lodge were steadily supplied with clothing brought in from the site that needed to be laundered and folded.

156

Plumb allied himself with the visiting avalanche experts. He arranged to have Jill Fredston come back to town some months later and put on an avalanche awareness program geared mostly at avalanche preparedness and mitigation—and zoning. He busied himself with ordering beacons and talking about this event with Points North and the Eyak Ski Patrol. He downloaded photos posted on the internet of the Mile 5 avalanche and collected maps and charts with run-off projections. He searched for documents about the area. He acted like he was on a secretive quest. Every week all summer long, he would go to the site and video it to record the rate of snow melt. I didn't understand the purpose for most of that. With all that focus, he never finished writing a simple narrative report on the event. Maybe Vicki was right.

What was most odd, in my mind, was that we have spent years wallowing in other people's blood or tiptoeing through their tragedies with little effect on us. But so many people got upset over this event—and it had a miraculous outcome. People are really goofy.

As an aside, what nearly took Jerry's life was "tourniquet shock." It is the second most critical reason that tourniquets are not used to control bleeding. The most critical reason is that tourniquets cut off all circulating blood and the oxygen it carries to all the cells distal to it. The cells then die and that portion has to be amputated. A secondary problem, "tourniquet shock," also occurs. When, for example, a tourniquet is applied to a bicep, the cells downstream begin to die from lack of oxygen. As a last-ditch effort to remain alive, the cells begin to burn their own fat. Whenever anything is combusted, there is an exhaust. When oxygen is combusted, the exhaust is carbon dioxide, which is transported back in the blood to the lungs where it is exhaled. But when fat is burned, the exhaust is lactic acid. This lactic acid, which is extremely toxic, is not circulated back out of the body but remains trapped in the limb by the tourniquet.

After a period of time, the lactic acid reaches lethal concentrations. A sudden release of the tourniquet allows the toxic substance to rush back up the veins and into one's system, often with lethal results. Actually, that's why the old Korean War first aid films instructed rescuers to write—in blood—on the victim's forehead, the time that the tourniquet was applied. That way, when the victim reached the MASH unit, the physicians could calculate the amount of lactic acid that was built up and estimate the amount of sodium bicarb that should be injected so that the metabolic acidosis could be neutralized as the tourniquet was slowly released.

Unless I have been reading the wrong EMS textbooks, the books are deficient in treatments for a not uncommon problem of patients being trapped or pinned under trees or debris for long periods of time. Perhaps, before lifting the object off of the patient, a tourniquet should be applied to keep the acid trapped, and not be released until the patient is in the emergency room and IVs and sodium barcarb are ready.

In classes that I've taught, students have asked me how long it takes for lactic acid to build up to dangerous levels, and I didn't know. But recently, my son Jason told me that it takes about 4 hours to build up to levels high enough to be concerned about. Apparently, that is not routinely taught in EMT courses, but Jason is a medic on the Anchorage PD SWAT team, and was sent to attend military medic training and that's where he heard that.

Jerry's upper arm was pinned for about six hours. As soon as enough weight had been lifted off the door that "tourniqueted" his arm, circulation returned and the acid sent him into cardiac arrest.

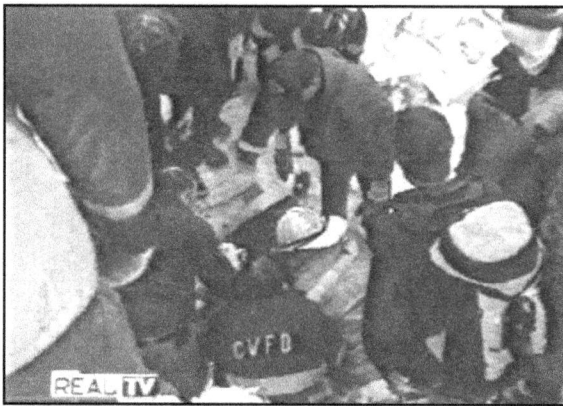

LOCATING JERRY

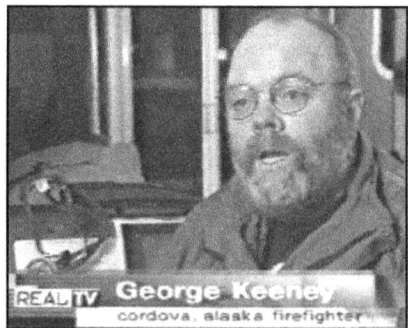

DEPUTY CHIEF BOB PLUMB (LEFT) AND EMS CAPTAIN GEORGE KEENEY (RIGHT) BEING INTERVIEWED BY REAL TV FOLLOWING THE RESCUE

Photos this page from video clip made available to CVFD by RealTV

SAR ANECDOTES

Before U.S. Senator Ted Stevens got tired of the constant requests that Bagron and I were sending him for a Coast Guard helo base in Cordova (we finally got one), CVFD would respond to unorthodox ocean calls during the two and a half hours it took for a Kodiak chopper to arrive. One of the funniest was for a capsizing in the breakers on the flats. The fisherman in the water was so fat, the most our guy in the Chisum helicopter could do was drop the end of a looped line, and drag him like a ski-less skier through the water and up onto Egg Island like a beached whale. They then sat down on the sand and waited with him until the H-3 got there. The Chisum chopper couldn't lift him.

Thank God, when Gary Davidson and I flew completely across the sound to within 20 minutes of Seward to rescue two guys from a floatplane that crashed in the water, they were already out. Their floatplane was bobbing upsidedown and the only equipment—other than medical stuff—that we took was a crow bar and a claw hammer.

In the early '90s, the "new kid on the block" in the SAR business in Alaska was the Alaska Air Guard and their para-rescue people with their olive-green Black Hawk helicopters. We had heard that after the Gulf War, it was deemed a good idea to have them keep their skills sharp by getting involved in civilian SAR operations. We'd never called them because we only like people we know…the Coast Guard. And that's exactly who we called for help when we got dispatched to a plane crash on a mountainside near a logging camp in Two-Moon Bay.

In the middle of our summer manpower problem, we sent EMT Mike Gundlach out on a floatplane to try to stabilize the three people that crashed. As soon as he was taking off, we called the Coast Guard. Mike and the pilot lugged all his equipment up the slope through the thick brush and he started tending to the crash victims. When he turned to the sound of helicopters, it wasn't just one, there were two. And they weren't white or orange. They were dark green. Apparently, the Air Guard and the Coast Guard agreed that if the incident were over the water, the CG would take it. If over terra firma, the Black Hawks would take it. The Coast Guard had called them.

The Black Hawk hovered and a para-rescueman zipped down a line, packaged up the patients, hoisted them all up, and they vanished. Gundlach stood there deserted and amazed. Then he scanned the ground around him and scowled. The oxygen unit, the IV equipment, splints and C-collars were gone with the patients. We never got any of it back.

Hi-tech never impressed me much at first, but when GPSs first came out… okay, when I first heard about them, I could immediately see their usefulness.

We used to use forestry maps which were gridded in 1-mile squares. We would then describe a spot by which quarter of the square the patient or victim was in. So GPSs which could pinpoint a spot within feet was an excellent invention. So, I bought a couple of them.

Within a few days we got a call about an accident in the Heney Range. A guy named Sepulvida was hiking and came upon a ravine that had a cable stretching across it. I didn't know what the cable was for. I still don't. But for some reason, Sepulvida decided that instead of hiking down to the bottom, then back up the other side, he would cross, hand-over-hand, on the cable. He fell and landed on the rocks far below. He really fucked himself up. That's what firefighters consider a detailed patient-condition report.

"Describe the patient's injuries."

"He's really fucked up."

"Thank you." Eyes roll.

Anyway, I sent a team of medics and firefighters and all the crap they would need to stretcher-carry the patient out of the ravine and through the nasty tangle of woods. But we also called the Coast Guard to see if a chopper were available that could bring him down.

Chopper pilots are leery of picking up patients in the woods or mountains. They prefer not to sling-load a litter out because if the cable gets snagged in a tree, they have a couple of options to deal with the situation: 1) Stay up in the air, tied to a tree until they run out of fuel and crash. That's near the bottom of the list of preferred options. 2) Sever the cable and let the patient fall crashing to the ground. They are prepared to do this, but patients aren't crazy about that option. So, if a sling load is the only technique available, they are very picky about the clearance of trees and such to make it safe. Their optimum choice is a landing zone (LZ), so the patient can be loaded into the chopper.

Was there a clearing up there to facilitate their operation? No one knew. So, I gave a GPS to one of our younger members, Josh Graham, and told him that while the others are tending to the patient and getting him out of the ravine, he—Josh—should try to find a suitable clearing and radio the latitude and longitude to the chopper, which was on its way.

Josh was the right guy for learning how to use this fancy thing. A high-schooler, he spent every waking hour, until the wee hours of the morning, studying computer programs. It was an obsession with him. In fact, he got so good at it he opened his own business in Cordova, then shortly after graduating, he moved to Oregon where he is now making about a thousand dollars an minute.

Anyway, as soon as the crew drove to the place they would start their climb up, Josh tore the plastic off the package, opened the instruction book and started reading. Since no one can read and hike up a rocky terrain at the same time, a firefighter got on each side and grabbed him under the armpits, and like cops hauling a prisoner away, started hiking up while Josh read. Up top, he located a LZ, radioed the coordinates to the chopper, and the rescue went fine.

PRIOR TO GETTING A COAST GUARD HELO BASE, THE CITY OF CORDOVA PAID CHISUM AIR FOR THE SAR AVAILABILITY OF THIS BELL RANGER CHOPPER UNTIL IT CRASHED IN A STORM, KILLING JOHN STIMSON

Photo: Cordova Times

SCRAMBLE FOLLOWING A SAR TONE-OUT

Photo: CVFD

161

THE SECRETS OF "SAR CENTRAL"

Running a Search and Rescue operation requires an efficient support group: the Emergency Operations Center (EOC). For SARs, the EOC is really a logistical support mechanism. Often the command of an operation takes place from a command post near the site of the incident. On the other hand, if no one knows exactly where the incident is, or if the incident is everywhere (like a massive earthquake or tsunami), the command post is actually placed in the EOC.

Take this scenario: A missing boater, whose empty boat is spotted in the northern portion of Orca Inlet one October morning. Trooper Jeff Babcock asks CVFD to put a SAR package together and while he (Jeff) will run air operations.

It will be a three-day search operation comprised of two planes and one chopper; several fishing boats (each with 3-person crews) for the search; and numerous skiffs transporting personnel and supplies to and from the search area. There will be several 6-man teams used for ground searching the islands nearby.

There will be not only numerous radio frequencies being used, but in this modern age, even cell phones will be used.

Here is the point of this story: After so many years of running or coordinating Search and Rescue operations of all sizes and types, CVFD members stress out more over planning a department picnic than a SAR operation.

* * * * * * * * * *

Here's how it will go: Someone alerts us, we get the basic information and tone out the department. The first person through the door grabs the 3-ring binder off the shelf that contains the lists of resources from just about anywhere on the planet, and our "so's-we-don't-forget-somethin'" checklists, and walks into the apparatus room. The next person to arrive, moves the apparatus outside. Now, our 4 by 8 foot white marker board is visible and ready for action. You don't think that's hi-tech enough? Just wait 'til you see our Secret Weapon. Ta-Daaaaah (trumpet fan fare): Stacks of various sized Post-its! That's it, Buckwheat. That is the nervecenter of SAR Central: A 3-ring binder, a marker board, and a stack of post-its.

Our Situation Unit Leader (the person that grabs the erasable marker pen) writes "EOC" in one corner of the marker board and each person who comes into the station has his name written on a single small Post-it, and that Post-it joins the list of other Post-its stuck under "EOC."

I stand on the third step of the stairway that leads to the training room, explain the situation to the crowd, and the group quickly brainstorms a plan of action (Incident Action Plan). That takes about 5 minutes, or in a really complex situation, maybe 15 minutes. Then, it's, "To the phones, men! Dial! Dial like the wind, you magnificent bastards!" Sorry. I went from being Chief John Belushi to Chief Errol Flynn. I have an image problem.

Anyway, shortly after, here come our resources. As in the above example, here they come: Maxwell, owner of the fishing vessel (F/V) Little Swede. Next, Bailey, owner of F/V Shiloh. Then, Kopchak who has F/V Gnarley Macho; Van Dyke who has Sarge; Johnson who has Commodore; Webber who has Westerly.

On the marker board, we draw a map of northern Orca Inlet, and decide to assign Maxwell with F/V Little Swede to search the area from Chugach Cannery up to Sheppard Point. Since Maxwell is alone, department members George Keeney and Brandi Keeney will go with him.

"G. Keeney" is written on a small Post-it and stuck to the larger F/V Little Swede Post-it. If he has a cell phone or a portable radio, that number or frequency is written on it, too. B. Keeney's little Post-it also gets stuck to the edge of the F/V Little Swede Post-it along with any communications data. Also, on the larger boat Post-it is its radio stand-by frequency.

As soon as Maxwell and the Keeneys leave, we put the F/V Little Swede Post-it over on that part of the map we drew. As soon as they radio us and tell us they are at their spot, we write what time it is on the Post-it. We also keep a running log, a narrative journal of everything that happens. That job is usually given to the newest person on the department.

Anyway, you get the idea. Our little EOC looks like the old war rooms you've seen in the movies about World War II. But instead of using sticks to push battle ships around on a table (as information is received), we move Post-its around on a marker board, writing little notations on it.

Okay, George Keeney doesn't want to miss a dental appointment. We send a replacement out on the skiff F/V Quantum Leap (run by owner Bob deVille) to meet up with the F/V Little Swede and bring George back. Quantum Leap gets its own Post-it, too. As soon as the switch is made, the little "G. Keeney" Post-it goes to the "unavailable" section of the marker board, and his replacement's little Post-it gets stuck on the F/V Little Swede Post-it.

If we have a 6-person ground search party, all six little Post-its are stuck together and designated by a division or group name. These groups can exchange members and all we do is swap Post-its.

One look at the marker board will tell the observer how many people, boats, aircraft, snow machines, air boats, pick-up trucks, fire apparatus, 4-wheelers, or whatever, are out there, where they are, and who the people are. Every so often, we slowly pan the marker board with a good video camera. That will help with a chronological record of the entire event. At the end of the day, everyone that was out there must return to the station to assure his/her Post-it gets put back under "EOC." If one Post-it doesn't get put back under "EOC", we gotta go find this guy.

Well, I will have to admit that, in addition to using GPSs, we did swallow our pride and use one more piece of hi-tech equipment. The city planner for the city of Cordova had a really neat CD-Rom. It does interesting things with maps. Slide that thing into the computer and type in your zip code and you see a map of your entire area. Keep zooming in to a smaller and smaller area. One time we just wanted a map from Mile 6 to Mile 9 of the highway and the area north of that up to the glacier and the mountains. Now, here's the cool part. By dragging and clicking the mouse, the map—which looked like any map drawn from overhead—could be seen, with each click, more and more from the horizon. The mountains became 3-dimensional, just like in a video game. All we needed to do was choose which angle we wanted to use, and print out a series of them to distribute to our search teams. The other thing we did was to designate the individual teams (divisions) not by numbers or letters, but by colors, i.e.; Red Team, Green Team, etc. When they came back to report what they had searched, they drew the area out on the same computer screen map and filled it in with their color. At the end of that operational period, we printed out the map with the swaths of red, green, yellow, etc and gave it to the trooper as part of our report, and as a tool for planning the next operational period. That's about as hi-tech as we got.

So, after Jeff Babcock flew the search area for the missing boater, he would walk into the fire station, look at the marker board and maybe peruse the hand-written journal and know everything he needed to know.

At the end of the operation, we filled out all the forms he needed, attached all the receipts, the maps and so forth. We would even type out a narrative report to fill in any holes. Yep, this is real rocket science.

The key to it all, is the resource list. Everybody who owns anything that might conceivably be useful at some time, needs to be in that book. That means statewide, too.

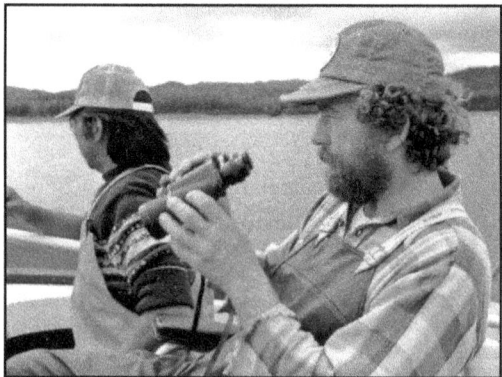

CORDOVA FISHERMEN SACRIFICED TIME AND INCOME TO ASSIST IN SEARCHING

Photo: CVFD

164

SECTION II

DISASTERS

DISASTERS AND THE RYAN AIR CRASH

It's a very short trip from running a SAR from an EOC to running a disaster response from an EOC. Let's make the transition from a single-site emergency to more complex incidents.

Let's take a hypothetical situation and put it in New York City. Let's say that at a given time, 100 people scattered around NYC get their arms broken. You have 100 ambulances and paramedics to tend to each of them simultaneously. That's not a disaster, that's a routine day in "The Big Apple." But put them all in the same building and cause a collapse of that building, and you call it a disaster. You still have 100 EMS crews converging on that site and each patient gets individual attention. It really isn't a disaster so much as just a major event. It becomes a coordination challenge for the IC there because he has so many cops, firefighters, medics; plus public works or even private contractors removing debris to access the patients. He has all the resources he needs, he just needs to choreograph their movements—you know, like a fire fight. Here is a disaster: You have fifteen seriously hurt people, but enough resources to tend to only ten of them. Five of them go unattended.

A disaster—by my definition—is when you have more emergency than you have resources to deal with it. Of course, poor coordination of resources can turn almost any event into a disaster. But back to my point. You can reduce the magnitude of a disaster by being able to acquire lots of resources (know where they are and how to get them quickly), and how to utilize them fully (eliminating duplication of effort, and tracking their every movement). Then, lastly, hope like hell that those organizations outside of your control have their shit together, too.

A case in point: The crash of Ryan Air.

In comparison to newsworthy natural disasters or human-caused catastrophic events, the 1987 crash of commuter airliner Ryan Air in Homer, Alaska, was simply another mass casualty incident.

It was after dark in November 1987 when the fully loaded Ryan Air twin-engine Beechcraft 1900, landing at Homer, Alaska, plowed through a hurricane fence and hit the ground 600 feet short and 250 feet left of the runway. The belly-skid sheared the seats from the floor and the sudden stop caused the 21 persons aboard to be piled atop each other in the front of the plane.

The plane looked intact as it sat on the snow and frozen ground. But the passengers, most of them high-school students returning home from a basketball tournament, were still strapped to their seats, dead or dying. Chief Robert Purcell and his EMS supervisor, Marge Tillion, directed the firefighters and medics who managed to have all the occupants extricated in about 90 minutes. Only seven were alive at that time. Four more were to die later.

The seven survivors were taken to South Peninsula General Hospital where hospital staff tried to stabilize them. And in accordance to the area's disaster plan, Providence and Humana (now Regional) hospitals in Anchorage were being alerted at the same time. The Anchorage police officers were notified and converged on the Alaska Blood Bank to donate blood.

The first 10 units of O negative blood (universal donor blood) were rushed to the Anchorage airport for one of the two planes sent to Homer. Two doctors and six nurses were on those planes, enroute to assist the seven doctors in Homer.

At Providence and Humana hospitals, dozens of doctors and nurses clustered in the emergency rooms waiting for the seven survivors who were to arrive back in Anchorage at about midnight. Everyone and all the equipment was readied. But there still wasn't enough blood from the blood bank. Messages were carried on radio and TV stations. More than 130 people responded quickly and then waited in lines for up to two hours. There wasn't enough staff at the blood bank. The bank is usually set up to handle a few donors at a time. Extra staff had to be called in.

As it turned out, two of the seven died on the flight back to Anchorage and one died after being admitted in Anchorage. But the lesson here is, a bottleneck anywhere in the system (like the blood bank in this incident) can make the process crash. Each detail in your contingency plan needs a contingency plan of its own.

When I would recount this incident during disaster management lectures in Cordova, I highlighted all the advantages that Homer has that Cordova lacks:

Homer is on the road system and can summon aid from nearby towns. Cordova is isolated. Homer had good weather that night. Suppose we had a similar incident in high winds and driving rain? Our airport is 13 miles from town, and the drive could be further complicated by deep snow or icy roads. Their plane landed off of the runway. However, if wreckage were strewn on a damaged runway, aid arriving by air would be hampered. Incidentally, most plane crashes occur while landing, which puts them on or near the runway. There are usually more survivors in crashes that occur during landings than during take-offs where air speeds are increasing rather than decreasing and height is often gained before plummeting down. But equally critical, plane failure on take-off puts the aircraft further away from airport fire/rescue personnel. The Cordova airport property is surrounded by miles of marsh where no vehicle could go. Accessing a wreckage site out there requires air boats or (in freezing temperatures) possibly snowmachines. And, even though Cordova gets a couple of commuter-sized aircraft a day, we also get two Boeing 737s a day. The passenger load on 737s is 85. A shortage of donor blood and inability to collect it timely is only one of numerous factors that could render even the best plans impotent.

Homer law enforcement restricted public convergence to the crash site, and Cordova's plan addresses that problem also. But a subsequent convergence of loved ones to the small hospital would tax our small police department. The Coast Guard would need to be requested to assist. Interested persons, agencies, and media would be plugging the phone lines to town. Loved ones would be flying to Cordova and would need transportation into town from the airport, lodging, counseling, etc. The airlines have plans and personnel trained to deal with that convergence, but the city must accommodate the NTSB, FBI, politicians, and media.

Each problem I just described needs to be considered and possible solutions created. During my career, I have seen so many communities that point to their Emergency Operations Plan (some dust-covered, generic, bureaucratic spew-out), and smugly assert that they have all their bases covered. They're full of shit.

One major problem that is seldom considered when developing a disaster plan is the problem of "convergence." My first experience with convergence wasn't too dramatic, but the fact that it involved the department and me made me aware of this previously unforeseen phenomenon.

RYAN AIR CRASH–NOVEMBER 1987, HOMER, ALASKA

Photo from video clip provided to CVFD by Homer VFD

168

THE CRASH OF COAST GUARD HELO #1471

August 7, 1981, local fisherman Skip Holden was fishing his little 26-foot bow picker Marlene outside the bar in the gulf when the weather began deteriorating quickly. The weather had been rough for a couple of days already. Most of the other fishermen in the area had noted this and, expecting the worst, had reeled in their nets and were well inside the bar, cruising around the flats for a place to anchor up.

By the time Skip had pulled in his gear, the swells had become very large and were breaking over the bar for miles in either direction. A friend of his was cruising inside the line of breakers searching for any "hole" that Skip could slip through. There was none so Skip was stuck out in the ocean in a small boat with the weather growing more ominous by the minute. His only communications was via a small CB radio.

As was recounted in Spike Walker's Coming Back Alive (2002, St. Martin's Griffin). The swells increased to 15 feet and the wind was now about 70 mph. Then the torrential rainfall began, which during the next 3-day period would break all previous records. It was at this time that the Marlene lost its steering. Skip let out about three-quarters of his net, which kept his bow into the waves, as assurance he would be carried by the tide away from the deadly breakers. He had taken on so much water, he was forced to shut his engine down to bail out the bilges.

His friend, who was monitoring the situation from the flats, called in a "Mayday" to the Coast Guard. It was now the middle of the night.

By the time the C-130 made it to the area and circled at 18,000 feet, the winds were up to 90 mph. The H-3 helicopter (#1471) took about 3 hours to reach the area and now the seas were 20 feet high and winds were spiking at 100 mph.

For one hour, Helo 1471 tried getting Skip off the boat, but the chopper was being pitched around the sky and finally out of control. Similar to the scene in The Perfect Storm, the chopper snipped the top of a huge wave and plunged down into the ocean. The C-130 knew what happened, and radioed their base with the alarming news.

By daylight, Skip got his engine started again and had survived the storm. The four crewmen on Helo 1471 had not.

I remember getting a phone call in the middle of the night saying the Coast Guard search and rescue teams were going to be arriving at the air station sometime around daylight to look for their chopper and crew.

The air station at that time consisted of only the hanger with some office space inside. No barracks. They needed sleeping facilities for about 25 persons. These 25 people would be crews and relief crews for arriving helicopters, C-130 airplanes, ground support and administrative personnel. Since the Cutter Sweetbrier was already underway to search the area, the Coast Guard called us.

I toned out the department members and they got on the phones in our part of the building. It was fishing season, but each cannery had a few beds from their bunkhouses available. The volunteers took pickup trucks and made the rounds to the different canneries, disassembled beds, hauled them down stairways and transported them to the airbase at Mile 13. We got some bedding from the canneries, from the hotels, and from the hospital. We woke up store owners who were gracious enough to drive to their stores and open them up. We also got towels and toiletries, coffee makers, disposable cups, hotplates and the sorts of things they would need to get started. We had a caravan of vehicles speeding out to the air station. There, we assembled cots, distributed bedding, set up tables, plugged in coffee pots, and such, and had everything ready to go by the time planes and choppers started arriving on that dismal morning.

I knew that their relief crews and support personnel would be able to provide all the logistics after that. Our involvement was pretty minor, about four hours of equipment gathering and, of course, documenting where we got all the stuff so it could be returned.

Even though the next ten days of searching for the chopper and the four airmen would involve Coast Guard aircraft from Kodiak and Sitka, also taking part were USAF aircraft from Anchorage, three Coast Guard cutters, two Exxon oil tankers, a cargo vessel the Aleutian Developer and the Civil Air Patrol. Logistics for the Mile 13 hanger was handled easily without our help. But think about this:

More often than you think, a disaster or significant emergency that occurs in your area and one that is not your responsibility to respond to, still impacts you. The impact is usually "convergence." The crash of Helo 1471 only involved the CVFD in a minor way. We handled some logistical support only. Things like that can certainly be handled by any number of other organizations within a community or within the local city government. But it won't be. A fire department—whether career or volunteer—comes complete with a readily available roster of members easily contacted and summoned, a membership bent of serving unselfishly any day and at any time of day and for any duration. It has an integral command structure designed specifically for accomplishing tasks under stress, and a radio communications network to pull it all together. It doesn't matter how simple or complex the needed task is, the local FD is the first organization that pops into the mind of someone who wants something done. When something odd happens around your little town, don't think for a minute that your phone won't ring. Be ready.

Now, let's suppose the event is really big. The "convergence" will be tremendous and overwhelming. You may be the first one to get the call, but soon after, your entire community or region may be dealing with the effects of an emergency. One of the best ways to prepare is to note what happened to other communities and how they coped.

CRASH OF SWISSAIR 111
AND CONVERGENCE ON NOVA SCOTIA

On September 2, 1998, the MacDonald-Douglas jumbo jet, SwissAir 111, with 213 people aboard spiraled down out of the sky at between 450-650 miles per hour and hit the Atlantic ocean 6 miles off shore from Peggy's Cove, Nova Scotia, and sank in 180 feet of water. Nova Scotia lies right under a major flight path from Washington DC to Europe, and 1200 flights a day pass overhead. What can be learned from Peggy's Cove, population 60, and Nova Scotia, an area the size of West Virginia and home of 900,000 people can be very enlightening.

When I was training the Red Dog Mine fire/rescue crew to respond to mass casualty incidents, three Nova Scotia responders were contracted to travel to this remote arctic site and give a presentation on this incident. Even though my 3-day session was completed, I stayed over to hear these Canadians.

Even though much of the official action and focus of the local response centered around Halifax, capital of the province, and Shearwater, the military base, Peggy's Cove was the least able to cope with the problems. The fishing fleet converged on the crash site, searching for debris and working under the direction of the military. Everyone with a TV saw that. What they did not see was the closing of the fishing industry in that part of the ocean. The impact of shutting down the primary industry for cities and towns would need to be addressed. There would be no livelihood for fishermen, for cannery workers, truckers, longshoremen, hell everybody gets hit in circumstances like this. The influx of people to this rough and nearly roadless rock outcropping had to be dealt with.

A "privacy tent" was set up so that converging loved ones of passengers, who felt the need to look out at the ocean could do that within the confines of this 3-sided tent. A microphone was set up near the tent in case any of them wanted to answer questions from the media, which were held at bay some distance away. And there were lots of media reps there: ABC, NBC, CBS, FOX, BBC, ATV, Global, CBC, CNN, and print media. These folks were aggressive and relentless. The cell phone site was overwhelmed, media folks were using cell phone scanners, and officials had to use regular phones to keep conversations confidential. One lady who owned a small restaurant in Peggy's Cove was startled to see a reporter walk in, pull several tables together in a corner, and inform her that he was "renting" this space.

Nova Scotia quickly installed 100 more phone lines and did everything to accommodate the media. When instructing officials how to deal with the media, supervisors used the terms "truth, clarity, courtesy." In the end, it was not only the right thing to do, it proved to be the smartest approach. Reporters digging for the truth uncover all sorts of horse shit. They will talk to anyone who looks official and conflicting statements abound. Later, much time will be wasted in confirming what is true and what is erroneous. "Truth, clarity and courtesy" ultimately saves time and diminishes aggravation.

Route 333, a coastal road, was a two lane highway to the site. There was limited parking, limited landing sites for boats, and logistical control for water and food was a nightmare.

Intruders (non-responders) at first found it easy to bluff their ways past local designated guards. After military guards were in place, their strictness maintained much tighter security.

On September 3 at 5:30 a.m. Public Works department received the following message for the Emergency Management Office: "Plane went down, no survivors...need Public Works to build temporary morgue...go to Hanger B, Shearwater Air Base, find Dr. Butt, Chief Medical Examiner...do whatever he wants." The medical examiner must confirm identification of each victim to issue a death certificate. The plane's manifest would not suffice. As far as Public Works goes, some provinces "privatized" that service and could not mandate a response like that to a public disaster.

When Public Works found Dr. Butts, who was wearing an orange immersion suit, he pointed to the hanger and told them he wanted six autopsy rooms, hot and cold water, waste water was very important, need electricity for each, and by the way, X-ray machines are on their way. There was lots of electricity in the building but the only plumbing was in two washrooms at the far end.

People and materials arrived en masse. Tradesmen dropped occupational lines: Electricians grabbed hammers, military doctors carried plywood. They had two functioning autopsy rooms by 4:30 p.m. and two more by next morning. There were eight dental records rooms in two days. And a 60-person lead-lined locker room right after that.

They managed to build or find wash stations, separation stations, shelves, easels, dividers, tables, built a laundry and a second morgue (even though finding building supplies was next to impossible on a Sunday). They remembered that the military can "feed an army," General Electric loaned portable x-ray machines, AT&T sent boxes of cell phones and batteries. Yet it was difficult to find contractors willing to clean refrigerator vans that had been used for storing body parts. Ultimately, what worked was empowering workers to do whatever is necessary without seeking supervisor approval.

The premier of Nova Scotia called upon all residents to help in any way they could and maintain the reputation as a friendly and helpful people.

As far as Critical Incident Stress Debriefing (CISD) goes, every agency seemed to have a team. There were almost too many and they did not coordinate with one another. It's the same situation that followed the famous Sioux City air crash. There was no place a responder or relative could go for privacy without a well-meaning psychologist or clergyman stopping to talk. Seemingly odd things would have bad impacts on workers: One person upon entering an unlit freezer van of tagged body parts; another person hit bottom when he saw a computer image of the plane's passenger seating plan. Those that coped the best were the ones that focused on the nuts and bolts of their jobs. The medical examiner read passages from a children's book on death.

When an "All Hands" emergency occurs, responder's regular work piles up during their absences. They feel useful, making a difference, adrenaline is their primary nourishment, but when it's over, there is a letdown.

In preparation for the arrival of family members, drivers were selected and briefed about mannerisms and decorum when picking up these people from the airport. They were to drive these people to a spot overlooking the ocean if they wished. The privacy tent was waiting. Small tables and chairs were put in hotel rooms because it was depressing for relatives to be in a room with no place to sit except on the edge of the bed. People need to sit around a table and face one another. Even the smallest hotel room was outfitted. The airline company generally plans to handle this.

SOME OF THE LESSONS LEARNED WERE:

- When you have resources and know how to contact them, you can compensate for lack of planning.
- Teams must be empowered.
- Don't allow organizational protocols (or their proponents) to get in your way.
- Keep good notes, designate a full-time note-taker.
- Cell phone batteries do go dead.
- Know how to contact local contractors and suppliers on holidays and weekends.

Learn from Nova Scotia; develop truly comprehensive plans ahead of time.

Here's what it took: 2,000 sheets of plywood, 90,000 feet of 2 x 4's, 30,000 feet of wire, 5,000 feet of piping, several tons of fasteners, bolts, clamps, connectors, pumps, tanks, furnaces, fans and ductwork, not to mention all the tools used, and—most important—25-30 talented, committed and inventive tradespeople willing to work 24 hours a day, seven days a week for as long as it takes. Remember that buildings, equipment and facilities will expand rapidly into the space available. In this case, two airplane hangers and half of the airbase tarmac.

Hanger B was the morgue with autopsy "suites" and cubicles for fingerprint and DNA tests (public service announcements should ask family members to bring things like victim's hair brushes with them, or send DNA kits to family members.) How about X-rays, service chases, mechanical services, uniform ID chart, dental X-rays and document driers? This place was manned by 60 medical staff. They estimated 200 victims at 200 pounds each would amount to 40,000 pounds of human remains. Waste water had to be disposed of, and plans for how to handle and dispose of fluids from body bags. The analysis of body parts was complicated by other debris: Tiny steel pellets were found in many body parts, because 2,300 pounds of tips for ball-point pens had been in the cargo hold. Also sticking to body parts were some of the 5,000 pounds of female receptacles of car cigarette lighters and particles of frozen pizza.

Mechanical ventilation for the hangers—as ingenious as it was—was not adequate enough for workers who, upon going home at night, undressed on their back porches, but upon getting back in their cars in the morning were hit with a wall of suffocating smell. Dental X-ray machines were run off a diesel generator outside. They had to create document driers for tons of documents and paper money. Some of the passengers were carrying papers from the U.N. In our own experience, we had to construct dryers for Jerry LeMaster's papers that we dug out of the avalanche for months afterwards. Two by fours, chicken wire and lots of space did the trick for us. Their money dryer had to be capable of drying $200,000 per hour. A lot of the money was suspected to be Mafia money bound for European banks. Bundles of money that were bound for Swiss banks were "weighted" to sink, but individual bags broke loose and floated. Four garbage bags of Croation money, which now won't buy a cup of coffee, were recovered.

Rubbermaid containers were used for "John Doe" parts and placed in the freezer vans. More wire mesh and lumber was used to construct the "fuselage" of the plane upon which plane debris would be mounted. When I listened to these Canadians, two years had passed, utilizing 23 people full-time to construct it. Another two and half years was planned for the operation. It was here that the horrendous experience the plane's passengers felt came to light. Most of the metal foot rests were bent by passengers pressing so hard on them as the plane plummeted. Investigators examined the plane's engines and found the tail engine had been torn off, one engine was at half throttle and one was at full throttle meaning that the plane spiraled straight down into the ocean.

Whalesback Cannery, since there would be no fishing season, was used for cleaning 300 workers. These workers repeatedly searched 50 miles of coast line. They often were inadvertently drenched in waste spilling from body bags.

Public Works had to design apparatus for removing sea water and jet fuel from seat cushions. They found and used old wringer washing machines.

They used shop vacs for removing sea water and contaminated fluids from body bags.

By mid-November, most flesh had been identified. Tons of "X-File" tissue was left: mostly fish. About 5,000 pounds of unidentifiable human tissue and remains were left for burial.

Nova Scotia began to demobilize. The hangers were washed down with a Clorox solution. Body parts were shipped home. Canada spent up $850 million in recovery costs. Some costs of building the morgues were recovered from the airlines.

A local memorial and internment service was planned for September 2, 1999. Three coffins were used to represent Christian, Jewish, and Muslim victims. There was lots of community involvement. Residents invited visitors to stay in their homes. The Red Cross provided kitchens and food for attendees. The service was delayed two hours because all the attendees wanted to stop and see the names on the memorial wall that was erected.

That plane never even came close to Nova Scotia.

Are you starting to get the idea? That cheesy, little disaster plan doesn't begin to address the impact a full-scale disaster can have on your town or on your people.

The Exxon-Valdez Oil Spill and Convergence

The seven oil companies (Arco, Exxon, Mobile, Amerada Hess, Unocal, Phillips, and British Petroleum) that had sunk nearly $3 billion into the North Slope oil discovery, and had formed the pipeline consortium "Alyeska," built the largest and most expensive private construction project in history—the $8 billion, 800-mile, Trans-Alaska pipeline. The Prince William Sound fishermen (in combination with national environmental organizations) protested the pipeline. Even though they won a temporary injunction against it, they could not prevent its construction in favor of an overland route through Canada. Starting in 1977, up to 2.1 million barrels a day would be shipped by oil tankers through the sound.

In retrospect, most Alaskan residents looked at the protests of the fishermen as silly. In Valdez, Alyeska became known as "Uncle Al." One hell of a tax base for the city, that just a year before the spill, 90 percent of Valdez's $33 million budget (to provide for its less-than-4,000 residents) came from Uncle Al and what is one of the largest oil storage tank farms anywhere: eighteen enormous tanks, five floating tanker berths and a huge support facility were built across the bay from the city.

At the time of this writing over 70 tankers visit Valdez each month accounting for 13 percent of the nation's tanker traffic. State coffers were swelling so much from oil tax dollars that state income tax was abolished and every resident receives an annual dividend check from the state that has never been less than $1,000.

* * * * * * * * * *

On March 23, 1989, no one knew that while the super tanker Exxon Valdez was getting loaded, so was Captain Hazelwood. Then, right after midnight, the 24th, while a harbor pilot guided the ship through the Valdez Narrows, Hazelwood retired to his cabin. He returned to the bridge after the pilot was dropped off, and to avoid icebergs, changed coarse to 180 degrees, switched on the autopilot, and told the third mate to steer back to the regular traffic lane when the ship came abeam of Busby Island. Hazelwood stumbled back to his cabin.

The correction in heading was not made and the ship struck Bligh Island reef 28 miles out of port. The reef's submerged pinnacles sliced open 8 of the ship's sixteen compartments with a gash that ran down 600 feet of the hull. Later, salvage divers videotaped boulders the size of pickup trucks jammed into the hull. Hazelwood radioed "…We've fetched up hard aground north of Goose Island off Bligh Reef, and evidently leaking some oil." The ship began listing to starboard while the eight cargo tanks gushed oil at 20,000 barrels an hour. The third mate feared the ship would capsize if it came off the reef. Nevertheless, Hazelwood tried to maneuver the ship free. The ship only swung and wrenched back and forth. Had she freed and sunk, the spill would have ultimately been an unbelievable 53 million gallons. But she was stuck so tightly that a subsequent brief hurricane-force blow of 70-knots did not budge her.

Alyeska Pipeline Company was responsible for making an immediate response to such an incident, but over the years the barge and booming equipment designated for the response had been neglected as was any training for personnel, and no response was made by Alyeska.

At 7 a.m. that morning, my stroll into my office was interrupted by the dispatcher telling me to call Michelle O'Leary right away. Michelle, a local commercial fisherman, informed me that a super tanker had run aground and was spilling oil. She went on to say that local fishermen had been trying to call Alyeska to offer help in booming, but that Alyeska was not answering their phones.

I tried calling there myself several times and found that to be true. So I called the Alaska Division of Emergency Services (ADES) and asked them to try contacting Alyeska with that offer from the local fishermen. I suggested that ADES call Michelle back with the answer to their offer. Back to work.

Before long, City Manager Don Moore was in my office, pacing (and hiding for a few moments of quiet). I hadn't realized how extensive the spill was (ultimately 11 million gallons) and what impact this was beginning to have on the whole town. We were "out of the loop," the manager complained. We didn't know what was going on over in Valdez or what remedial measures were being taken out in the sound. His phone was ringing off the hook, and elected officials, commercial fishermen, and cannery managers were cramming the offices in city hall. They were all convinced that the fishing season for this year would be shot.

THE IMPACT OF THE SPILL ON CORDOVA

Cordova's economy is a house of cards. The bottom layer is the fishing industry. One third to one half of the city's operating budget came directly from the fishing industry. The previous year, Cordova received $575,000 from the raw-fish tax and $400,000 from harbor fees for fishing vessels. The city's general operating budget was $4 million. In fact—jumping ahead just two weeks—the pot shrimp fishery was closed by emergency order, followed by black cod fishery. Then the purse seine for herring, gillnet for herring, then herring roe on kelp were closed. The previous year the herring fisheries were valued at $11 million. In total, commercial fishing was a $36 million industry for Cordova and $120 million industry for the sound. Much of the salmon fishery was to be closed next. Some remained open, but the price dropped from $1.25 a pound to 55 cents a pound. Besides with the fishermen not fishing, the four fish processing plants

178

would not be able to pay property tax nor would the cannery workers be able to pay city sales tax. Cannery workers are our primary employment base during the summer. Many stores in Cordova issue credit to fishermen during the winter, expecting to be reimbursed at the end of the summer fishing season.

The house of cards began to collapse. Contractors stopped applying for building permits, major building renovations already underway were cancelled, so carpenters, plumbers, and such were now out of work. The school district cancelled its plans for a vote on a major bond issue for school renovations. Without expectations of exporting fish, longshoremen expected a very lean year. All of this was accurately predicted in the manager's mind that morning.

What was most traumatic for Cordovans was not knowing what was happening. For the next few days people sat glued to CNN and network news. Before long, it was impossible to get anything done in city hall because the cable company ran TV cables into the city administration offices, the police department squad room, and the fire department training room. If it weren't for CNN, Cordovans would not have any idea what was happening.

I tried to turn a deaf ear to the groups of outraged people milling around city hall asking questions. The staff—interrupted from their duties gave up and flipped through TV channels trying in vain to find some positive, proactive approach to this problem. It was completely disruptive to our operations. Our training room became a shambles with groups of people meeting or just hanging out.

The fishermen were the least patient of all the groups. The fleet began leaving the harbor, and the president of the fishermen's union, Jack Lamb, chartered a flight to Valdez. They learned that nothing was being done to attack or contain the spill. One fisherman never got to Valdez, he stopped at the site and put his crew to work picking up oil in 5 gallon buckets and dumping it into his fish hold.

The fishermen set up an office in Valdez and started ordering boom and calling the City of Cordova for purchase orders to pay for it. They started "borrowing" unguarded boom to close off the bays leading to fish hatcheries on the western side of the sound. They acquired planes to fly the area to track the spread of the oil, and no one was keeping track of expenditures. It seemed like the only people on the face of the earth willing to put up a fight were Cordovans. Others only gave lip service, gestures of outrage, and lots of hand-wringing. The fishermen would locate boom and sorbent material in the continental U.S. and in Europe, hoping to get it before the oil reached the hatcheries. The city council appropriated $200,000 from its own treasury to pay for it. You gotta fight.

Exxon officials set up a town meeting in Cordova at the high school gym. The place was packed. One fisherman was so angry, he kicked and shattered the glass in one of the main doors of the school. This was going to be ugly, and the local cops were edgy. Up on the stage sat Coast Guard and Alaska Department of Environmental Conservation (ADEC) officials and Exxon executives. As Exxon officials stood up, one at a time, to assure that the spill would be handled professionally and that we should thank God that it was Exxon, with its resources, that would handle the problem and make everything right, the crowd grew more and more confrontational. Local police and state troopers in the room became very uneasy. I felt sorry for Frank Iarossi, President of Exxon Shipping. Threatened and overwhelmed, he looked pale. This was no place for him. His PR people, when they spoke, used such patronizing and condescending tones to these "hicks in hip boots," they seemed at first, oblivious to their surroundings. The members of the audience, when responding to these folks, never called out to the dais using titles, or last names—first names only. The audience told them, in no uncertain terms, that they were full of shit, and there was no way they could contain or clean up a spill of that magnitude, and that it was a "death knell for this fishing town, Frank!" We were ruined. The meeting ended only by ending.

About 4 days into the spill, the city manager and mayor told me to be at a special council meeting. There I learned that ADES had called and wanted me in Valdez to work with them (I'd conducted disaster management classes for them in the past). The mayor also gave me some objectives:

1) Examine how that whole operation was organized

2) See if I could assist organizationally

3) Keep Cordova's Oil Impact Committee (which we had just formed) informed about the spill response

4) Keep those agencies in Valdez, and especially Exxon, informed of Cordova's needs and interests

The pilot of the plane I chartered to go to Valdez asked me if I wanted to fly over the spill area, but since that route would have been longer, I said, "No. Let's just go over the mountains, I'm in a hurry." An ADES rep picked me up at the Valdez airport, which had more aircraft there than I had ever seen there before or since. Civilian, Coast Guard, fixed-wing, and helicopters were everywhere.

Valdez is a suburb-looking town. It has no center. Without a center, you have no forefathers. You drive around looking for it but feel unfulfilled because you never find it. Its courthouse is not tall nor made of brick. It has no old statue or cannon in its yard. It has no yard. Its floors are not clacking-loud marble. They're carpet. That doesn't affect me, because I know those other things exist. Young Valdezians don't.

We went straight to the community college, which is where ADES was setting up offices. The place was a flurry of activity setting up computers, banks of phones, maps of the Prince William Sound, paper and people everywhere—completely typical of every other public building I saw. Since there were no hotel

180

rooms available, I was to stay at the home of Fire Chief Tom McAlister. But going there would have to wait. I dropped my bags on the floor and walked to the spill Command Post, half a block away.

The federal building was in chaos. Every room was in use. CDFU had an office there, as did ADEC, National Oceanic and Atmospheric Administration (NOAA), Exxon, the Coast Guard, the media, Department of Fish & Game biologists, VECO (oil spill clean-up contractors for Exxon), and VRCA (oil spill clean-up contractors for the state). Representatives of other agencies either had office space there, or were hanging out there: Prince William Sound Aquaculture (PSWAC), U.S. Forest Service, Representatives of area cities (like me), Alaska Department of Fish & Game, Alaska Air National Guard, State Department of Labor, and State Department of Public Safety. Of the entire bunch, the only ones who would take the time to stop and explain what was going on to me were the Cordova fishermen. Whenever I tried visiting Exxon reps, they were in meetings. I couldn't talk to anyone of authority with the Coast Guard or ADEC because they were in the same meetings. But the briefing I got from CDFU was in-depth and clear. In short, there was no incident action plan, and there was no coordination in acquiring or using resources.

Back at the ADES office, I asked Pete Wuerple to get a DNR Overhead team to come to Valdez. The State Department of Natural Resources Forestry Division dispatches Overhead Teams every spring to fight wildland fires well into the fall. These guys are organizing geniuses. I was to learn later, that Dave Liebersbach, from the federal Bureau of Land Management (BLM), one of the Incident Commanders of the Yellowstone fires (who ultimately commanded 14,000 people) had already been to Valdez and offered his services. But this "Top Gun" of wild, screaming emergencies was brushed off by the agencies and industry that had no time to talk to him. So he traveled on to the Seward at the request of the National Park Service and organized local agencies there in preparation for the spill that would eventually reach those shores.

When the Overhead Team—nine of them—arrived the next day, I explained how we might get agencies to re-organize. I asked them to divide themselves up and visit each of the offices in the Federal Building and start explaining the Incident Command System to whomever would listen, but preferably to rank-ing personnel. In the meantime, I contacted Cordova VFD to put my chief's vehicle on the ferry for Valdez and to throw in some extra portable radios so I could network with folks around town. In those pre-cell phone days, the local phone system was impossible to use. Even payphones had long lines of people waiting their turns.

One morning, I was with ADES, watching in awe as Pete Wuerple spun in circles in the middle of the room, barking out orders. Resources were arriving from across the nation and around the world: Boom from Norway flying into

Anchorage on Northern Air Cargo aircraft, and caravans of ATCO trailers coming into Valdez down the highway. The place was a frenzy of activity. Then some little guy walked in and spoke loudly, everyone stopped and turned to him. I can't remember his name, but he was from the governor's OMB (budget/finance) office.

"I'm ordering all acquisitions to stop."

You could hear a pin drop

He continued, "From now on, I will be the only one issuing purchase orders. Anything you want will be properly listed on numbered requisitions, and the requisitions will include the cost of the items and the estimated cost of shipping to Valdez. Some items may require competitive bids from suppliers, and that will be up to you to get them. Turn those requisitions over to me for approval. Then, I will issue the purchase orders."

Pete, stood stunned. His shirt was rumpled, he had been sweating and his hair was sticking up. His staff all looked at him.

"Oh, great, " he said. "The elephants are stampeding, and we're tripping over piss-ants!"

The OMB piss-ant never acknowledge the statement, he just acquired a desk and sat down.

I always admired Pete. I admire anyone who will work himself into such a lather that he has to be hauled away in an ambulance, which is exactly what happened about a week later. He recovered of course.

* * * * * * * * * *

Within a week of the spill, the population of Valdez doubled. And even though the rates for hotel rooms had more than doubled, no rooms were available. No rental cars either. The Valdez airport went from it's typical handful of flights a day to nearly—and sometimes exceeding—600 a day. The federal government had to send in a force of air traffic controllers to handle it. By mid-summer, the population hit 12,000 (*In the Wake of the Exxon Valdez*, Art Davidson, Sierra Club Books, San Francisco, 1990)

Much of the new population included folks who were not there to handle the problem but to express opinions. Boy, there's something sorely needed at this time. I heard it described as a media carnival and mental killing field of exasperation and emotional excesses. Jean Cousteau was there, being followed around by a TV camera crew from Channel 2, Paris, France. He had a Cordova fisherman point to a wall map of the Prince William Sound, while he—Cousteau—grasped his chin with his hand as though internally deliberating a tactical decision while being briefed by a subordinate staff member.

182

When I finally got to talk to Frank Iarossi and his assistant Dan Paul, I tried to explain the Incident Command System (ICS) to them, but they just didn't get it. Hoping in the meantime that the Overhead Team would have better luck with the Coast Guard or state regulatory agencies, I went on to explain the economic disaster that was about to occur in Cordova. As a result, much of the resources arriving in Valdez, due to their smaller airport runway, was being trucked down the highway, or being barged over from Whittier. They agreed to bring some through Cordova instead. Our runway, 7,500 feet long, accommodates larger jets (before Alaska Airlines bought a fleet of small 737s, we used to get 707s and 727s daily), our dock was very big and our longshoremen needed the work. Exxon would also set up an office in Cordova to run that part of their operation. Having gained that commitment—to send some work Cordova's way—was the stepping off place to my complying with objective #4. Objective #3, keeping Cordova current on events would follow my attendance at the daily debriefings held late each evening.

Securing an invitation to attend command meetings with the ADES reps, I was introduced as the representative of the City of Cordova and took my seat against the wall with the other "observers" who were not in authority to make any decisions, and watched the "triumvirate" beat the shit out of each other. Representing the Federal Government was Coast Guard Rear Admiral Ed Nelson of the 17th Coast Guard district and his aid. Representing the state government was Commissioner Dennis Kelso, Alaska Department of Environmental Conservation, and his aid Larry Deitrick, and also for Alaska, the Department of Emergency Service's Pete Wuerple and Pete Petrum. Exxon was represented by Frank Iarossi and Dan Paul. ADEC's loose cannon, Dan Lawn (later, the character-made-hero in the HBO movie "Dead Ahead"), was always present to insure that Exxon or Veco would not be bullshitting people at the table. He would validate his statements by showing aerial-shot videos of ineffective booming attempts; boats adrift, patiently awaiting orders for some action, and other indicators of complete lack of direction and planning.

Iarossi—intelligent, dedicated and dignified—had been taking a beating from the media and faced the muzzle blast of public outrage from Alaskans since he got here. And this room was no sanctuary for him. Each proposed plan of action from him was in the form of a question. Each would be shot down by someone. He sat overwhelmed. He was a victim who could do nothing right. Seeing him like that stopped my natural tendency to obliquely blame him for his position of power and wealth. And my first impression of Dan Paul as the thin-lipped corporate Quasimoto soon softened as well. The pressure was on everyone at the table.

At the head of the table was Admiral Nelson with his piercing ice-blue eyes, white hair, and cutting East Coast accent. It seemed he felt his role was that of a moderator. It was obvious he was used to giving orders, and it was a strain for him to refrain from that. ADEC's Kelso felt his role should have been representing the public's outrage, the person to whom you make excuses, except that his subordinate, Dan Lawn, nightly erupted in genuine outrage, and not only at Exxon, but at the Coast Guard's and ADEC's loosening of standards that contributed to this. This robbed Kelso of his preferred role and put him on the defensive. Kelso's open, college-boy face was looking strained. Lawn had been stationed in Valdez and over the years chronically complained about the industry's slip-shod practices and the regulators' indifference to his complaints. No one basked in illusions of control, and everyone knew that in this room there was enough ineptitude to go around. Everyone seemed to focus on blame, but no one focused on an attack plan. There was not enough equipment in Alaska for an attack, nor an organizational command structure capable of coordinating it. The whole thing was painful to watch.

At these meetings, NOAA would present spill trajectory predictions and weather forecasts, which were followed by hesitantly suggested plans of action. The meetings traditionally would last until well after midnight. I would then head for the bar and have a drink (or two) while I composed a lengthy, handwritten report to be faxed next morning to the Oil Spill Response Committee in Cordova. My reports also contained crude hand-drawn maps of islands and coastlines showing oil deposits and trajectories. I discovered later that they never typed my reports, corrected the spelling or grammar, or omitted expletives—at least for the first week. They just made stacks of copies of my handwritten reports and distributed them around town. It was kind of fun. I didn't know they were doing this, otherwise I would not have written in such an informal—sometimes sarcastic, sometimes flippant—manner. I felt like early war correspondent Hemingway, the biggest difference being, he was a good writer. Shut up.

* * * * * * * * * *

One quiet young man in the briefing room introduced himself to me as Dean Monterey from Vancouver, British Columbia. His government, because they had experienced an oil spill the previous December, sent him to Valdez to see how one was handled in America. Pretty much they same way, unfortunately. Dean was also a volunteer firefighter in Vancouver, and when I explained that we had a DNR Overhead Team here trying to convert these people to the Incident Command System (ICS), he was impressed.

Dean was a little less impressed, later, when the team reported to me that no one was interested in reorganizing. ADES sent the team home.

About that same time, Jim Sellers, aid to Governor Steve Cowper (pronounced "Cooper"), walked up to me on the street and said that Governor Cowper told him to go to Valdez. Jim laughed as he recounted that Cowper told him to look up the fire chief from Cordova. "Dewey something. I can't remember his last name, but he's there somewhere." That was the extent of his orders. The governor and I became acquainted when he was running for that position and I was president of the Alaska Fire Chiefs Association. We had a long discussion about the selection of the Commissioner of Public Safety, if he (Cowper) won the election. Then four months before the oil spill, I called him and suggested he send search dogs to Armenia following an earthquake there, which he did. Anyway, Sellers had also been a career firefighter with the Anchorage Fire Department and we'd met when he worked in Cowper's campaign headquarters.

Dean, Jim, and I agreed that I would try to get the floor at the next briefing and push ICS in front of all the major players. There were always local U.S. Forestry people in the room, so I was certain that they would be concurring with the idea.

"You guys back me up on this." They agreed. Jim also promised that if the triumvirate were not convinced, he would call the governor. If the governor agreed, he (the governor) could direct DEC to go along. That would not be necessary for ADES, because they were in favor already. In fact, in my opinion, and not only mine, ADES should have been the lead state agency anyway. ADES is staffed with emergency managers, DEC with biologists, environmentalists, and a staff of starry-eyed college kids. The only reason DEC got it is because it was viewed primarily as "pollution," forgetting the primary problem was managing a disaster.

About an hour into the typically unproductive meeting, I asked if I might make a suggestion. I said, "The inability of Exxon to handle this problem is very understandable. You people are in the business of finding and selling oil. You never expected that at some point in your career you would be expected to handle an emergency of this magnitude. Expecting that of you is simply expecting too much." (Years later, at a chance phone conversation with Dan Paul, he told me how much that statement meant to him and Frank) I went on, "But there is a much better and easier way to handle this spill. What you have out there is really the world's slowest-moving forest fire, and it can be handled the same way. In fact, if the U.S. Forest Service or DNR were handling this event, we probably wouldn't even be having this meeting. It's simple." The term "simple" made Iarossi, Paul, and Nelson literally sit up straight in their chairs and lean forward. On the marker board, I sketched the ICS organization's general staff: just the Incident Commander and the Section Chiefs. I explained that thousands of people working on a disaster would ultimately fall under one of the Section Chiefs: the Plans Chief, the Operations Chief, the Logistics Chief, or the Finance Chief. Period. "With everyone working in one of the sections, the Incident Commander could have total control."

Wuerple was winking and the triumvirate was salivating. Sellers and Monterey would tactfully mumble "yep," or "That's right," and it was effective punctuating. The affirming body language of the forestry guys did not go unnoticed at the table. Kelso was noticeably and predictably negative about this. The Overhead Team told me that DEC only wanted to cruise around the sound and find fault and stop operations and swing their weight around, but did not want to be saddled with accountable responsibilities. They viewed themselves as regulators and analyzers only, not team players. So I knew they would hate the next part. "Anyone who wants to be part of the operation becomes a definable part of the organization. No freelancing. And when you become part of this team, the first thing you leave behind is your agency affiliation. You no longer belong to the agency that sent you here; you belong, ultimately to one of the four section chiefs or one of his staff members." A quick flick of expression in Admiral Nelson's eyes showed some reluctance to go that far. "The beauty of this system is eliminating duplication of effort. That makes it cheaper and effective."

As the listeners started asking questions, I moved from speech to lecture. I then distributed handouts of ICS organizational charts and samples of daily Incident Action Plans. As I sat down, I remarked how unfortunate it was that the DNR Overhead Team—who could have quickly straightened out this whole mess—had not, er…ah…had the …um…opportunity to help. And one last stab, another mention of the 14,000 firefighters on the Yellowstone fires were astonishingly well choreographed and were fighting a fire that moved a hell of a lot faster than 5 knots. Too bad Yellowstone IC Leibersbach didn't have the opportunity to help, either. Neither Iarossi nor Nelson truly understood those statements.

Leaving the building that night, on my way to the bar to write my report, I think I saw Weurple cartwheel across the parking lot. What I did not know until the next night was that Admiral Nelson and Frank Iarossi did not call it a night. They headed to the hotel, at Nelson's insistence, and practiced drawing ICS organizational charts incorporating agency representatives into specified sections. It was a noble attempt, but one doesn't suddenly ignore political appointments, jurisdictional authorities, or legal responsibilities after a lifetime of being acutely aware of them.

* * * * * * * * * *

After Logistics was consolidated into one location, as a test, I went in and asked some guy who obviously worked for Exxon (by the way he was dressed and his hair was combed) if he were to get a request for some supplies, how would he provide them? The first thing out of his mouth was that he would have to check with his supervisor (corporate supervisor). I explained to him that his supervisor was the Incident Commander. He argued the point and I left looking for Iarossi. I found Dan Paul and told him these folks needed to be briefed better.

186

Deitrick asked me to attend a small meeting at the Valdez fire station. There, he asked me for a list of all certified advanced-level EMTs in the state. I told him I gave that list to Logistics (pending the appointment of a Resource Unit Leader under the Plans section). He said he knew that, but he also wanted the list. He also said during the meeting something about "Operation Bartlett." When I started to ask about it, he gave me a look to hush before others in the room realized he'd slipped about something.

Out in the parking lot he told me in confidence that the governor was going to support the state handling it's own operation by assigning a state ferry as a mother ship for clean up of a portion of the sound. Running a separate operation was why Deitrick wanted a list of EMTs, and—I imagine—a lot more resources. He said to keep it quiet because the governor was going to make the announcement this coming weekend. I told him that was completely contrary to what I had been trying to promote and I could not support it and would not help. But I kept my mouth shut about it.

But I wonder if Deitrick believes that. Because the next night, Admiral Nelson made a statement about working as a team. Then he looked at Kelso and asked him about "Operation Bartlett." Kelso turned pale and stammered. Nelson blew up and slammed his fist on the table. It wasn't until I was writing this that it occurred to me I might have been suspected of telling Nelson. But I hadn't. I also wonder how Sellers, a firefighter, felt about this "end-run" or if he even knew. I know Kelso didn't like me; we never got along from the very beginning.

President George H.W. Bush was under a lot of pressure over this spill and felt he had to do something. As a result, Nelson was to be relieved of command by Admiral Yost, commandant of the U.S. Coast Guard, who flew into Valdez in a white and orange Lear Jet. With Yost was his boss, U.S. Secretary of Transportation, Samuel Skinner. At President Bush's (and Skinner's) direction, Yost stepped off the plane like a stern father-figure who'd had enough of this bumbling, squabbling, and bickering. He intended to lay down the law. But at the same time, Bush had declined Iarossi's request to federalize the spill, so what did all this posturing really mean? Leadership image; something completely lacking, no discredit to Nelson, who I really like. I thought that sending Yost was a slap in the face to Nelson. I really think it was just a way of showing the world that the federal government was not shrugging off this spill. Even after we'd left Valdez and corresponded a bit, I never said that to Nelson—I didn't bring it up at all. Hell, for all I know, maybe he wanted to pass this on up a notch.

Well, reader, if you think I'm writing all this to hear myself talk, you're wrong. I'll spell it out for you before going on. Somebody needs to take the reigns, take absolute command of the entire incident. Making bad decisions is better than indecisiveness. Any plan of action is better than inaction. And, people are

187

waiting—poised—for direction. These people, and all spectators, want to see a real face at the helm: leadership image. If you choose to take on a leadership role in this business, then learn to swim in the deep end of the pool with the rest of us or get out with the kids. Okay, back to the story.

* * * * * * * * * *

I returned to Cordova in mid-April and prepared for a Thursday night training session with the CVFD. It was a coincidence that the evening's training was scheduled to be on the Incident Command System. Usually our ICS training is centered on search and rescue operations since we perform those routinely and not wildland fires, because they are so rare in the rainy area. But rather than a SAR scenario, I began this evening by saying we will pretend that we've been granted the adjuvant authority to reorganize the oil spill response.

I distributed generic ICS organization charts and job definitions. I listed all the "players" that I could remember from Valdez. The objective was to organize this attack all the way down to task forces consisting of, for example, fishing boats with booms, a skimmer and air recon, to strike teams consisting of two boats with booms that might either deflect and direct flowing oil or corral it in a sacrificial bay.

We were to divide the sound into geographic "Divisions" for catching and scooping oil, and develop "Groups" of roving boats or barges to pick up used sorbent boom and deliver food and supplies to the different divisions. The 30 volunteers in the room that night ranged in tenure and background from veteran firefighters with backgrounds in wildland firefighting from other agencies, to 17-year-old high-schoolers. In two and a half hours we had organized several thousand responders into a smooth-running machine. Then we put the newly developed stack of flipcharts away and played pool.

One week later, a couple of guys walked into my office and introduced themselves as Mark Hutton and Mike Williams. Mark Hutton was a tall, pleasantly assertive, and robust commercial fisherman hired by Alyeska to work on this response "problem." Mike Williams was smaller, slightly older, Welsh, and Master Mariner (which means he can skipper ships of any size, anywhere in the world). Previously, he had transferred from BP London to BP America and was living in Cleveland. They wanted to know more about this response system the Valdez folks were now talking about. Alyeska, incidentally, was never part of the oil spill response organization in Valdez, since almost everyone forgot about them in their rage against Exxon. It made Alyeska happy to see that the world had momentarily forgotten that Alyeska was responsible for making the initial response to the spill—and they'd dropped the ball totally. They hadn't done shit. They spotted this opportunity of deflected rage to buy them time to develop a workable response plan.

I led Hutton and Williams to the training room, where I pulled out the flipcharts. I delivered a more detailed explanation about the whole system, its history, and its potential uses on all projects—emergency or non-emergency—to the two of them. Mike told me later that he lay awake that night and mentally examined the system until—like a flash—he saw it in its entirety and knew that's what they needed. To him, ICS was like finding the Holy Grail. The next morning he reported to his headquarters in Anchorage that he had their new response plan.

Also, the three of us met with some folks from our Oil Spill Response Committee and from CDFU for lunch. While Hutton was talking about the spill with the others, I pulled Williams into a conversation about Welsh poet Dylan Thomas, my favorite poet. As a coincidence, it turns out that Mike's uncle Frank Beese (a.k.a. Buzzer Beese), a teacher in Swansea, Wales, taught Latin and Greek to young Dylan Thomas. Mike's father was a teacher in a different school. Mike's father and Frank were in a pub some time before WWII when Dylan walked in and joined them. Mike's father asked Dylan where his inspiration to write poetry came from. Dylan's reply was, "Mr. Williams, what do you hear when you hear a clock ticking?" He was referring to an old grandfather clock in the corner of the bar.

Mr. Williams listened and replied "tick-tock."

"No, no," said Dylan. "Listen carefully. It says 'Death knocks, Death knocks.'"

Mr. Williams would not have a working grandfather clock in the house after that, and was known to stop other people's clocks when he visited them. Boy, Dylan was a laugh a minute.

I even asked Williams about a documentary movie I saw once in Detroit. I had walked into a dingy movie theater, not knowing or caring what was playing. I was shit-faced and simply looking for a place to sit down. But there was Richard Burton on the screen, walking around some town in Wales talking about Dylan Thomas and quoting lots of stuff from Thomas' "Quite Early One Morning." I still can't find that movie. I eventually gave up sleeping in cheap movie theaters, because too often I'd be awakened by some stranger trying to stroke my shlong. I didn't mention that to Williams. And, no, he was not familiar with that movie.

Anyway, after they left town, they contacted me again to provide some ICS training for their people, and I agreed. I got Chief Charlie Lundfeld to accompany me to the Clarion Hotel in Anchorage for the training session, which was attended not only by Alyeska people, specifically from BP, but by representatives of the Coast Guard and DEC. About mid-morning, as I was speaking, the door opened and an entourage of gray suits walked in—two of them were Hutton and Williams. The "alpha male" of the group was Jim Hermiller, the new president of Alyeska. They all sat. Having previously made a comparison between attacking an oil spill and attacking a wildland fire, I was being specific about developing tactics and strategy when Hermiller asked what the two events had in common. I explained it to him, he thought for a moment and said "okay," and left with his entourage.

I learned, some months later, that Hermiller went to Houston to meet with the other oil company partners and said that the new response plan will be an ICS plan. Those others had never heard of ICS and were raising hell. They'd always simply contracted out a clean-up campaign to an outfit like Veco, exchanging millions of dollars, and went on about their business. They wanted to know where Hermiller got such a hare-brained idea.

Hermiller started to explain, but stopped himself and shortened the explanation by reminding all of them that British Petroleum owns 51 percent of the stock in Alyeska, that he (Hermiller) was with BP and is now president of the company; that his vice president in charge of oil spills will be Mike Williams, and that ICS will be the basis of the new plan….period. I understand that the screaming from the board room could be heard throughout Houston.

By mid-summer the pressure was on. The state (ADEC) wanted to see Alyeska's new written response plan. I was so fed up with spill stuff that I stopped answering my phone. Finally, Tom Casey, from Alyeska flew to town to see me. Casey, a mammoth, sweaty, hulk of a man came to my house and said that Alyeska needed a formal, written response plan which had to be approved by the state by the end of August. He explained that one million dollars worth of oil goes through that pipeline every hour. That's $24 million a day, and that if DEC doesn't see a workable plan by the end of August, the state will shut down the pipeline. I explained that I'm not very good at writing that sort of document. Besides, I was trying to run a fire department. He offered to hire me, and I could have charged a gazillion dollars to write the plan, but that type of work would be torturous for me. He stepped back and started telling me the story of BP, and its history of having been a major player in the oil industry many decades ago—thanks to the finds in the Middle East. He rambled on about the Ottoman Empire, and I think he even said something about Lawrence of Arabia, but I couldn't swear to it. Anyway, after British Petroleum peaked in the late '20s, the decades that followed were fraught with one problem after another, including the nationalizing of the oil fields in Iran and other places, and BP became a bit-player in the business until the North Slope discovery. And now they were a major player again. But if the pipeline gets shut down, all that disappears and it's gonna be all my fault!

I was astounded at how Casey could take this entire scenario and drop in into my in-basket. I told him I'd think about it.

Right after he left town, a fisherman friend of mine called to state that one of his crewmembers had to get home and would I like to work on his boat over in the western part of the sound. He had a contract to operate and maintain an oil boom near one of the hatcheries. I said okay.

While on the boat, there wasn't too much activity in that location so I had plenty of time to talk to folks about boom operations and to scribble some notes for Alyeska on their new plan. I would write in longhand on tablet paper, and draw illustrations. Whenever I was done with portions of it, and saw a float plane land in the area, I would take a skiff to the plane and ask the pilot to deliver the packet to the Cordova Fire Department where someone would fax it to Alyeska. Those documents became the foundation of their new plan. Between those moments of inspiration I would read Hemingway in an attempt to maintain a proper view of the world. It worked and I did not become a dweeb.

Afterwards, Lundfeld and I discussed Alyeska's predicament and recommended to Alyeska that they hire Fred Bethune and Don Fuller, who were willing to give up their jobs and their retirement package with the U.S. Forest Service, and—since both were qualified as members of wildland fire Overhead Teams—to write the plan. They did that, and the only other thing I wrote for the plan was the Annex E, on the use of local fishing boats for oil spill operations. DEC approved the plan, did not shut down the pipeline, and everybody was happy.

However, having a plan is only the first step. Training the plan's operators must follow quickly. Alyeska wanted me to do that. I wanted to be left alone. Compromise was in order. Lundfeld was now the president of the Alaska Fire Chiefs Association (AFCA). So at our fall conference, I proposed that the AFCA provide training for Alyeska, and they agreed. Over the next few years, AFCA would hire qualified instructors from Alaska and other parts of the country, skim some administrative costs off the top and managed to get some money into the bank. Still, there were problems.

ADEC, which approved the response plan, still believed they had the authority to monitor activities during a spill, give some orders, yet would not be part of the team nor be accountable for their decisions. I remarked numerous times that you can't fight fires that way and you can't attack a spill that way. You are either part of the team, or you don't play. No "interested party" is going to walk up to a commander of an incident and start telling him how command that incident. That "interested party" would be hauled away in handcuffs. Yet DEC held to this position, stating that they would run a "parallel organization." I got tired of talking to them, so I called State Senator Jay Kertulla who subsequently had a legislative bill passed stating any state agency taking part in an oil spill operation would operate under the proper ICS organization. End of debate. Dennis Kelso, commissioner of DEC, was pissed off at me and would not return my calls when I was calling to offer training.

The one thing about the plan that Alyeska insisted upon but which I was completely opposed to was to establish a Unified Command consisting of three persons: Alyeska representative (who later would pass the position to a rep of the spiller company), a Coast Guard official representing the federal government, and an ADEC official representing the state. I was picturing minute-by-minute conflicts of ideas over the action plan and no consensus ever being reached. I was surprised to see the Coast Guard and DEC willing to participate at this level, but they were. My skepticism was dead wrong.

After providing classroom training for some months, we organized a 3-day drill in February. Within a few hours of the starting gun, the Valdez convention center was packed with 300 people handling written "inputs" from us facilitators. We threw in everything from vessel collisions and fire, to a boat crewmember who had forgotten to pack enough insulin with him. We had people acting like out-of-control politicians storming into the Emergency Operations Center (EOC) and environmental activists interrupting press conferences. But always, developing Action Plans based on new and unforeseen circumstances, and the managing and tracking resources equivalent to a military operation. When I was walking quickly through the EOC, I was stopped by Senator Kertulla who flew up from Juneau because he wanted to see how all of this was going to work. He was pleased and said it's nice to see everyone fitting into the system so well. I started to answer him but got tugged away by an urgent request, and got so busy at it I left the area forgetting to talk to Kertulla. Later, Charlie said, "I always wanted to be too busy to talk to a senator."

Anyway, the Unified Command worked so well, I was a bit embarrassed by having opposed it. However, it was easy to see why it worked so well, and Alyeska had accurately predicted it. Since so many of the circumstances regarding tactics and strategy have been predicted (with the help of hindsight) and like any good firefighting pre-plan, decisions have already been agreed upon and documented. The only thing the Unified Command must do is agree upon basic tactics, based on good data, and direct the Plans Chief to develop the Incident Action Plan, then pass it on the Operations Chief. Primarily, here is why the concept works: All three fingerprints are on the knife. Each is equally responsible for the outcome of the operation. Cooperation, rather than taking pot-shots at one another, is the order of the day. They win or they lose together. It doesn't take long for them to forget they are from separate agencies and they become a confederacy. That is one of the primary objectives of the Incident Command System.

Incidentally, Hermiller did something that few emergency responders would ever do. He invited everyone to come and watch the drill. That's ballsy. Letting people watch an operation that could make you look inept. He did it time after time. Funny, that, even though each drill cost over a million dollars….making it a major event in this state, there was very little attendance by either the press and or by TV news. It was almost like the oil industry was being boycotted whenever they did something positive, because the political correctness at the time was to only publicize the negative things one could find about them. It wasn't until 1993 that one of these drills was properly covered by the media. By that time, my ire was up. I felt paternal about this whole thing and was taking media snubbing personally. I spent two weeks calling newspaper and TV news reporters to tell them that pending drill would involve 300 people from not only the oil industry, but numerous agencies and would cost $1.2 million dollars.

I made several trips to the Alyeska headquarters in Anchorage and attended meetings of the different sections—primarily the Logistics Section and the Plans Section. They were attended by huge numbers of people, but each one knowing exactly what was expected of them. When Tom Casey took me in to the meetings he would introduce me and add that Hermiller said I had a favorite son status and "carte blanche" invitations to all of their meetings and training. I couldn't wait to leave so I could look up what "carte blanche" meant. Upon closing the dictionary, I felt like a real big shot.

Later that spring, Alyeska called me and wanted me to go to Valdez prepared to reiterate all of the advantages of using the Incident Command System on oils spills—"and be convincing." I didn't think anybody else needed convincing. I was wrong. I was to speak to the presidents and vice presidents of shipping of the other oil company partners.

On the charter flight over, I'll have to admit I was getting a nervous stomach. Comes from being steeped in the world view of hierarchies. You know, peasants and aristocracies. We peasants are supposed to feel a fearful awe for the world's shakers and movers, the power mongers, the exclusive club of gray suits. I'd known a handful of people that honestly never felt intimidated by that. Jack Lamm was one. For him, it was not "Mr. Iarossi," it was "Frank." And that was just fine with Mr. Iarossi. I could be that way, but at that time, was usually faking it. Anyway, I used one of my head games I'm so proud of to ease my nerves. First, I convinced myself to remember that I am going to do these guys a favor. I have knowledge that they sorely need. They should be and will be grateful. I would have no need to, nor impulses to, put on airs. I am not pretending to be a world-renowned expert in oil spills. I'm only a small-town fire chief with some information that can save them millions and millions of dollars. I also happen to know that—like it or not—if they spill oil in Alaskan waters, they will handle it the way I will be outlining it. Alyeska, the initial responder, and the government agencies will have the command organization already established by the time the spiller's reps show up (about 72 hours later) and move into the IC's chair. My task will simply be to brief them on what it will look like and convince them to be grateful for that.

My presentation went fine, but their facial expressions were totally non-committal. The barrage of questions that followed caught me completely unprepared. With few exceptions, they were all legal questions—mostly about exposure to litigations. Some of the questions, I was able to draw similarities with fire service operations, but being too sensible to try to fake it, I advised them that attorneys for British Petroleum and Alyeska could probably explain that aspect of these operations better than I. Later, the feedback that I got from Alyeska was that the gray suits (and, incidentally, they really didn't wear gray suits) felt that ICS was the right tool and some had implied that it would be used by their companies wherever they spill oil.

Meanwhile, back in Vancouver, British Columbia, Dean Monterey returned and got himself transferred from the Ministry of Emergency Services, to the Ministry of Environmental Protection. He made the change from their disaster office to their version of DEC so he could make identical changes in their spill response protocols. Having done that, he represented British Columbia by assuming the vice-chair position on the International Oil Spill Task Force, which met regularly in Seattle. It really only encompassed Alaska, British Columbia, Washington, Oregon and California.

Canadians sound just like Americans except that they articulate better and don't say the word "motherfucker" nearly enough to make themselves credible; but even with this flaw, Dean was able to convince the Task Force that the rest of the Pacific Coast should follow Alaska's example. He told me that he gave them copies of my magazine article on the conversion of the Alyeska owner companies to ICS and they were willing to take a look at it. The toughest nut to crack was the Coast Guard. But I think the commander from the 17th Coast Guard District (Alaska) was able to convince the commander of the 13th District—headquartered in Seattle—to go along. Anyway, it took a while, but it was done. One has to remember that the oil companies weren't the only ones that needed to change. State governments and the federal government had to commit to being part of it.

This stuff really started to spread. During the '91 Persian Gulf War, when Saddam Hussein spilled all that oil into the gulf, the provincial premier of B.C. sent Dean and another guy over there. They organized Bahrain and Qatar into the system to deal with the spill.

In addition, the Task Force got a chance to mobilize by responding to a two-ship collision that occurred near B.C. and Washington State. They were pleased with how smoothly the multi-agency response went.

When an oil tanker had a spill near Morocco, Lundfelt called me to say he'd heard from some oil folks in Valdez that he and I were being considered to be sent there. "You wanna go?" he asked. "Not on your life," I replied. I never heard anything more about it, so Charlie could have just been pulling my leg.

* * * * * * * * * *

The training and the quarterly drills continued, using personnel from the other oil companies. When it was BP's turn again the following February (1990), a bunch showed up from BP American out of Cleveland. They went back home after the three-day exercise. Ten days later, a BP tanker ran over its own anchor near Huntington Beach, at Newport California. The crew from Cleveland was sent. It went so well that everybody loved them: the press, the state politicians, even the greenies. I got a note from Mike Williams afterward with a glowing statement that ended with, "You see, Dewey, the gospel has been spread even unto BP."

I was busy, so Charlie got Chief Mike McGowen from Fairbanks and met with Marathon Oil's vice president of emergency management in Cook Inlet and pulled a surprise drill on their facility there. It was a 9.0 earthquake followed by a 50,000 barrel spill and ruptured gas pipe. The VP was impressed and sent Charlie to Finley, Ohio, to conduct training for the International Emergency Managers of Marathon.

None of those changes in oil spill response in Alaska would have taken place if Mike Williams had not advocated for it and subsequently Jim Hermiller hadn't bucked the powerful forces of other oil company executives. The Alaska Fire Chiefs Association developed such affection for Williams they awarded him as an honorary fire chief by presenting him with a white helmet at one of our annual conferences.

Hermiller visited and hosted receptions regularly in Cordova to mingle with locals whose first names he remembered easily and spoke openly on any topic they chose. Only one time did I ask him for a favor.

For decades, the Cordova High School boys and girl's varsity and junior varsity basketball teams have played against rival teams of the Valdez High School. Valdez players would come here for homecoming games, Elks Tip-Off tournaments and so forth. CHS, in turn would travel to Valdez for their tournaments. Even when the ferry schedule made it difficult, phone calls to the state capital in Juneau usually resulted in modifications to the sailing schedule to accommodate us. That did not happen for the Valdez Christmas Tournament in 1991. The state simply could not alter its schedule and the cost of air fare for that many people meant CHS would not be going. Both high schools were terribly disappointed. The Coast Guard Cutter Sweetbrier— which we had used once in the past was not available, either. So I called Jim Hermiller.

I explained the situation and asked if one of the 210-foot Ship Escort/Response Vessels (SERVs) might be available to haul the ball teams to Valdez. Jim called the SERVs office in Valdez and advised them to make it happen. I had no idea what was involved in that. The head of SERVs called the Coast Guard for a permit to carry students. They said, "No way." Every laden tanker leaving the Alyeska terminal is accompanied through the 60-mile trip through the sound by two escort vessels. At least one of these vessels had to be an Escort/Response Vessel (the other could be a tug or another ERV). An ERV carries 4,600 feet of rapid deployment containment boom, two portable surface skimmers and a 20-foot workboat. Mike Williams called me and said he called the Coast Guard and told them to find a way…dig into their regs. Later, SERVs said they could not get a permit for 50 students, but maybe an "excursion permit" (special one-time) for 25. That meant the J.V., the cheerleaders, and the band would not be able to go. To get the rest of the students over there, a second ship would be necessary. Alaska Department of Environmental Conservation (ADEC) would have to issue permission for SERVs to be released from the job to do that. Also, Hermiller had scheduled—for the day before the tournament—a big open-water boom deployment demonstration near Knowles Head for the mayors of the coastal cities to watch.

The phone calls continued. SERVs acquired many cases of life jackets, more life rafts and life boats and found a way to store them on the ships. The ships were big but most of the space on board was for their massive engines and the oil recovery equipment, not for passengers and extra gear. But they managed to do it.

Eventually, they got all the permits and extra gear necessary and two ships arrived in Cordova Friday morning and loaded all the students on board for the five-hour trip to Valdez. Sunday morning, the two ships brought them all back.

I took photos and videos of the kids boarding the ships, wrote a description of what had occurred and why and sent them to the Anchorage newspapers and TV news after calling them on the phone. I don't know how many tens of thousands of dollars and man hours of footwork it took to do that, but the media wasn't interested in airing or printing the event. However, Hermiller did fly in to Cordova for a city council meeting where he was given letter of appreciation for what he did.

BACK TO CORDOVA'S RESPONSE TO THE SPILL:

People all around town had been working on this disaster from the very first day. Within a couple of weeks, Cordova was busy trying to transform its disaster response and recovery actions into "routine" daily activities. The town did that very well. The first axiom of Zen is that "life is difficult." Once that is finally accepted, the energy wasted on rage can be used to defend, attack, or cope. Pick one of the three and get busy. There were numerous aspects of the life and economy of this community that needed addressing. Much of what I saw happen was ingenious and inspiring….and damned effective.

I imagine there are communities and regions out there that have suffered disasters so regularly (those in the flood plains along the Mississippi River, or the hurricane belt, for example) that they are very familiar with various agency responsibilities from local to federal. The Exxon Valdez oil spill, not being a "natural disaster" and occurring in a very remote and inaccessible place, made this incident very unique.

Besides those who "went and fought" were those who stayed and tended to the community. I am always amazed at the ingenuity of people. People formed themselves into groups to deal with the issues presented to them. And if I were to get the position of Chief of a volunteer fire department in some other small community somewhere else, I would start drafting up organizational charts for the local community based on what happened in Cordova and other locations.

* * * * * * * * * *

On May 4, Vice President Dan Quayle (and wife Marilyn) flew to Cordova. A little before they arrived on Air Force 2, Quayle's advance team came in on Wilburs Airlines (a small commuter airline). I don't remember what kind of aircraft they flew, but it was along the same lines as a Convair. Accompanying the advance team was our mayor, Erling Johansen. Their flight was so bumpy that one man's seat broke loose and he was rocking back and forth on one bolt. The plane landed so hard that the front landing gear collapsed and the plane skidded on its nose to a stop. Local DOT employee, Robert Cunningham rolled the fire department crash truck out of the station and pulled up to the wrecked plane, which was not on fire. Mayor Johansen stepped from the plane and Cunningham remarked, "God, you're everywhere." Johansen smiled whimsically and walked off. Shortly thereafter, Air Force 2 landed, rolling passed the skid marks, the wrecked plane, the fire engine, and the sign saying "Welcome to the Mudhole Smith Airport" for all the passengers to see. Quayle, wearing his James Dean "Rebel Without a Cause" red nylon windbreaker and his wife deplaned as though they were not at all surprised. He then flew from the airport to the sound on a helicopter to see the spill, then returned and met with the locals. R. J. Kopchak, who had advanced from hippy to vice-mayor, but still had those wide and jittery eyes behind his wire framed glasses, shook Quayle's hand, introduced himself as vice mayor and remarked, "We both have a lot in common."

"Really? What's that?" the VP asked.

"We're both just a heartbeat away," R. J. explained.

"How's it feel?" Quayle asked.

Before Quayle left, he was asked to support the establishment of a Prince William Sound Science Center in Cordova which could turn this spill and it's aquatic effects in to the most studied spill in history. Quayle assured the locals that he would personally tell President H. W. Bush about the need. The science center later was established and is still here today.

Margy Johnson promoted Alaskan salmon and assured him that none of the salmon caught and processed would be oil contaminated. When I was in Valdez, I ran into a fish broker who had come up to try to find a way to save the salmon industry from financial ruin since salmon buyers were already canceling orders. Some salmon was later sent to the White House and was served for a dinner.

Mayor Johansen accompanied Quayle on Air Force 2 when he left Cordova for Anchorage to meet with Governor Cowper, mayors of the sound, and local media. Secretary of Transportation Samuel Skinner, Admiral Yost, and Presidential Assistant Richard Breeden were also in Anchorage to brief him.

* * * * * * * * * *

The Cordova Chamber of Commerce's efforts after the oil spill were directed at assuring that the economic damaged suffered by Cordova businesses were compensated, hoping to save the economic future of the community. Exxon then recognized that we were uniquely dependent solely on the fishing industry and all businesses were declared eligible to file claims based on lost net income. Cordova was the only community in the sound to receive this recognition.

The standard disaster policy provides for paying claims to the "directly affected" parties, in this case the fishermen and processors. The idea is that as the claim settlements are spent, there is a normal cash flow circulating through the city. But the Chamber wanted to show that this standard policy would not suffice for us. Accepting the invitation, senior Exxon officials met with local business owners and managers and saw what made Cordova's economy tick.

The fish canneries were the biggest customers of our Cordova Electric Cooperative. With the canceling of the herring fishery, revenues from the canneries were down $35,000 just for the month of April. It was estimated that those revenue losses would be up to $120,000 for each of the next four months. The electric co-op already decided to stop plans for their traditional $300,000–$600,000 annual system improvement projects. And the losses also jeopardized a planned $2 million hydro-electric project.

Exxon Treasurer Curtis Fitzgerald and Senior Vice President Ulysses LeGrange saw what a house-of-cards our economy was. LeGrange stated at a town meeting April 27, "....we agree with you, you do a have a unique situation here, it is a one-product economy." Consequently, detailed guidelines for claims procedures were drawn up and refined during the summer of '89. Exxon's Cordova business claims policy was to compensate businesses for lost net income as a result of the spill. And Exxon Claims Manager Dick Harvin said, "Exxon will evaluate and process valid claims as long as there are losses that are a direct result of the oil spill." Boy, that was an open-ended promise. I wonder if anyone previously had ever tried to calculate legitimate long-term economic effects of a spill like this.

Exxon and Veco were encouraged by the Chamber to purchase supplies from local merchants. The Chamber also produced a Cordova products and services guide with funding assistance from Exxon. On the short term, the '89 second quarter Cordova sales tax collections were about $70,000 higher than the previous year, indicating higher-than-usual sales in the summer following the spill. Additional visitors to Cordova kept local hotels fully booked.

Since many locals went to work on the spill, mostly on boats contracted to operate boom and support services for the fleet, 20 percent of the town's labor force relocated to the sound. Labor costs in town increased substantially. It was difficult for local businesses to pay these higher costs. The Chamber brought this problem to the attention of Exxon, who then implemented a labor search and housing program. The Exxon Community Assist Program paid for turning the Bidarki recreation center gymnasium into a bunkhouse-style dormitory by constructing cubicles with plywood and 2 x 4s. Funds were also provided for expansion of a trailer park and provisions for tents down by one of the canneries. A new workforce was brought in by Exxon's labor search.

The Chamber also continued to promote the use of our airport for support of the clean-up operations reiterating that Cordova had the best all-weather airport in the sound. They also explained how important it was that our truckers and longshoremen be employed to transport spill equipment from the airport to vessels enroute to the spill site.

* * * * * * * * * *

Oil Spill Disaster Response Committee, officially formed by city ordinance on April 17, was to be the local focal point for us. Prior, city employees were spending every minute of the day dealing with people, calls, and faxes about the spill. People were working well into the night and, at times, all night just on the spill, and the regular business of the city was being ignored. Remember that I described what happened in Nova Scotia? City staff will work round the clock for as long as it takes and will run on adrenaline. In the meantime, their regular workload piles up. Their in-baskets look like the Library of Congress. Well, this appointed committee was the answer.

The committee was given office space upstairs in city hall and given their own phone and fax numbers. It was to the committee that I faxed my reports every night from Valdez. Since the media did not have access to the nightly meetings in Valdez, and I did, Cordova got immediate briefings on the situation.

The committee (with it's staff of three people) was the contact point for any resident to find out what was happening out there and was also the contact point for outside agencies and persons. The office received calls offering time, money, and expertise. It was the contact place for our congressional delegation and the governor's office.

The Oil Spill Response Committee eventually became the model for other communities affected by the spill. Later, when I re-wrote the city's Emergency Response Plan, I made a point of quickly establishing such a committee for this type of event. Even before the committee was established by resolution on April 2, eight days after the spill, the unofficial committee had a newsletter, the "Cordova Fact Sheet," which was involved in information exchange with national media and with state and federal agencies. At first, neither the staff, nor anything produced by the committee had the polish and professionalism one might hope. And you know, nobody cared or probably even noticed. These folks were just folks. I was not Hemingway, the war correspondent, there were no professional journalists writing the Fact Sheet; no one was trained as a Public Information Officer or media liaison. Nobody cared. It was cool. It didn't take as long as one might think for the Cordova Fact Sheet to look polished and professional, and the committee obtained the bulk mailing stamps to mail it to all local post office boxes. Shortly afterwards, people across the state and outside Alaska were placed on the mailing list.

Staff member Susan Ogle organized the town meetings; Linden O'Toole worked with the Prince William Sound Aquaculture Corporation, which sponsored daily news conference; and Monica Reidel acted as the press liaison (teaching the media about the fishing industry) as she took them around town daily.

Including reports from Fish and Game, the fishermen's union, PWSAC, and sometimes Exxon, the committee updated Cordovans and the international press. Looking at it clearly, the information disseminated from the committee did less to change or improve oil spill operations as it did the community's shock, anger, and fear of losing a way of life.

Don't view a "loss of a way of life" as simply unfortunate change; like how life must change—evolve—in this country moving from an industrial based economy to an information based economy. No, this change would result in completely erasing a town from the map. Katalla, once a booming oil town on the Gulf of Alaska, no longer exists. There are no remnants, no indications that it ever existed. And there are others. A town does not exist for its own sake. Without the fishing industry, the deserted buildings would decay and disappear and be taken over again by the elements.

The office continued its role as host to dignitaries and media. Alaska's state and federal legislators, Congressional committees, scientists, high-ranking federal agency officials, international and national conservation organizations, and media visited Cordova to learn and report the impact of the oil spill on the daily lives of its citizens. Oddly, one of the greatest impacts on the town was the visitors converging on us to study the impact. Of course, that was nothing compared to the convergence on Valdez which is on the road system

GOVERNOR'S OFFICE COMES TO CORDOVA

Of course the Governor sent a "fact-finder" to Valdez early in the event. But the first of April, two persons arrived in Cordova to set up the Governor's Oil Spill Coordination Office and also needed office space in city hall. They were the first liaisons to arrive in town with the intentions of staying more than a few hours. They were to set up communications links with spill cleanup assessment and impact operations, and address local impact reimbursement concerns. They were hit with a deluge of requests for assistance and information. During their first days, they worked from about 8 a.m. to 11 p.m. On April 3, more people arrived to help them. From Anchorage came the Division of Governmental Co-ordination and the Department of Natural Resources. The office was the focal point for community groups and individuals wanting information from the state or wanting specific action from the state. A State Economic Impact Response Workshop was held so that local organizations could discuss how state agencies could help relieve Cordova's' economic and other impacts. A Disaster Assistance Center to help residents was established. The same was done for the village of Tatitlek out of this office.

The state instituted a Tax Obligation Loan Program for fishermen who'd lost their income, to borrow money from the state to prevent losing their boats or fishing permits (and consequently their livelihood) to the IRS for delinquent taxes.

* * * * * * * * * *

The Cordova District Fishermen's United (CDFU) was the first non-oil industry or non-government agency to mobilize against the spill on March 24, 1989. CDFU began the first day with just a handful of people in their offices getting information and assisting local fishermen who intended on going out there and scooping up oil. By the beginning of April, the staff had grown to 32 people working around the clock dispatching and coordinating the work of volunteer fishing boats, acquiring boom, making over-flights and dealing with massive number of media from around the world reporting on the spill. From the first day of the spill, when about ten fishermen sailed to Valdez and met up with and worked with persons from Exxon, DEC, and other agencies, they relentlessly pressed for action to be taken.

As time wore on, CDFU continued to press the government agencies and Exxon to continue clean up efforts throughout the winter and to be sure that local commercial fishermen were the ones employed to do this. CDFU wanted a better review process on Alyeska's Oil Spill Response Plan, and when Alyeska formed a citizen's advisory committee in July, CDFU took a seat on that committee.

CDFU, replacing the fire department in this operation, worked directly with Alyeska to formulate a local response program using both a winter and summer inventory of boats in Cordova, Valdez, Tatitlek and Chenega. Massive storage for boom and other spill equipment has been established and maintained in Cordova. Starting in mid-April, CDFU sent members to Juneau and Washington DC to testify in congressional hearings and had considerable input into state and federal legislation.

CDFU obtained funding from Exxon for these and many other expenditures. They received donations from all over the world. One of their better-known fund raisers was the high-quality videotape entitled "Voices of the Sound" made in the early days of the spill, contrasting scenes of the sound before the spill with graphic footage of the effects of a 240,000 barrel spill on the beaches.

* * * * * * * * * *

Mayor Erling Johansen arranged for hosting a visit from Soviet super-skimmer the Vaydaghubsky. The ship—classified as not only an oil super-skimmer, but also as a dredger, tanker and cargo vessel—arrived in Alaska on April 21 to assist Exxon in the oil recovery operation. The vessel had been under contract with Alyeska Pipeline Service Company as part of its interim contingency plan against any future oil spills. In late June, Cordova residents Ken Adams and Deborah Buchanan pulled along side the Vaydaghubsky and traded some fish for Russian bread and plum preserves. Ken and Deborah returned to Cordova to report that the 41 Russian crew members were eager for a visit ashore—they had not had shore liberty since the skimmer joined the clean-up effort two months previously. Cordova Mayor Erling Johansen extended an invitation to them.

After numerous phone calls and with the assistance from Alyeska, state, and federal officials, permission was received for the 425-foot vessel to travel to Nelson Bay, at the head of Orca Inlet to anchor.

A flotilla of local boats, led by the locally-based Coast Guard buoy tender Sweetbrier, which carried the mayor and other local dignitaries, went out to meet the Soviet ship and presented them with an Alaskan flag and flowers. A third of the crew was allowed off the ship each day for their visits ashore. They were everywhere. And Cordova residents were allowed to see the ship, literally from top to bottom. Few of the crew members spoke English and there was no formal tour, but everyone was heartily welcomed aboard and allowed to roam throughout the vessel. By the time the vessel left Nelson Bay, about 350 residents had found their way to the vessel.

The skimmer's visit here was the farthest it had traveled from its home port in the Sakhalin Islands, just north of Japan. The Russians had a great time. I don't know how much impact the skimmer had on the spill.

* * * * * * * * * *

The Prince William Sound Community College (PWSCC) usually closed during May, June, and July remained opened due to many requests. Several classes were provided in bookkeeping and computer skills. The college responded to numerous requests from visiting journalists, researchers, televisions crews, NOAA, Forest Service, university professors, audio-conference calls, and many others. Rooms were lent, equipment was used, and supplies were consumed. The college also acted as a public shower service to many who arrived on the ferry in the early morning hours and could find "no room at the inn."

* * * * * * * * * *

Duayne Gill sociologist from the University of Mississippi showed up in tow of Mark Johnson, state EMS director. His objective was to study and write a report on the effects the spill had on the collective psyche of Cordova. He had done similar studies in other areas of the country. So he spent a great deal of time interviewing residents about how the event affected them. He told me that the aspect that made this event different from others that he had researched was that since this was not a natural disaster, local anger could be directed at a specific entity, whereas "Acts of God" left people feeling helpless and victimized. I never asked him which he preferred. Maybe that would be like choosing between a heart attack and a stroke.

As an aside, on one of his follow-up visits, ten years later, he and a colleague climbed up Eyak Mountain and got lost when the fog rolled in. He was very happy to see a couple of CVFD searchers the next morning.

STATEWIDE EMS GOES TO THE SOUND

When EMT-III Pam Weaver from the Chugiak Volunteer Fire Department just happened to be in the area and got shanghaied by VECO to set up the EMS system for the clean-up effort, she hired Debbie Schneider, another EMT-III from Chugiak to help her. The difficult job was made even more difficult by how spread out the "housing" of spill responders was. To house 10,000 workers and managers and EMS personnel, berthing vessels (ships) were employed. The U.S. Navy was called to help, Exxon leased and purchased ocean liners and large barges with housing structures stacked on top. The sound was divided into numerous divisions and assigned task forces. Each task force consisted of a combination of berthing vessels, barges, and "taxi boats." Medics—working 12-hour shifts—were assigned by Weaver and Schneider to those task forces, according to how many people would be working the area.

Medics were stationed on the berthing ships or cruised the beach line in "taxi boats" to watch the oil spill workers.

There was one time, however, when medics didn't have to be so concerned about a particular crew on the beach. When Vice President Dan Quayle went to observe the clean-up effort, he took along his own walking hospital. Accompanying him were his personal physician, nurse, and anesthesiologist—not to mention a supply of his own blood.

Thousands of dollars worth of supplies were ordered for the task forces, exhausting medical supply warehouses in Alaska, most of Washington and much of California.

Medics complained that—operationally—communications sucked. Incidentally, I have never seen, heard of, or read about an emergency operation where communications inadequacies were not at the top of the "suck" list. But that eventually was improved substantially. The problem of weather and transportation was impossible to control. The long time it took to get to the site, combined with high waves made transportation often difficult or impossible. Communicating operational changes was a constant source of problems for medic supervisors. A short notice that a vessel was being sent out and needed medics resulted in a scramble to staff the vessel or to shift personnel from another vessel.

But, no matter what the problems were, it was a job. Most medics in the country—like firefighters—are volunteers. This experience was unique....actually getting paid to be an EMT. But volunteer ambulance services throughout Alaska suffered. Kevin Koechlein, EMS coordinator for the Matanuska-Susitna Borough, said the results were devastating for that area near Anchorage. He said many EMTs left their areas for one of those $16.69 an hour jobs as a medic or oil cleaner. "Overall, we probably lost one-sixth of our EMTs in the valley. I know some of our smaller squads lost as much as 50-60 percent of their volunteers." The loss resulted in volunteers being on call seven days a week, instead of one or two. Plus, response times increased. Schedules were also confusing because people working on the sound would come home for one week, only to get called back in two or three days. In Cordova, many members of our squad were on call round the clock, seven days a week.

Weaver and Schneider could speak at great length about the project's implementation, the constant changes, the ongoing effort, and demobilizing of the program or the fact that there were no guidelines to go by. That last one, I believe, is the one that should require our greatest attention.

204

* * * * * * * * * *

Jim Hermiller, who became the new president of Alyeska after the spill, moved from Ohio to Alaska and was the welcomed new image of the oil industry in Alaska. He put on a fresh new face that the Cordova residents for the next four years felt they could deal with. That was until the "Wackenhut Affair" sent him down in flames. The story began nearly ten years before when Chuck Hamel began investigating practices of Exxon and Alyeska. Hamel was not an environmentalist nor an investigator for any regulatory agency. He was an oil shipping broker—small time compared to most. He believed that Exxon and Alyeska tried to run him out of business by diluting the oil he was shipping with water. He constantly received complaints from his clients about it. He eventually lost $12 million and his business in 1980. Hamel surmised that they ran him out of business not because he presented any real competition to major shippers but because he shipped oil at five dollars a barrel cheaper than, for example, Exxon. He believed that his shipping costs openly demonstrated that the major shippers were inflating their shipping costs for tax purposes. And that by doing so, they had managed to divert about $1 billion of what should have been state and federal tax dollars into their own pockets. So, not wanting to draw attention to this, they ran him out of business. Hamel then began investigating other practices of theirs by secretly talking to Alyeska whistleblowers.

Hamel also alleged that he was being threatened by the oil industry and stonewalled by regulatory agencies and the legal system in Alaska. He gathered a lot of information during the nine years prior to the spill but was unable to satisfactorily pursue the issue until 1989. After the spill, he moved back to Valdez and opened up an "office" with multiple phone lines, fax machines and became a resource library of supposed violations that Alyeska had been committing all those years. He was forwarding leaked information and documents to the media and politicians.

Later, when Hermiller was appointed to run Alyeska, he tried to put a stop to the leaks from who he believed were disgruntled Alyeska employees. In January 1990, after the contents of an internal legal memo had been shown on a foreign television program, Hermiller hired the Wackenhut Security Agency to identify the whistle blowers and recover the stolen documents. Wackenhut has contracts all over the world and is on Fortune magazine's list of "America's Most Admired Companies of 2001" and Forbes magazine's "Platinum 400" list of "America's Best Big Companies." It is the world's second largest security agency and employs 40,000 people. I don't know if their methods have always deserved scrutiny, if their methods were officially sanctioned by the company, or if it is simply impossible to monitor the activities of that many employees, but in this case of industrial espionage, their dirty tricks were spotlighted eventually in a congressional sub-committee investigation.

Wackenhut set up a bogus organization, "Ecolit" (as in "ecological litigation") and lured Hamel to their office in an attempt to get him to divulge his sources. In addition, he was tailed when in Anchorage, his phone was tapped, his mail was stolen, and his trash was taken and inspected. Hamel's home was wired for eavesdropping and a Wackenhut employee parked a full-sized RV, equipped with sophisticated surveillance equipment outside Hamel's home. In November, on the advice of Alyeska attorneys, Jim stopped the operation and fired Wackenhut.

Late the next year, 1992, Wackenhut's methods were revealed during California's Representative George Miller's sub-committee investigation and Hermiller was summoned to Washington D.C. to be questioned by them. When he returned to Anchorage, I called him and told him how good he looked on TV. "Up yours, Whetsell," was all he could say. It was time for him to retire. I always liked Jim Hermiller. He spent about $50 million a year on oil spill prevention and response and set the stage for what I believe is the most proficient emergency response team of its type in the world. Alyeska could only be rivaled by wildland firefighters in large-scale disaster management proficiency, and our own state and federal emergency agencies could learn from them. The U.S. government would be well advised to send Alyeska staff to national disasters instead of FEMA. And, in fact, I said so when employees from Alyeska's Anchorage office were compiling a book of memorabilia for him for his retirement celebration. In response to a memo from Alyeska to me, I simply typed out a note to him:

> Retiring? No shit? And do <u>what</u>, pray tell? Okay, so we won't have old Hermiller to kick around any more. Well, look at the bright side. Now you can tell the world to kiss your ass without it being aired on national TV. The Alaska Fire Chiefs' Association will be very disappointed when they sober up and find out about it; but not as disappointed as I am. On behalf of them, and the residents of Cordova, I'd like to tell you what a pleasure it has been dealing with you over the years. I especially want to thank you for assisting our high school basketball team in getting to Valdez so they could play in the tournament over there. And as far as preparing your people for disaster response, FEMA could learn by your example, and I'm not kidding. You really did it. You should be proud.
>
> Your pal,
> Dewey

Incidentally, that note was copied into evidence by Exxon attorneys during Cordova's law suit against them. I don't know what it was supposed to be evidence of.

Of course, Alyeska was as much sinner as saint. Aren't we all. You know, we all have to do what we have to do. Environmentalist have to put up a fight. Industry has to put up a fight. They keep each other somewhat straight. Just do what you have to do and quit anguishing over it. Combat is an integral part of life, pick a side, and go for it. It's only a fight, don't take it personally for Christ's sake. After all, losing is only painful when you consider it a personal affront.

Personally, I thought the entire Wackenhut scandal was funny. It must be a hold-over from my fire service gallows humor. Maybe I need therapy.

So, besides learning—from this section—that someone needs to take command and others need to know exactly who that someone is, don't succumb to "analysis paralysis." Make a decision and move on it, and, lastly if you make a bad decision, don't try to weasel out of it; take the criticism like a man.

THE ALYESKA OIL SPILL EOC IN VALDEZ
FILLING UP TO RUN ONE OF THEIR MILLION $$$ DRILLS

Photo by CVFD

A Word About FEMA

Now, a word about FEMA and about your expectations: Well, hell….you know better than that. Here's who you can depend on in a pinch: those guys riding on your engines, trucks, or ambulances; those guys in the cop cars—local, county, or state. You can count on the military. But if you think you can count on civilians to swoop in and save the day, you're living in La La Land. And, of all of the above, the fire service is the only organization that has the frame of mind and broad understanding to prepare a community for a disaster. Period.

A case in point: When FEMA botched their response to Hurricane Hugo and the San Francisco earthquake in 1989, they were told to get their shit together. Then three years later they botched their response to Hurricane Andrew. Everybody went after them. Well, I wrote several suggestions to them to improve the effectiveness of their organization. First, I suggested that they appoint fire service personnel to key positions. Second, I suggested that these new appointees organize "Overhead Teams" just like the wildland guys have. These teams could be "first-in" to a damaged area, liaison with either local or state officials and help walk them through the disaster.

I got a rather snotty letter back letting me know—in so many words—that I was free to apply for a job with them if I wanted.

Okay, so I sent copies of my letters (recommendations) to Maryland Senator Barbara Mikulski because she was on FEMA's appropriations subcommittee. In return, I got a glowing letter back from FEMA (3-page letter and 4 pages of attachments) detailing all the improvements they were making, including the formation of "Federal Overhead Teams." They sent me that letter in October, 1993. Any questions?

Think about the dynamics of the fire service. Major city or small town, the local fire department must take the lead in disaster preparation—for the entire city. It should be part of the department's enabling ordinance.

SECTION III

MANAGEMENT

SCABS GALORE

Okay, now that we're tired of running everybody else's business, let's concentrate on running ours.

You can buy numerous books on the management of fire departments. Go for it. But what follows in this section goes beyond what you will find written by normal authors. Just consider this stuff as… well…ah…extra.

Maybe it is only because I live in Alaska, but most of the career firefighters I know began as volunteers in the fire service. I was a volunteer before I was offered and took the first paid position in Cordova. I wasn't a paid chief, I just happened to be elected as chief and was re-elected by the membership each year for 28 years. I was a paid firefighter. I also remained an uncompensated volunteer for anything that occurred outside of my normal 40 hours of work per week. So, I was a volunteer for 34 years. Therefore, in the eyes of the International Association of Fire Fighters (union), I must have been a scab of immense proportions.

I had not been a scab all my life, however. As a factory worker in Muncie, Indiana, I was a member of the United Auto Worker's union (UAW) and also member of the musicians union. My brother-in-law was a career-long union member in the Muncie Fire Department, as was his father before him, and my mother's two cousins.

In Cordova, fate had me negotiating with unions from the other side of the table from time to time. On behalf of the Cordova School District, I negotiated several years with the NEA (teacher's union). And on behalf of the City of Cordova, I negotiated with the IBEW. The NEA was straight-forward and candid. The IBEW reps acted like they were in the middle of a Cold War spy intrigue. Completely untrustworthy.

In the fire service, I never liked the "us and them" attitude of animosity I had seen between volunteers and career firefighters and was glad to see it was not as prevalent in Alaska as in some other places. Most civilians don't even know such a mind-set exists. That is why I was not surprised by a question that was asked of me by a TV news reporter back in the mid-'80s.

I was president of the Alaska Fire Chiefs Association and was in the middle of an Association conference in Juneau, when during a break I was being interviewed by a TV news crew. After a few questions about the events and topics of the conference, the reporter asked my views on some recent newsworthy problems that were occurring in Fairbanks (Alaska's second largest city).

A bitter dispute arose between the Fairbanks city fathers (mayor, city manager, and city council members) and the firefighter's union. I doubt that the reporter knew I was a volunteer, and that volunteers don't always see eye-to-eye with career firefighters. But he knew I was a chief, and perhaps he thought that I might make an anti-union statement. I knew that those firefighters up there were making lots of money, and apparently the public—upon learning how much—got a bit frazzled. But I knew a bit more about Fairbanks, too. I knew that when the Trans-Alaskan Oil Pipeline was being built, people across the state were leaving their regular jobs to make big bucks constructing that pipeline. I knew that Fairbanks had dramatically increased firefighter's wages to keep them there. For the most part, it worked. But there was more.

Without sounding too scolding, I explained to the reporter and camera crew that Fairbanks was a city that was "built to burn." That city was "the last bastion of the Great American Libertarian whose anti-government views fought every attempt at common sense fire code enforcement. Because of this anti-regulation mind-set, the city fathers created a tinder-box city. Then they have the gall to resent paying firefighters who risk their necks dealing with that mess."

I ended by saying that "the manning of their department is so low, it's irresponsible. It's almost criminal, in my opinion." And it was. At that time, they had enough manning for an engine company and a truck company.

By the time it was aired, most of my statements had been edited except "Fairbanks was built to burn"; they "fought every attempted at code enforcement"; and "fire department manning is low."

But even after the TV homogenized my statements that much, I was still surprised when I received a letter from the Fairbanks City Manager some time later, asking a series of questions about volunteer fire departments. Fairbanks was toying with the idea of supplementing their fire department with volunteers. My guess is that this guy was not aware of my TV statements and chose to query me because of the reputation the CVFD had. He wrote:

Dear Chief Whetsell:

I am writing to you in hopes that you will be able to provide information on your volunteer firefighter personnel. Our city is looking at the feasibility of volunteer fire fighters. Members of the city council have asked that I provide this information as soon as possible. Please send as much of the following information that you have:

1. How many volunteers do you employ?
2. What is the working agreement with the volunteers and their regular employers?

3. The number of injuries sustained by the volunteers?
4. The breakdown of cost of injuries, i.e., workers comp, benefits, or any other liabilities or costs associated with training or volunteer-related injuries?
5. Are there any insurance benefits allowed volunteers?
6. Any costs associated with volunteers, to include training, equipment, or any other costs to your city?
7. The service miles covered by your volunteer firefighters.
8. The population of your service area.

If you can, please include any other information that you feel would be pertinent, or that might give our Council a better idea of the total picture of the complete costs of a volunteer fire service.

Thank you very much for your assistance.

Sincerely,
City Manager
City of Fairbanks

It took me several hours to compose my response to him. Why am I recounting all this data to you when, I'm sure, you bought this book expecting it would be one long series of war stories? First of all, you are not likely to find this stuff in one, single publication anywhere else. Second, it resolves a lot of misinformation floating around about us scabs in the fire service. Third, it makes you smarter than the guy that didn't read it.

My reply to him contains numerous management principles that explain why some communities must have volunteers and other must have full-time career. Just like fighting fires, it's all math. What is most unfortunate is that many career firefighters do not realize that they have their jobs, not in spite of volunteers, but because volunteer firefighters insisted that full-time personnel be hired. I answered:

Dear (city manager):

This letter contains some of the information you requested regarding volunteer firefighters and the service provided to the City of Cordova. The city limits takes in a little over seven and a half square miles and a population of about 3,000 persons (this was before Cordova annexed the surrounding rural area). However, we provide a secondary service to about 400 persons and the buildings outside the city limits in exchange for state revenue-sharing money (which doesn't amount to much). This department has 40 volunteer firefighters and one paid person. That is less than average since generally there is one paid person per 1,000 population in most towns. My annual budget is about $151,000. Converting to the equivalent of mills, my budget is less than the usual 2.3 mills I have received in the past. Most departments (regardless of size) operate on an equivalent of about 2.5-to-3 mills. Major capital purchases and large inter-fund transfers often inflate that figure.

Quick and dirty—The average cost of emergency services throughout the '80s was:

1.5 mills for EMS

2.0 mills for fire service

3.0 mills for combined (or $2.5 - $3.00 per $1,000 of taxable property)

You asked about what working agreement volunteers have with their regular employers. In rural communities it is generally understood that for the benefit of all residents and for the protection of the businesses and economic backbone of a community, that employers must be civic-minded enough to allow employees to respond during the workday with no loss of pay. That doesn't seem to adversely affect employers in rural towns that have 200 calls a year or less. As towns become bigger in population, the run volume increases proportionately and employers become more and more resentful of making the sacrifice of their businesses in the name of community-spirit. In addition, most volunteers start getting a little resentful as well, and start demanding that the city hire more full-time firefighters to handle the low-consequence calls. In Cordova, we have kept the run volume down through fire prevention and code enforcement.

I then gave him the figures for our insurance and number of injuries per year and the cost of providing gear for the volunteers. I explained that our training costs were low because we use in-house instructors even though we require all our firefighters complete the state certified Firefighter-1 course (approaching, at times, 160 hours) and our engineers complete our own 40-hour Driver/Engineer course (strongly encouraged by our insurance company). Company officer training and numerous other courses necessary for professionalism are also provided in-house. I then explained other (following) options that he had not asked about, but I was sure at some point, he would ponder.

I continued:

In looking into your dilemma, I've noticed several circumstances that would need to be addressed and solutions found to be able to provide a good level of service using volunteers in a town the size of Fairbanks. For example: Some communities experimented with

213

privatizing or contracting for fire protection. In Alaska, the City of Homer contracted with the 'Homer Volunteer Fire Department, Inc.' The city just recently took that back over. Their budget was virtually the same either way ($500,000) with 4 paid staff and 64 volunteers. Homer's population is about 4,000. The city wanted to have governmental authority over the fire department's activities, which was difficult under private contract.

Billings, Montana, tried on two occasions to contract fire protection and went back to having it a city function. I believe it was for the same reason: city control. One thing to consider is that the authority to enforce codes is difficult to extend to a private company. Scottsdale, Arizona, contracts with Rural/Metro for private fire protection. It has been a number of years since I read up on their arrangements. But if I remember right, some of the things they did to make it work was all city employees were required to "volunteer" to be on the department. That was before the passage of the Fair Labor Standards Act which states that a person cannot be required to volunteer as a stipulation of employment. So, if they must make themselves available during off-duty hours, I assume they must be paid stand-by time while at home, and must be paid overtime pay during call-outs with a minimum of 2-hour or 4-hour call-out pay. I'm only guessing about this. Rural/Metro did design and purchase innovative equipment to allow them to flow more water with fewer people than traditional equipment. I assume that prototypes are more expensive. Scottsdale was built from the ground up to be fire safe: Low profile buildings, lots of sheetrock and compartmentalization to reduce fire flow requirements, and lots of sprinkler systems. Also, Scottsdale, from what I understand, is a "bedroom community." It doesn't have a large industrial section or those types of high-risk neighborhoods that most other towns have.

Another problem you would have to resolve is the availability of volunteers. Even though about 85 percent of American communities rely on volunteers, it's much less than it was decades ago, because of the way we live now. In my community, the volunteers work and live in the same geographic area. But in many places, residents work in the city but travel back to the suburbs in the evening, making them unavailable 75 percent of the time. Recruitment would have to target those who live and work within the city itself, really reducing the 'bank' of people you could draw from.

The response time for volunteers, of course, is longer, but there are ways of dealing with that. But first, let me give you some details about the effects of response time. Fighting fires is nothing more than mathematics. It starts with what is called Fire Flow Requirements. That means the rate at which water must be applied to a building when it is fully involved. There is a very complicated way of figuring it out, of course, but also a simple 'rule of thumb' for doing is faster. The square footage of the building divided by 3, then multiplying that by the number of stories that are burning will tell you the rate at which water must be applied to put the fire out. If you apply water at a lesser rate, the fire will just eat the water and continue to burn. However, if firefighters arrive before a small fire 'flashes over' and starts consuming the entire place, it can be extinguished by a handful of firefighters applying a small amount of water. 'Flashover' generally occurs in less than 5 minutes after the fire starts. Smoke alarms allow the fire to be discovered soon after it starts. And if the alarm can be turned in quickly and firefighters can arrive before flashover occurs, they stand a better chance of beating the fire. Alarm systems may be the answer to the response time of volunteers.

Should firefighters arrive after flashover, then the water must be applied in accordance with the calculated Fire Flow Requirement. Here is another rule of thumb: A wood-frame building 50 feet by 50 feet, 3 stories high will require 2,500 gallons per minute to be applied. An NFPA rule of thumb is that every firefighter on the scene is worth 50 gallons per minute (that figure includes engineers and officers). That fire is a 50-man fire. It takes that many persons to handle and move all the hoses that would deliver that much water. It takes two persons to handle a 1½-inch hose that delivers 100 gallons per minute. It takes 3 persons to handle a 2½-inch hose that delivers 250 gpms.

Most departments have now purchased 1¾-inch hoses with high-flow nozzles to allow two persons to deliver 200 gpms. So now, that 2,500 gpm fire we're talking about takes 25 persons, not 50. I assume that Fairbanks has done the same. For this scenario, you can double the size of your fire department (as it were) by making this one-time purchase.

Regarding this topic of Fire Flow Requirements, it's that very thing the ISO looks at in determining a city's insurance rating. The Fire Flow Requirement of your community determines how many engine companies and truck companies you need on a first alarm, second alarm, etc. In Cordova, since we have more than 5 buildings with a flow requirement greater than 3,500 gpm, we are required 3 engine companies and 1 truck company if we want a good rating. Here's one place where volunteers are looked at differently than paid. ISO says that they consider no less than 8 volunteers assigned to on a rig to be a company, but 4 paid firefighters assigned to a rig is a company. It's the old problem of 24-hour availability again.

ISO, of course, has no authority over how a department is run. But the cost of insurance really encourages communities to try to reach those standards. For example: a number of years ago I got some figures that might interest you. In a Class 6 city, insuring a wood-frame house cost $312 for $40,000 of insurance. A Class 5 city would cost $285 for the same insurance. If you were to take the appraised value of the building in your community and add another 50 percent for its contents it would be very simple to get a rough idea on the overall impact ISO ratings have on your residents. Actually, you would need to inflate that figure a bit since on average, the insured value of buildings and contents is about 56 percent higher than the appraised value. In 1987, when we improved from a Class 6 to a Class 4, we saved our residents about $100,000 a year in insurance premiums. That made them very happy since that amounted to about half of our annual budget.

Insurance companies, being private businesses, are not required to be reasonable in the premiums they charge and, in fact, are not required to do business in a town at all. If you can show them that using volunteers can still be effective, no problem. If you cannot convince them of that, they may decide that they would be at too great a risk to charge reasonable rates. Not only would that cost all residents more money, it may make it very difficult for businessmen to get business loans for buying or improving their businesses, since those loans must have insured buildings as collateral and the monthly premiums may be more than the businessman can afford. The impact of that might be that business property in your city may deteriorate, and people trying to sell out may not be able to find buyers.

So, it's good to keep in line with the factors that ISO uses. There are two other things that occurred to me while thinking about your project. First: 'Mutual Aid' agreements with surrounding communities. But the one factor there that must be remembered is that Mutual Aid is just that…Mutual. You might be able to count on outside help, but at the same time, you would be required to assist them. Ordinarily, no problem, I suppose. Except during wildland fire season when neighboring departments spend so much time fighting those fires. Second: Consolidation of some functions. I'm only guessing because I've never spent much time in the interior. But if there are 11 fire departments in your area, then there are 11 chiefs, 11 training officers, 11 maintenance facilities, and 11 dispatch centers. Perhaps you can look at consolidation similar to what Anchorage and Juneau did.

I realize how long this letter is and that it contains more than just the things you asked for. But fire protection is a complex thing and I found it hard to just answer the questions without offering some background information. I hope you find this useful, and if I can be of further help, don't hesitate to contact me.

I chose not to complicate the letter further by explaining that I was speaking of towns with normal tax bases. Valdez with its tank farm or Kenai with its refinery, can generate fortunes with a low mill rate. Also, there are large cities with total volunteer manning. But these are exceptions.

As you just saw, there was not one glimmer of hope in that letter for the use of volunteers. There are a million of us in this country, and there's lots for us to do. We don't need to allow ourselves to be used in such a way. That would be a disservice to the fire service and the citizens.

Incidentally, the city manager never thanked me for the effort it took to compose that 2,000-word letter, nor even acknowledged receiving it. I guess he expected more support from me for the use of volunteers.

AMMO FOR THE UNION

A few days after I'd sent that letter, I was chatting on the phone with Anchorage Fire Chief Larry Langston. I told him about some of the data I put in the letter and he told me that he had gone to school with the president of the Fairbanks fire fighters union, Dave Rockney. He thought Rockney, who had been under tremendous pressure lately, could use some of the information that I'd sent to the city manager. I told Larry that since I was answering direct questions from the manager, and didn't "cc" anybody, it would not be proper for me to sneak a copy to the union. "Un huh," he said.

"Ah, screw it," I said. "If I sent you a copy, maybe you could forward one to Rockney."

He agreed and soon got that letter to Rockney.

As it turned out, the Fairbanks city fathers and city administration were still not through bending to the whims of profiteering builders and building owners. The next thing they wanted to consider was just abolishing the fire department altogether and paying the increase in insurance premiums. Now, I know you don't believe this, but it's true. So, Dave Rockney called me a few weeks later from his house. I don't know why he wouldn't call me from work. I hope it wasn't because he didn't want his buddies to know that he was talking to a chief or a volunteer. But he hoped I had some figures he could use in the next stage of this battle. I did.

I sent the following figures to give him some hard financial data to support the firefighters in their struggles. Later, Dave said he found the information useful and appreciated the help. Of course, the dollar figures listed have changed a bit since then.

Dave, here are those insurance calculations you requested. But first, let me tell you the criteria I used for them;

1- This is the most current assessment in Cordova of all structures. I included exempt structures, because even though they may not pay taxes, I'm sure they are insured.

2- National average is that contents of a building equal about 50 percent again as much as the structure (A little less in residents, but often much more in commercial buildings).

3- Insured values are usually 56 percent higher than assessed value (by national average).

4- My price quotes are from Nationwide and State Farm Insurance Companies. They are current quotes.

218

5- The figures are based on insured residential structures, about 10 years old. So they are conservative.

6- Most fire department budgets include much more than just the cost of fire protection. They include things like Rescue Services, Public Education programs, Fire Prevention, etc. However, for this illustration, I wrote the figures as though the entire budget was for fire protection. So the figures are REALLY conservative.

ASSESSED STRUCTURES AND
ESTIMATED CONTENTS IN CORDOVA

Class 10 city means no fire protection at all

Class 4 city is Cordova's current ISO rating (upon which the insurance premiums are based).

$ 100,000 POLICY:	Nationwide	State Farm	Average
Class 10	$ 754	$ 925	$ 842
Class 4	$ 320	$ 376	$ 348

Average savings per $ 100,000 unit: $ 500 rounded

In Cordova the Assessed Value of all structures was $112,307,413. Based on national averages, building contents were valued at an additional $56,153,706. So the total value of insured property was $168,461,119.

Insurance costs without fire protection would have been $842 for every $100,000 of insurance, or $1,417,928 total, city-wide. Insurance costs for a class 4 city at $348 per $100,000 of insurance was $586,896 total, city-wide. Just in insurance costs alone, our fire department saves the entire community $831,896. Our budget is $151,000. A return of $831,896 on an investment of $151,000 (budget) is 555%. So, one-tenth of one percent of a property's value goes toward its protection in Cordova.

Even more dramatic is that—national average—people insure at about 56 percent higher than assessed values. That means that the insured value in Cordova is more like $262,799,160. At the same rate of savings, city-wide savings are $1,298,232. That's an 859% return on their fire protection investment (our budget). Without the fire department, city-wide insurance would cost the residents $2,212,776. Only an idiot would spend $2,212,776 in insurance to save $151,000 in tax dollars.

Quicker and Dirtier—For different ratings:

Class-10 ISO	$11.00 per $1,000 of insurance
Class- 8 ISO	$ 7.00 per $1,000 of insurance
Class- 5 ISO	$ 4.00 per $1,000 of insurance

The neat thing about writing these figures down and presenting it to elected officials is they cannot later plead ignorance.

Anyway, for the career vs. volunteer arguments, here it is: You will have as big a paid staff as you can buy with 2.5–3 mills, whether it's one or 1,200. Period.

* * * * * * * * * *

Before I move on, there is a little more data I would like to share with you. First, I never had the opportunity for formal education in this field, although I did take one correspondence course in the late '70s. But I did order and read some text books from a Fire Science course and found invaluable information about Fire Loss Management. Actually, I read everything I could get my hands on.

Sequestered voluntarily in Cordova, it took me several years to come out of my shell and travel to gatherings in Anchorage of other fire officers. I didn't want to talk with real fire chiefs and have them discover what a hick I was. But when I finally went, the first thing that I discovered was that I was commanding more fires and bigger fires than most of them at that time…..even the paid guys. CVFD was driving up and down the streets, covered with soot, soaking wet, on almost a daily basis. Hell, the paid guys were responding to calls just within the neighborhoods covered by their companies. And, they were responding only 56 hours a week. I was responding 168 hours a week.

I discovered that I had a knack for tactics and strategy. Funny, I couldn't—and still can't—fix anything mechanical. Clerical work or filling out official forms always made me feel retarded. In fact, other than music, or reading literature or philosophy, this was the only thing I ever found that I had a knack for. So I worried for nothing. I also discovered that on the executive level, all my reading had prepared me for discussions with the most senior chiefs I'd met.

MILL-EQUIVALENTS

Now, here's why I use mill-equivalents in talking about budgets or fire losses. In 1980 I was asked to deliver a lecture to other fire officers regarding an administrative-level topic. I had read a comparison of national fire losses from 1925 to 1977. Looking at the dollar losses, the startling increases would make you think that fire protection was getting worse instead of better. In 1925 fire loss nationally was only $ 559.4 million, but that was equivalent to .61 percent of the gross national product (GNP); in 1977 fire loss was $ 3.76 billion, but that was .19 percent of the GNP. That comparison factored in inflation and actual physical growth of the nation's industry. It's easy to see that if you burned down a $100,000 house 20 years ago, and burned down the exact same house this year, it might be a $200,000 loss. So, I began talking about comparisons to "mills" of a town: 1/1000 of the assessed value of the town. That $100,000 loss 20 years ago may have been the equivalent of half-a-mill. And that $200,000 loss this year might also be worth half-a-mill in the same small town. That way, I could compare fire losses without worrying about the distorting effects of inflation. I could also compare budgets the same way over the years. And as it turned out, I could compare fire losses or budgets with other communities that were of different sizes. Otherwise, I could state that the Juneau Fire Department must be pretty lousy because they have ten times the fire loss of Cordova, not mentioning that they are ten-times bigger. So, I used mill-equivalents in my lecture and explained why.

Volunteer fire departments might find the next equation interesting. This has not been researched as far as I know. I just noticed it by checking out a bunch of VFDs around this state, then later, noticed the same trend whenever I read about other departments around the country. This is not solid, but most VFDs revealing their membership and the population of their towns (or coverage area), memberships are generally about 1 percent of the population served. A town of 3,000 usually generates about 30 volunteers. 5,000 residents generally generates about 50 volunteers. So, if your membership is better than 1 percent, you can be pleased.

Here's something I picked up from the International City/County Management Association (ICMA) a number of years ago. It was from a survey of 50 large cities; the largest city in each state. These are averages from the survey. There is one cop per 600 population served. There is one paid firefighter per 1,000 population served. Remember, a cop works 40 hours a week and a firefighter on the Kelly Schedule works 56 hours per week. Combine all public safety functions (law enforcement, fire protection, emergency medical services) into a budget category and the figure will be about 26 percent of the city's operating budget. Public Works consumes about 45 percent of it. City administration takes about 8 percent. The rest is doled out to Parks and Recreation, and other services. If, for example, you discover that city admin is sucking up 12 percent or 15 percent, and the blue-collar, direct-service-to-the-public departments are scrimping for bullets, nomex hoods, or shovels, then you know that pencil-necked, number-crunching geeks have taken over city administration and it has become top-heavy. Conversion to mills? Police Departments generally need the equivalent of 10-12 mills; local contribution to the school district is also about 10 mills. Incidentally, I'm talking about towns with a normal tax base, not the Valdezes or Kenais of the world with their money-generating oil industries within their tax areas.

Naturally, I am talking about equivalents to mills, not actual property tax mills, unless you're looking for armed revolution. City income not only comes from property taxes, it may come from sales tax, fees and licenses, and so on. In the old days in Alaska, lots of money came from the state in the form of revenue sharing money on a per capita basis. Then, as the oil revenues began to drop and

Alaska went from being filthy rich to just obscenely wealthy, the state stopped all revenue sharing.

Look at the figures in this chapter, make a comparison with current figures in your area, and put them in your arsenal of information. I used to quip to the city council as I walked in for the annual beg-and-grovel budget hearings, "Why don't you just abolish fire/rescue entirely, then each taxpayer can get a refund of a whopping 2 percent." They always smiled and rolled their eyes, then went through the motions of reviewing my budget requests, but in the end, were always very fair.

As an aside, on the topic of budget hearings with your city council, most chiefs hate the idea of the council assuming authority over individual items in your budget (i.e. "line-item options"). But here is one benefit: If you want to expand your services beyond what your enabling ordinance describes, you can—for example—budget for Water Rescue gear. If your Council has line-item options, and they approve the purchase of that equipment, then they have authorized you to start providing that service. You're now legal.

FIRE SERVICE SAFETY: FACT VERSUS HYPE

Managing your meager budget gets more and more difficult as outside agencies or organizations assume the authority of telling you what you need to buy.

The most dramatic impact any outside agency ever had on the fire service was the National Fire Protection Association (NFPA) and their compilation of "Standard 1500"—the Safety Standard. The NFPA has been writing standards for the fire service for decades. But what made this standard different was that since it was a safety standard, OSHA saw the opportunity to expand its jurisdictional authority. Since 1500 was a recognized consensus standard, OSHA could quote it on any investigation of an accident or complaint, and issue a violation. They now had a ready-made voluminous reference whenever they felt the urge to monitor the fire service. Now, federal OSHA has no jurisdiction over the municipalities, including fire departments, but any state that has its own state OSHA, and that OSHA does have jurisdiction. About 50 percent of the states have state OSHAs. Even though virtually no fire department in the country complied with this visionary, all-inclusive standard, a department wouldn't stand a chance in court.

Even though exposure to litigation became tremendous, no one in the fire service would openly criticize the standard and appear to be a dinosaur, or worse, be viewed by his people as being indifferent to their welfare.

All the chief officers became worried about this Sword of Damocles dangling over their heads. I asked the Alaska Fire Chiefs Association (AFCA) to write a resolution (protest) to NFPA and ask them to change 1500 from a "consensus standard" to a "recommended practice." Even if the resolution didn't cause a change, the association would be on record as opposing it. They were too timid.

Instead, they chose to jump on the bandwagon of hyperbole, denouncing themselves as the source of needless deaths and injuries, and "By God—we've got to stop this senselessness." It was a shameless show of political correctness. I explained that there were two aspects to 1500: 1) Performing our duties safely and prudently; and 2) Allowing ourselves to be exposed needlessly to potential litigation. A department could still strive to meet the intent of 1500 while not being threatened with law suits or be open to endless complaints to OSHA by disgruntled department members. For years past, my opinion had influence in the AFCA, but not this time.

So I took a different approach. After the third law enforcement officer in Alaska had been killed on the job in just a few months, I wrote to OSHA. I stated that the CVFD was reviewing its protocols for responses we make in conjunction with police officers. I asked them what their investigations of these work-related police fatalities had revealed. I suspected the incidents were not investigated by OSHA, and I just wanted a record of that in writing. Not surprisingly, they did not answer my letter. After several weeks, I wrote and asked them again. There was no return letter. But an agent did call me on the phone and said they had not investigated those fatalities. I logged that in my journal as a way of stating—later on, if I needed to in court—that the fire service was singled out by OSHA. I did step into the police chief's office one morning and asked him if he had a policy requiring his officers to wear their bullet proof vests. He said he issued them, but it was up to the individual officer to decide if he wants to wear them. I asked if it were not an OSHA regulation. He said no. Further, he said some departments do require wearing the vests and some leave it optional. There was no regulation. Shortly after that, a fourth officer was killed in the state in a 14-month time period.

* * * * * * * * * *

Anyway, since the AFCA would not write to NFPA, I did. I had been researching firefighter fatalities and injuries and NIOSH reports for some time to find out what the real risks were and to mitigate them in my department. My motivation was not the 1500 standard, but the reality of making a dangerous job safer. As it turned out, for my department, most of 1500 was irrelevant, and many of the real risks we faced were never addressed in 1500.

Let me preface this by saying that for years we used to call firefighting "the most dangerous occupation in America." We were bragging because it was macho. It was also not true. We also spoke about the Line of Duty Deaths (LODD) in the service.

There are approximately 100 fire service LODDs per year—every year—in this country. Of those, about half are heart attacks. I'll mention some studies done on that later, but the remaining 50 annual LODDs were firefighters who were actually *killed* on the job. It is those 50 deaths that would warrant OSHA scrutiny, if each one occurred in a state that had its own state OSHA. But OSHA aside, each of those tragic incidents are investigated to mitigate such occurrences in the future. However, 50 LODDs out of 1.2 million firefighters in the country statistically pales in comparison to the LODDs of construction workers, commercial fishermen, loggers, or police officers.

LODDs in the fire service are declining steadily. Related to this, according to NFPA statistics, fire incidents have decreased 45 percent over a 20-year period. In 1977, 3.3 million fires were reported, but by 1997, the number lowered to 1.8 million. During the same period, the number of deaths in residential fires also decreased 45 percent, and civilian injuries in residential fires decreased about 24 percent. It is also my understanding that even though the number of firefighter LODDs has decreased, those LODDs listed are not entirely structural firefighters (which are the focus of the standards), but include wildland firefighters, which is a separate issue. As an aside, this again affirms that preventing fires is the single best way to prevent fire-related deaths and injuries. Fire Prevention gives this country the best return on effort and resources in reducing our problems. This should be our focus. Even though this should be abundantly clear, even to the most casual observer, it seems to be a point nationally missed in our efforts to provide a safer environment for citizens and firefighters.

Putting LODDs aside for a moment, look at the injuries in the fire service. The injury statistics in the November/December 1999 NFPA Journal were extremely enlightening. Of the 87,500 injuries, less than 50 percent occur on the fireground. Fireground injuries dropped 34 percent in 10 years, even though the *rate* of injuries per 1,000 calls has not changed, which means—again—the 34 percent drop coincides with the drop in fire calls. The majority of injuries are "muscle" injuries, strains from over exertion, followed by the category of cuts and bruises from slips and falls. The smallest percentage of all is due to inhaled smoke or gasses.

Firefighters from small towns in Alaska find the next data most interesting: The largest rates of injuries per 100 calls—and those requiring hospitalization—occur in large cities (particularly one-half to one million populations). The largest number of injuries per 100 calls occurs in large cities of the Northeast (over twice the number of those large cities of the West). Apparently, firefighters from New York City or Boston will have twice as many injuries per 100 runs than firefighters from L.A. or Seattle. Where does that rank a little Alaskan town of 3,000? According to the article, towns with a population of 500,000 to 999,999 have 45 times as many firefighters as towns of 2,500 populations, yet they attend more than 300 times as many fires. Why should we be mandated to implement the same standards as them? Small towns have virtually no effect on the national statistics.

Look at California: In 1974 there were 216,528 fire incidents and 2,031 firefighter injuries. In 1994 fire incidents dropped to 111,193 and firefighter injuries to 1,369. In 1995 there were 72,702 fire incidents and 751 firefighter injuries.

The safety record for firefighters in Alaska is quite good, especially compared to other occupations. There were four law enforcement officers killed in the line of duty in Alaska in a 14-month period. In the fire service in this state, we have recorded 6 LODDs in the last ninety years. Nationally, law enforcement LODDs average 150 annually. With only 700,000 law enforcement officers, that means their LODDs are 50 percent higher than ours and they are only half our size. Why has the fire service attracted so much attention and law enforcement attracted none? It is because we have attempted to improve our own operations "in house" with the compiling of "consensus standards." In doing this, we have inadvertently and unintentionally provided to OSHA a document that they could use as mandated regulations against us.

But check this out: Accidents while responding to/from emergencies account for less than 1 %, and has not decreased, but injuries from them have decreased, possibly due to seat belts. Accidents responding to/from in personally owned vehicles (POVs), and resulting injuries, have *increased*: 1,350 accidents and 315 injuries. I didn't need national data to tell me that. My own data showed me years before, that POV accidents were our greatest exposure to problems.

Bottom line, the fire service works tirelessly to improve its safety while its safety record is quite good, particularly for smaller towns.

Eventually, I was able to get the Alaska State Firefighters Association (ASFA) to pass and forward to NFPA a resolution to have NFPA inform OSHA that 1500 was not intended to be used as a regulation. NFPA chose to ignore the request of the resolution. Total compliance with NFPA is an absolute financial impossibility.

One would have to read the entire 1500 standard just to understand what full compliance would cost a fire department, not only in equipment, but in developing systems necessary to assure compliance. The subtle effect is that this monumental effort insidiously dominates the focus of nearly all department members. The focus turns inward.

By the way, unlike other fire service injuries, the majority of heart-related problems occurred on the fireground (the other 50 annual LODDs). The factor that proved to be the constant was stress. The April 2000 "Fire-Rescue" Magazine published findings from several studies that revealed some startling physiological responses in firefighters responding to alarms.

In the year of one study, the 50 fatal heart attacks were among 715 heart-related problems in the fire service. Without going into too much detail about the Los Angeles or Helsinki tests, the heart rates of firefighters increased an average of 61 beats per minute (bpm) within 15 seconds of an alarm. With an average heartbeat of 60-70 bpm, doubling that rate in 15 seconds is something that is not demanded even of our professional athletes who get 15 minutes of warm-ups to raise their heart rates that high. The average rate was 157 bpm, or 88 percent of the maximum sustainable rate. And the volume of blood pumped was related as well. Ischemia (inadequate return of blood) resulted. The firefighters began at a high capacity, then added further encumbrances like heavy lifting and heat. They were then at 80-100 percent capacity until the fire was out. Since the initial increase occurred with false alarms as well, it was hard to tell how much was a result of exertion and how much was psychological.

Also, with a succession of several wake-ups throughout the night, the rate increases graduated with each successive alarm.

226

I often wondered if air pack design might not be related to cardiac problems. Numerous times during strenuous work on the fireground, I couldn't suck air in fast enough. My BA couldn't deliver it in the volumes I wanted. I wondered if I might be actually "running away" from my air supply. It was so painful it almost made me give up smoking. But back when I was wearing them, the regulators delivered air at 40 liters per minute. And the regulators were the "demand" type. I would be very surprised if the new design that now delivers 100 liters per minute under pressure, might not alleviate some of the problems of ischemia by allowing more of the oxygen into your system to keep up with the rate you are burning it.

As an aside, I did alleviate one problem the Cordova firefighters had with wearing airpacks. A couple of decades ago, I couldn't figure out why—after 10 years of wearing BAs—I still felt fearful every time I put one. I thought perhaps I was claustrophobic. Out of curiosity, one meeting night, I asked how many other firefighters felt claustrophobic while wearing BAs. I was startled at the number of hands that went up. Oddly, they looked relieved to see the room was full of sissies. No one wants to think of himself as the lone coward in an organization.

So, for remedial therapy, inside one room of an abandoned house, I had all of us hand-raisers don our BAs and sit along the walls in a blackened room. I lit off a smoke bomb and told everyone to just sit there and chat with each other about anything they wanted. We all went through a couple of tanks each. It helped a bit, but not that much.

SOLUTION: THE HIDDEN ENEMY

Then one night at a Fire Chiefs Association conference, several of us were drinking beer and I brought this up. Finally, one of the guys came up with a hypothesis. The problem was not fear or claustrophobia, but carbon dioxide. When exhaling your carbon dioxide into the old-style masks, some of the exhaled air exits the exhalation valve, but a lot of it remains in the mask for you to inhale on the next breath. And what is the physiological response to breathing carbon dioxide? You got it—increased respiratory and pulse rates—the same feeling you get when you are afraid. Some of the guys at the table had airpack masks with exhalation cups (directing exhaled air out of the mask) and some didn't. The next day, I found a vendor who said those exhalation cups were available for the type of airpacks I had, and I ordered a bunch immediately. It worked like a charm.

WELL-MEANING OUTSIDE AGENCIES

Rest Assured, NFPA is not the only organization that is trying to assume authority over your department. The Environmental Protection Agency (EPA) for months had been sending me letters regarding abandoned underground fuel storage tanks in Cordova, wanting to know their locations and sizes and what they had contained. I would glance at those letters then throw them away. One day EPA called me from Anchorage and said that "according to our records, you have not responded to our letters. Did you not get them?"

"Oh, those. Yeah, I got 'em but I just pitched them."

A momentary silence, then he continued, "We need to have that information, because when those tanks develop leaks, they cause pollution to underground water."

"They do?"

"Yes. So we need to know about them so that we can have them removed. It's now the law."

Feeling giddy, I replied, "Well, you better get down here then, and find them."

"Aren't you going to do this for us?"

"No," I replied.

Another pause, then, "Why not?"

"Because I don't work for you." And I added, "But if one of them catches fire, call me." Then I snickered a little.

I never heard from them again.

Some months later, while visiting in Valdez, I stopped at the fire station to say hello to the chief. I noticed several thick, black 3-ring binders on his shelf labeled "Underground Fuel Tanks." There were similar video tapes next to them. Subsequently, I noticed the same thing in Kenai.

I guess we all have this tendency to believe that if a state or federal agency tells us to do something, like their jobs, we feel obliged to do it. Those agencies, in their exuberance to be gain more power and regulatory authority, strive to have laws and codes passed to do it. Then, to meet their new responsibilities, they sucker local agencies in to doing the work for them. Fuck 'em.

* * * * * * * * * *

The Coast Guard required commercial vessels to have Emergency Locator Beacons (ELTs) just like aircraft. After the John Stimson tragedy, we used the money that his widow, Trish Stimson, had people donate to us—in lieu of flowers—to buy survival gear for our members. One of the items purchased was a couple of Emergency Locator Transmitters (ELTs) so that we could be found if our aircraft went down. We also purchased a hand-held ELT locator, which was about three feet tall with wire antennas branching out horizontally. The problem with the thing was that it picked up signals not only from the ELT but the echoes of the signals from surrounding mountains. It was so bad, that once you were within 5 miles of the source, the sound could come from almost anyplace. We tried everything in the book of directions to be able to make it less sensitive and better pinpoint the source. I called the manufacturer and they could offer nothing new.

What made the situation really bad was that fishing boats were now carrying them by Coast Guard regulation, and whenever one of the sensitive ELTs would get knocked over or moved and set on its side, it would send out a signal that was picked up by a satellite, and the Coast Guard would launch a helicopter to find it. Oftentimes we would hear or see a chopper flying over the boat harbor, followed by a phone call from their headquarters saying that an ELT was going off there, and asked if we could find it and turn it off.

The duty officer would literally spend hours searching for the thing. Our boat harbor has 12 floats, one of which is nearly a quarter mile long. There are nearly 900 slips there, and in the middle of summer, some slips have boats tied two or three abreast. With many others anchored in the inlet or tied to the pylons of canneries, we can have about a thousand boats in the area. After a year of that, we convinced the harbormaster to purchase an ELT locator, and his crew could locate the things because they were being paid to be there. After they found out how hard it was, they "lost" their locator. So the Coast Guard started calling us again, until it occurred to me to "lose" ours. The Coast Guard then bought one for their staff at the air station, who would have to drive into town to locate the things. Don't get me wrong, ELTs are extremely valuable to have, but it was a Coast Guard regulation requiring them, it should not be the fire department's responsibility to dedicate so many hours because of an associated problem. Their people were also being paid to work; our people are volunteers.

SOLUTION: TRY LOW TECH

And by the way, after several years of trying to use that expensive piece of equipment at close range, pawning it off on to the harbormaster, then to the Coast Guard, I found the answer to the problem. Local electronics guy, Teeny Anderson, said to buy the cheapest, plastic, portable FM radio he had (about $5.00). Turn it to the lowest frequency on it, and I would hear the beeping of an ELT. Hold the radio next to my chest and slowly spin in a circle. When the ELT is behind me, the signal—being blocked by my body—will disappear. Walk toward it for a short distance, spin around again, and keep going. A really cheap radio has a less-than-sensitive antenna, so it never "heard" the beeps ricocheting off near-by buildings, or hillsides.

We tried it using one of our personal ELTs in our survival suits. It worked perfectly.

*　*　*　*　*　*　*　*　*　*

After a Union Carbide plant in Bhopal, India, blew up and the chemical release killed about 2,500 people, the U.S. passed a law called SARA Title III which, in effect, stated that workers had the right to know what dangerous chemicals are at the worksite, what happens if you get exposed to them, and a plan to deal with releases. Not a bad idea. Hell, even in Pennsylvania (I think) an engine company responded to "smoke showing" from a small tool shed at a construction site. They rolled up on this nuisance fire, stretched a line to it, and the explosives in it blew up, killing all of them.

The federal government put this new regulation under the authority EPA, which in Alaska then came down to Alaska Department of Environmental Conservation (ADEC). To be able to comply with this, these agencies decreed that every community needs to establish a Local Emergency Planning Committee (LEPC) made up of a cross section of local agencies and industry. They were very specific about who should be on this LEPC. Petersburg VFD has always tried to do things properly. Knowing they had been working on establishing an LEPC, I called them. They had been trying for two years to get their committee approved by ADEC, but ADEC was never completely satisfied with the make-up of the committee. So, I decided I wouldn't do it.

What I did was, I inventoried the few hazardous materials we had in Cordova. I knew where all the ammonia was and how it was to be purged if there were a release in the canneries. The chlorine, too. I checked out the bulk storage of LPG. Checking with the shippers, I logged how often that stuff was shipped in to town. We had no highway or rail system going through town, so our exposure to hazardous materials was pretty simple. I researched the properties of these materials and worked with the shippers and plant managers about a plan of action. I met the intent of SARA Title III, and most important, how to deal with the problem of a release. Good enough.

ADEC didn't agree. They kept calling wanting a list of the LEPC committee members, our meeting schedules, and our written plan. I explain what I had done and said that was good enough. Not satisfied, they wrote to the city manager and the mayor. I explained to the manager and mayor what I had done and said it was adequate for our town. They thought it was good enough, too. ADEC didn't agree. They suggested to Valdez, who was also trying to do it, that Valdez should be the hub of a regional LEPC, and get my plan as a part of theirs. I told

Valdez no. After a year or so, they gave up. I feel the need to brag a little about my approach. We knew where the stuff was and knew exactly what we would do about a release. Those other communities spent years paper-whipping their approach, and I doubt if they were more proficient on a call-out than we were. And mine was done in a week.

After September 11th, the U.S. Department of Homeland Security came knocking at our doors and insisted we prepare to work hand-in-glove with other agencies in responding to these terrorist attacks: biological, nuclear, incendiary, chemical, and explosive. What makes Homeland Security different, is that they showed up with a suitcase full of money to fund our attempts to improve our "interoperability" with these other agencies. With that kind of financial carrot, we try to muck our ways through the never-ending list of acronyms. But they don't assume to enforce regulations upon you.

OSHA, on the other hand, is a different story. They have the ability to inspect and levy fines. I can really suck up to outfits like that. When the law was passed requiring departments to protect their people from infectious diseases like HIV and hepatitis by providing training, inoculations, and a written plan, we did that. Just like the SARA III, it was a good idea. I just don't like people telling me how to do things. But with OSHA, I bit the bullet.

Also, OSHA can be used as a tool by bitter employees. The International Association Fire Fighters (IAFF), the major union representing firefighters, has used OSHA numerous times to "punish" fire department administrations. But the IAFF's most successful ploy was the passing of a law called "2 in–2 out." It was inserted (basically hidden) in a broader law, but was the final victory for the union in their attempt to make increased company manning a law.

When the IAFF was unsuccessful in getting NFPA to declare that a minimum of a 4-man engine company was a safety standard, the union withdrew their membership in NFPA in protest. They eventually went back, but they were able to get part of a new law to say that you need to have two firefighters standing-by outside of a building when you have a 2-man attack team inside. Unless a rescue needs to be performed, the attack team is not allowed to make entry until the stand-by team is available. That law sent people scrambling for solutions.

It didn't bother CVFD because we routinely had a back-up line stretched out and plenty of guys in airpacks hanging around outside anyway. But departments whose members converged from longer distances away, and never sure of who would be responding, had to wait precious minutes before they could enter. The law just said, in effect, that if the building's interior was so smoky that you couldn't see well, the team could not go in without a back-up team on site. The law didn't care if a house was a single-story house with a fire in the kitchen ten feet away from the back door. The law prohibited a full-grown, macho firefighter in full protective gear, to walk three steps into the smoke, extinguish the fire, and walk back out. That was far too dangerous for "America's Bravest." I have been waiting to see that very same scenario of two firefighters standing at that doorway with a charged hose line, waiting for a second engine company to arrive, while a 14-year-old neighbor boy in a tank top and cut-offs, walks right past them—holding his breath—into the kitchen and snuffs the fire out with a portable extinguisher. He would then walk back past them again and say, "Anything else I can do for you guys?"

That never happened here, but what did happen in Anchorage was an engine company stood by a small, ranch-style home which was in the incipient phase of burning while waiting for the arrival of another engine company. While waiting, the small fire flashed over and destroyed everything inside. A TV news reporter, looking for an explanation, stuck a microphone up to the face of a chief officer who could only say "It's a new OSHA law. We had no choice." Not completely satisfied, the TV news reporter went to the state OSHA office to interview an official there. This official, with his eyes wide and eyebrows arched way up, sputtered "We didn't know anything like this would happen." I don't know what the hell he thought would happen.

But the loss of that family's worldly goods, important documents, and irreplaceable family heirlooms was worth it, because the city agreed that next year they would hire enough firefighters to be able to comply with the new union-backed law. Apart from my snide remarks on that scenario, having a back-up crew and hose is a hell of a good idea in many circumstances. Even on a ground-level fire, if the fire is far enough inside, having a back-up is prudent. But on a small ranch-style home, where you can access the involved room by crawling 12 feet, it's ridiculous and embarrassing to have to wait. But that's the problem with laws. They are anal and bring out the analness in us all.

EMERGENCY MEDICAL SERVICES
FOSTER PARENTING TO ADOPTION

Changes are more justifiable when they come directly from the needs of your taxpayers. But they can make dramatic changes in the scope of your operations. No one thinks about EMS any more. But I was present when EMS first came knocking at the firehouse door. Everyone knows of stories when some middle-aged man opens his front door to see a young face claiming to be the child the man never knew he had. "Can I come in, so you can take care of me forever?"

Most fire chiefs in the early '70s never heard of organized EMS, but were certain that they were handling about all the emergency services they could. They weren't clear about why he (the fire chief) needs to adopt this new face. Yet, here that young face stood in front of him. "Daddy." Oh, man.

* * * * * * * * * *

About a decade after EMS became an integral part of the service fire departments provided, the term "burnout" emerged in lectures and magazine articles throughout the fire service. Whenever the members of the Alaska Fire Chiefs Association met, the after-hours conversation invariable centered not just on paramedic/EMT burnout but on the variety of personnel conflicts that arose between firefighters and medics. Conflicts between state EMS regulatory agencies, hospital groups and physicians as sponsors, and fire department administration staff seemed insurmountable. Those problems were territorial. The state: Those people who were writing regulations for administering state EMS training and protocols and certification requirements were non-providers; in other words those people were not saddled with budgeting, training, equipping, rostering, and all-in-all, providing the service to the local citizens. We fire chiefs were doing that, and quite frankly resented the subordinate role the state bureaucrats seemed to place us in. We were constantly filling out their forms and explaining ourselves to them. Yet when the tones went off, they were not rolling out to do the job.

Physician sponsors: A local physician who helped local medics develop protocols and authorized them to do advanced care in the street following strict guidelines, understandably didn't want any screw-ups to place them in legally precarious positions. So they, too, were telling our department members what to do. Suddenly, we chiefs weren't the sole bosses any more, and in some instances, were just in the way. Yet, we were to make everything happen.

Most of us chiefs are arrogant, overbearing, and domineering tyrants. And we like it that way. So, there were some—shall we say—adjustments we had to make. But EMS created tension within the line ranks as well. There were conflicts over the use of training facilities and duties around the station. That "I'm better than you, you…axe-swinging Neanderthal" perceived demeanor of this new group of people occupying the fire station contrasted with the grizzled, scarred, invading-Viking-looking veteran firefighter or fire officer with his not-so-subtle remarks about pencil-necked bookworms showed that there was a distinctly different temperament in the members of the two groups.

In fact, it was a much greater cultural shock then when women came into the service. After a while, we saw that women were spitting, tattooed, beer-swilling trailer trash like the rest of us; only with higher voices. But these medics, most of whom came in separately for their separate set of duties, had a whole different persona about them. Chief offices were constantly occupied by complainers from both disciplines. We weren't sure what to do to create a little harmony in the station.

Then, came burnout. Medics rolled out of the station twice as often as firefighters. When returning, often they would be sullen, and silently get back in service. And while firefighters at the station were doing their usual grab-ass, bulls-in-a-china-shop antics, the medics would drink some coffee and talk quietly to one another. Or more often, not speak at all. This further separated them from the firefighters. Now, it usually took several months for this behavior to surface, but most all departments noticed it. Attrition in EMS ranks began to exceed that of the firefighters. So did the rate of sick leave taken. This whole thing only validated what the grizzled Vikings had said all along about this new breed of member. Things were not good.

* * * * * * * * * *

Fire Chief Gene Fisher of Ketchikan told me a story that I found very enlightening. It wasn't about medics, but about cops. He had recently taken that Ketchikan job having previously been the chief in a town near Portland, Oregon. When down there, he and his crew were dispatched to assist the department in Portland with a crash of a commercial airliner. He said that it was a surreal scene, there at night, all the red lights flashing in the rain and firefighters, medics, and cops stuffing bodies and body parts into body bags through the whole night. When they were done, they left.

He went on to explain that the police chief noticed more personnel problems than was normal for his department. And a lot of sick leave was being used. Things just inexplicably turned sour. He summoned a psychologist who began a series of interviews with police officers. Out of curiosity, the psychologist went to the fire department, checked things out, interviewed firefighters and medics alike.

Here's what he found: There was a similar, yet not as extensive, morale problem with the medics as there was with the cops. But oddly, not much of a problem with the firefighters. His conclusion really struck me. After the incident and the gruesome job, the firefighters jumped back on their engines in groups of four and five, drove back to their fire houses, sat around in the kitchens and drank coffee together….talking. The medics did pretty much the same, but in groups of twos and threes. By contrast, the cops got into their vehicles—one-at-a-time, alone—and drove off into the night.

MESSAGE:

THE ONLY WAY TO CARRY A HEAVY LOAD IS TO SPREAD IT OUT.

At a working fire, when things go well, the high-fiving and praise is spread out 20 or 30 ways. When things go poorly, the blame is spread out equally among all those present. When training, after the individual skills are mastered (coupling hose, nozzle work, wearing airpacks), members are trained in "team plays" (hose evolutions, ladder raises, building searches), while all team functions are coordinated by the on-scene coordinator, or the incident commander. It is a choreography of separate team dances into one overall operation. It becomes a mind-set. This is a mind-set that is never completely understood or appreciated by members of sister agencies like police or EMS.

By contrast, watch EMS trainees. Their focus never moves beyond developing individual skills. Read, study, learn…read, study, learn promoted by the often repeated statement of the instructor, "Don't screw this up. If you make a mistake here, your patient will probably die." The burden of this level of personal accountability never wanes. Yet the new graduate struts with his new skills, a prima-donna. The volunteer medic carries the Dyna Med catalog around with him everywhere. He sticks star-of-life insignias on everything he owns. He loves the opportunity to use his new terminology. He becomes more of a blue-light "geek" than the new firefighters. Six months later, he has bags under his eyes. The fear of injuring a patient, the paranoia of exposure to liability, the scrutiny of patient's friends and relatives and on-lookers creates a level of stress unfamiliar to the firefighter. Firefighters act like construction workers during break, but medics are more like quiet college students….by and large, humorless. I believe that burnout is not so much a result of a sequence of tragic runs as it is the relentless feeling of personal accountability and fear of screwing up. There is not enough emphasis on "team" concepts.

Cops are like soldiers in their demeanor, but on the job, often dealing with calls alone, their level of personal accountability is higher yet, when factoring in the additional problem of personal threat. The only time cops act as members of a choreographed team are the non-routine calls like barricaded gunman, hostage situation, civil riots, and the like.

Even though medics are usually part of fire departments, they are more like cops than firefighters in nature and mind-set. Thankfully, most of them are affiliated with firefighters, because I think it's good for them to be around boisterous, guffawing "construction workers." Otherwise, I'm sure they would suffer morale problems of the magnitude one sees in so many police departments.

I will never believe that cross-trained firefighter/medics are as skilled in either function as those who are either one or the other, but it not only saves a few bucks for the department, I'm sure it prevents a lot of EMS burnout. It's much better than a firehouse full of "us versus them" attitudes. When North Star Borough's fire chief, Charlie Lundfeld, took his paramedic training in Los Angeles, he did his internship with the L.A. City Fire Department. At a supermarket fire, Charlie and his paramedic partner were leaning up against their squad watching the truck company firefighters from the station they were assigned going up on the roof of the supermarket to ventilate, when a battalion chief walked up to them and asked them what they were doing? "Standing by," they replied. "Get your turn-outs on and get your asses up on the roof and help," he ordered.

Up on the roof, Charlie was watching the truckies prepare to cut ventilation holes. He panned the roof to get a full view of the circumstances. While looking off in a different direction, he felt the entire truss roof drop suddenly about six inches. It was heart-stopping. He asked as he was turning around, "Hey, Cap, you think we ought to get off this……" He turned to see he was alone. "I guess that meant yes," he recounted to me. When he'd reached the ladder, he saw the others were already on or near the ground. Having a separate, exclusive skill can create a feeling of elitism, but right behind it comes isolation.

Anyway, in the early days most paramedics or EMTs were also firefighters. But it didn't take long for fire department administration to realize how much simpler it would be to have members who were either—or.

Anchorage Fire Department, for most of the years I remember, would hire persons for either firefighter positions or paramedic positions. Things went smoothly for years until greed and the "us/them" problem took hold. All line personnel worked the traditional Kelly schedule of 24 hours on, 24 hours off, 24 hours on, 48 hours off, etc., or an average of 56 hours straight-time per week. Then, one day it occurred to the paramedics that the department contract said that the firefighters would work the Kelly schedule. But, wait!—they thought—we're not firefighters. We're paramedics. So they brought suit against the city saying that since paramedics were not firefighters, they (the paramedics) had been working an additional 16 hours a week without being paid overtime wages for it. Nothing would dissuade them from pursuing this overtime money, retroactive for many, many years. Well, the city eventually reached some monetary settlement with them, but all the while the negotiating was underway for this settlement, the fire department was changing its structure.

Paramedics were now required to also be firefighters which included passing the physical agility test (a real ass-buster, too) and attend and pass the 160 hour Firefighter Academy. They would then rotate between assignments on suppression companies and ambulance assignments. Some very skilled, yet overweight paramedics washed out of the firefighter training programs and lost their jobs. Many female paramedics simply did not have the upper-body strength to keep their jobs either. It was tragic the way a ploy to pick up some easy money ended the careers for some paramedics. However, once the dust settled and everyone was used to the system, there were few complaints. But most important, I expect there will be much less burnout—comparatively—and virtually none of the decades-old bitterness between fire suppression and EMS.

Volunteer departments, mine included, can also be affected by this dichotomy. We began our EMS service in 1975, one year after the police department got an ambulance through a grant and sent a couple of police officers to the State Trooper Academy's EMT school. It didn't take long for them to regret ever taking this on. So they asked us to do it. Police officer Stan Shafer, also a member of our fire department, enjoyed doing it and would be glad to help run the service if we would take it on. City employee and local salvage diver, Don Endicott, thought that if he were sent to EMT school, he would like to do it was well. So, we said okay.

Don and Stan worked 12-hour shifts: Don made all the day runs and Stan all the night runs. First-Aid trained firefighters drove the rig and aided the EMT with equipment and loading and unloading patients. After about a year, they were ready to be committed to an asylum. Don was getting crap from his public works supervisor for being gone so often. Stan had quit the police department and got a civilian job and then was not allowed to leave the job for ambulance runs after a couple of lengthy runs. One evening Stan flew out on a float plane to get an injured fisherman, but the plane got stranded on a sand bar when the tide went out, and he never made it back early enough to make it to work on time. The other time, he picked up a critically injured fisherman way out in Prince William Sound. The man was so seriously injured, Stan made the decision to fly straight to Anchorage. He missed the next day's work.

Following a successful fire fight, grateful cannery owners gave us a sizeable donation. We contracted to have an EMT class brought to Cordova. Don and Stan recruited people specifically for this new division, even though some of us in the department took it, too. After some struggle, we decided we could simplify things by having medics meet on a different night than the Thursday firefighter's meeting. I was so rummy from trying to upgrade the department to professional standards (we wrote, had approved by the state, and were the first department in Alaska to conduct a 3-month-long Firefighter-1 course) that I could not keep up my involvement in our EMS. Stan took over the entire thing. But problems began to immerge. He would tell me about new people who had joined the ambulance "department"…without going through the normal department process. At some call-outs, a firefighter might ask who someone was on the fireground because this new person was unknown to most of us.

When someone is leading a group of people, taking on all the responsibility and headaches, and highly regarded by all those in his group, he resents having to justify himself to someone who apparently has no "stake" in the operation. After about a year, any time I disagreed with Stan on an issue, he wanted to talk about splitting with the fire department and possibly operating under the hospital. Many times, I found that option very attractive.

There were even those times when I would walk into the room of EMTs and the place would go uncomfortably silent while they struggled to make a comment that I was to believe they were talking about when I walked in. I noticed morale was extremely low. I would never hear laughing from them, no grab-ass, only frowns. So one day, I informed Stan that I was going back on the response roster and that EMT training nights will be on Thursdays again and that there would be a lot of cross-training from now on. All new applicants for membership will go through the normal channels as part of our monthly business meetings which must be attended by all members. This will be one department, not two.

Of course it's more complicated accommodating two functions on the same night, sharing the training room and training equipment. But, as we found out, it could be done. I was surprised to see that it only took a few months for the two groups to feel a part of one another. But that isn't the point of the story.

This is the point: It was about this time that AFCA after-hours discussions revealed how wide-spread the rivalry was. That's when Gene Fisher told me that story about the Oregon plane crash. Shortly after that I decided that being *excessively cerebral* was the biggest cause of EMS morale problems. I remembered that when I was a garbage man or a ditch digger, I busted my ass, but I was never tired. But if I spend a day working on the budget (lifting nothing heavier than a pencil) I would go home exhausted.

So, "Close the text books. Turn off the VCR. Put your coats on and get outside and have a drill," was my next directive at a squad training night. "Get physical," I said. "Work with the equipment, don't worry about didactics. No memorizing protocols. Hook up the suction unit while blindfolded or something."

One project was born of necessity and was very successful in improving their morale. The Insurance Services Office (ISO) would require that Cordova provide three engine companies because of the town's fire flow requirement. Well, I had a third fully qualified engine, but not enough manpower to call it a company. At that time, ISO wanted eight persons to be assigned as a company

238

if they were volunteers. I explained the problem to the medics and told them I needed their help. I needed them to be trained to use that engine for—at least—a defensive operation.

Hell, before we even actually did anything, I outfitted them with bunker gear and they were ecstatic. They felt like breaking things with axes. Most ecstatic were the female members.

We were out in the dark, in the rain, learning how to hook up to a hydrant, how to blow a thousand gallons a minute through the top-mounted deluge gun, but most important how to ride around town hanging off the back of the engine bellowing. Back in the station, soaked, panting, and grungy, they were giddy and bellowing and wondering why they'd never done that before.

Besides just standing-by during fire calls, the medics were trained by the firefighters to change out tanks on the breathing apparatus. They kept the tool tarp tidy, helped our safety officer with accounting for the whereabouts and welfare of our firefighters. They became part of the team and showed true caring for the firefighters. Rescue operations became a perfect blend of technical rescue—in things like extrication or high-angle—and patient care. Technical rescuers and medics became like the left hand and right hand of the same person. Medics trained the firefighters in CPR and First Aid, qualified them to drive the ambulances (if necessary), and identify and locate equipment carried on the ambulance in case the need arose. Training for disasters and mass casualty incidents further unified the groups.

Now, our medics are as rowdy as the firefighters and their behavior often startles itinerant EMT instructors who come to town to teach. Our medics are highly skilled and dedicated to professional patient care, but thank God they don't act like it.

"FIREMARK"
AND MANAGEMENT FUN

As you can see, this adoption process involved more than just money. It was the first glaring example I saw about the damage that "internalizing" (focusing on your "needs" rather than on the needs of the people you serve) can do to an organization. Well, if getting physical isn't enough, or re-directing attention back to the public needs isn't enough, then create a project...and the more outlandish, the better.

* * * * * * * * * *

One of the wildest ideas we ever had was in an attempt to raise money so we could buy our own heavy rescue truck. We got our first full-sized rig from the Juneau Rescue Council. We could never get the very costly item into the city's rather frugal budget on top of a new fire station and newer firefighting gear, but hauling our rescue gear around in a small, sagging van had to be addressed. So, Juneau had a full-sized, heavy-duty vehicle used as a command post for search and rescue operations and for carrying lots of equipment in the numerous compartments accessed from either inside or outside. It was a 25-year-old, 1953 Reo. That's all right, I never heard of one, either. It was like driving an old logging truck. But they were surplusing it out and we bought it for a dollar. We rebuilt the engine after buying a short block, and repainted it. It served us extremely well for many years. But after about 10 years, we had milked it for about as much as we could.

At that time, a new heavy rescue truck would have cost about $85,000. Still big bucks for our little department. Where, oh where, could we find that kind of money? Well, a lightning bolt of an idea came from the most unlikely place. The city had it's own refuse department for years. Contracting out the garbage collection to a private contractor was something the city had done in recent years and was simple and the costs were reasonable. However, the current contractor defaulted on the contract, grabbed his stuff and left town. The city went back to picking up garbage and was unhappy about it. So, the city planned to ask for bids again from private contractors.

After our training was completed one training night, and the beer was flowing, so were the ideas. "Let's bid on the garbage contract!" I blurted out. Lots of laughter, then Groff said, "We should do that, just because it's absolutely unheard of. We haven't done anything noticeably insane for years. We are starting to be and act like every other fire department....boring." Members were leaning forward, intrigued. We pay the workers, and the profits go toward buying the

240

rescue truck. Later, profits could go for anything we want that is abnormal for budget requests....like a condo in Hawaii.

As per our usual process, I went to the marker board for the brainstorming session.

First was the schtick that goes with such a concept: Change our name to "We be fires and shit." And designing our patch: A Maltese cross and in the center is a banana peel and egg shells with flies hovering above. A garbage truck with lights, sirens, and fire department paint job. Etc, etc, etc.

It was decided that a city department legally could not bid on a contract with normal bidders from private enterprise (private sector). To bid on this we would have to create a corporation for profit,. After a long list of hilarious proposed corporation names, Groff's name won. Based on the insignias borne by insured buildings in the 1700s and 1800s, we would be Fire Mark, Inc. So, Groff and an assistant would contact an attorney, draw up articles of incorporation, a list of officers (the same officers as in the department) and membership (the same members as in the department).

Sooner than we expected, we had become a corporation, complete with a seal. Others were gathering information about the costs of buying the dumpsters and the garbage trucks. Still others were determining which of our department's unemployed would most like to have the jobs, and determining employee costs. While all the information was being gathered that was needed for us to submit a bid, well, one grandiose idea generates another: We could branch out and get into other businesses as well...house and trailer rentals. The Kotzebue Volunteer Fire Department currently owns and operates the primary video rental store in that city and uses the money for uniforms and for sending members to training programs almost anywhere they want. I don't think they incorporated to do that, however. One of our female members suggested we could have our own whorehouse. She would be glad to design our new patch. That idea did not go up on the marker board, and we decided she'd had enough beer for the night.

After the money we paid to get incorporated and the time spent to put together the bid proposal, we were underbid by someone else. Incidentally, the low bidder went back to the city a few months later for a rate change because of unforeseen additional expenses. Funny, we foresaw them.

We kept our incorporation for several years afterwards "just in case" we came up with another idea, and it didn't cost much to renew it. We eventually got a new rescue truck with a lease-purchase deal, and never needed to buy it ourselves. But I'm a little sad that we never did anything as outlandish as that. Nevertheless I feel better knowing that it is doable, and the only thing that stops any fire department (or the members in it) to be major players in local businesses is lack of crazy ideas.

Photo: CVFD

CORDOVA VOLUNTEER FIRE DEPARTMENT FAMILY PORTRAIT
(SOME MEMBERS MISSING)

242

EMS Division 2002

Back row from left: Rob Mattson, Vicki Hall, Penny Oswalt, Hawk Turman, Dixie Lambert, Oscar Delpino, Brandon Doig

Front row from left: Mark Kirko, Joanie Behrends, Seawan Gelbach, Melanie O'Brien, George Mundy, Kyle Marshal, Melissa Grant, Dewey Whetsell.

Photo: CVFD

243

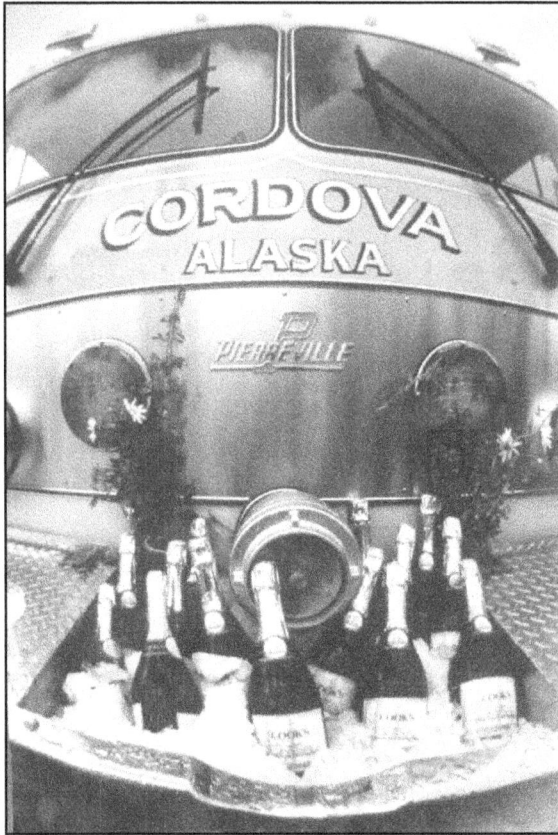

CVFD HOSTED A VISITING SOVIET BALLET TROOP THAT ARRIVED IN CORDOVA BY SHIP IN THE SUMMER OF 1991, JUST WEEKS PRIOR TO THE COLLAPSE OF THE SOVIET UNION. THIS PHOTO APPEARED ON THE FRONT PAGE OF THE LAST ISSUE OF THE SOVIET TIMES

Photo: CVFD

PART IV

LEADERSHIP

CULTURAL BASIS FOR GROUP CONFLICTS

All organizations are fraught with personnel problems that interfere—sometimes drastically—with the organization's ability to function. I've read tons of articles and books about dealing with personnel problems. But there is material beyond the usual approaches that is valuable. For example: Using Dr. Eric Berne's approach from his book *Games People Play* (Grove Press, Inc., NY 1964), I will explain the most common game played in the workplace that causes strife and conflict among the employees. Berne is quick to explain that he describes these interactions as "games" not because they are fun, but because they have a winner and a loser. These games can result in dissension, altercations, terminations, law suits, divorces, hospital treatments and can end up with funeral services.

TRIBAL AFFILIATIONS

This most common game—played at the worksite, street corner, bar room, or tea party, is the one Berne called "Ain't Men Awful." This game is clearly identifiable. The example goes like this: A group of women are chatting when the topic of "men" comes up. To illustrate a flaw in men, one woman laughs a bit and wonders why men will drive around forever before stopping to ask someone for directions. They all nod and laugh. That's a cue for another woman to mention a universal male flaw: Men strut and brag about tough manliness until they get the sniffles. Then they lay down like they are on their deathbeds, moaning, and sniveling for wives to tend to them like army nurses in the war. They all laugh and nod. Each takes a turn validating the flaws in men. Like I said, games have winners and losers. In this case the winners are the women and the losers—conspicuous by their absence—are men. Here, this version of "Ain't Men Awful" is harmless enough. Change groups for a minute. Try conservatives playing "Ain't Liberals Awful." Or "Ain't Teenagers Awful," or "Ain't Blacks Awful," or "Ain't Crackers Awful," "Ain't Chiefs Awful,".

What makes this game relatively harmless is that the loser is generally not there to hear it. Oftentimes the validating statements are humorous wisecracks intended to evoke laughter. At other times, the statements are dark or vicious and rather than evoke snickers, the winner's disdain or resentment of the loser group grows with each new validation. Only the most strong-willed can resist assimilating the effects of negative, defacing comments. I'm not even going to go into examples of the loser not being a group but a particular individual (the name of that game is "Bloodspot on a Chicken"), because I want to talk about "cultural diversity."

Some years ago, on TV, I saw a black gentleman lecturing an audience. He asked them what they saw when they looked at him? Like Samuel L. Jackson said

246

in an interview, "When someone meets me (meaning someone white), the first thing they see is that I'm black." The lecturer said, "The first thing you probably notice is that I'm black. That makes me different than a white." He was using examples of how professional advertisers might try to appeal their products to him. He went on. "Of course, you notice that I'm a man. That makes me different then a woman. I'm 56 years old, and that makes me different than a teenager. I'm from the West Coast and that makes me a little different than folks from the East Coast. I have a PhD, which makes me different than a high-school drop-out. My income is in the six-figures which makes me different than someone earning minimum wage. I own my own company employing 300 people, which makes me different than someone who works under supervision. I am a political conservative which makes me (and my world-view) different than a liberal. So, which 'category' do you put me in? Is it racial, or gender, or regional, age, education, economic, or political? Which pigeon-hole do you conveniently slide me in to? Who am I?"

What was significant about his statement was that he compared each "dimension" of himself with an opposing dimension of someone else. Instead of saying that he was an educated, wealthy, black, male conservative. He mentioned that each dimension made him different than people who were uneducated, or female, or young, or less affluent, or had different political views. Hold that thought, I'll be right back.

Years later on TV, a famous prize fighter, Rubin "Hurricane" Carter, was being interviewed after he had been released from prison after serving 18 years for a murder he said he was wrongly convicted of. The interviewer asked if Carter believed he had been convicted by the all-white jury because of racism. He said, "No—it wasn't racism. It was 'tribalism.' 'Racism' is just the most obvious form of 'tribalism'." Holy shit! That statement hurled me back to the affluent, black lecturer. It wasn't "categories," or "pigeon holes" he was describing, it was "tribes."

We are all members of tribes. Most significant, is that we are members not of a tribe, but of many tribes. When we play "Ain't Men Awful," it is not gender discrimination, or racism, it's "tribalism," and each of us shifts from one tribe to another, depending on the topic of conversation at that moment.

Picture this: We are all standing around, attending a big gathering in a banquet hall. It is a joint conference of firefighters and fire chiefs. Watch with amusement this categorizing by tribes flow around like mixing chemicals. Within the two main groups—chiefs and firefighters—there were scads of sub-groups: the volunteers, paid guys—union and non-union—urban firefighters, rural firefighters, and wildland firefighters. All of these tribes existed in the fire chiefs association as well. To make things really interesting, age and tenure are tribal factors as well as political and social views. So is gender. So is race.

247

What about the black female firefighter? As she strolls around the room, clutching her glass of punch, does she join the group of white, female paramedics, or move on to the group of black, male firefighters? What if the black, male firefighters are lieutenants or captains? Well, she could join a group of lower-ranking firefighters, but discovers they are all volunteer firefighters, and she is a full-time career firefighter. Her decision will ultimately depend on the topics of conversations within each of these tribes.

Anyway, tribal affiliation become more minute, until the invited state legislators walk into the reception, then the fire service personnel would all become one big tribe again.

Tribal affiliations become destructive games very quickly. A sure sign is when tribal members become "reductionists." Like when environmentalists play "Ain't industrialists awful" managing to "prove" that industrialists are the lowest common denominator responsible for all the ills of the world.

Focusing on differences (diversity) rather than on similarities promotes and feeds tribalism. Ross Perot put it well when he described how political campaigns try appealing to different groups, "You spend all that time trying to separate one group from another, trying to appeal to the uniqueness of them, then get frustrated when you can't pull them back together again later."

Each tribe makes demeaning remarks about the "opposing" tribe: rivalries between high schools or neighborhoods; southern hillbillies versus northern urbanites (who change their ties daily but couldn't change a tire if their lives depended on it); liberals versus conservatives; Protestants versus Catholics versus Jews versus Muslims; rich versus the poor. Hell, skin color is only the tip of the ice berg. These tribes of ours buzz like bees in a constant drone of belittling remarks keeping us separate until some foreigner intentionally flies a jet into the World Trade Center buildings. Then we all become one tribe again—Americans. White and black, rich and poor, liberal and conservative, all now arm in arm, swaying back and forth singing songs. The human race is one scary sonofabitch.

Incidentally, Mr. Samuel L. Jackson, if, when meeting you, the first thing I notice is that you are black, remember, the first thing you notice about me is that I'm not.

Oftentimes, going hand-in-glove with tribalism is showing your affiliation with it. People feel compelled to wear their uniforms to show their affiliation. The term "uniform" doesn't necessarily mean clothing. Take the slack-jawed, tobacco-chewing, red-neck hillbilly. He is a guy that is fully capable of saying "It doesn't matter to me." But prefers to say, "Hit don't make no never-mind (or 'never-mand') to me." Or, "Well, I may not be very smart, 'cause I ain't got no book-larnin', but I do know one thing......" Hillbillies are the only people who brag about being stupid. That's not the only thing they brag about. They go out of their way to sound like hayseeds. They brag about having never owned a tie, or worn one. "Chokes me." They don't own a pair of pants with a crease in them, and remark about never wanting to go into a restaurant where they wouldn't be accepted wearing greasy coveralls. They don't want to be around those fancy-pants, big city, sissy-boys anyway.

Less extreme, and more common are the ruralites, steadfastly disassociating themselves with the urbanites. There was one department member named Phil who was a ballsy, aggressive firefighter. He was good humored and well liked, but he had a problem. The prospect of being at a formal, social gathering of any size made him nearly catatonic. He would withdraw, stare at the floor and break out in a sweat. This would happen even knowing that he would be at a table with other department members. So, once, before a banquet at a state fire conference, I took him off to the side and talked to him. The other members would be wearing suits and ties, highly polished shoes, and had been ironing and grooming all afternoon. This 23 year old only felt natural in jeans or Carhartts, work boots or cowboy boots. He hated "big-city" clothes and just knew they would make him look like a sissy. He almost shook, thinking about it. I think his greatest fear was that some of his buddies from work might see him all dressed up and make fun of him for switching "tribes." His tribe was rugged, outdoorsy, and manly.

In a stroke of genius, I reminded him of the Indiana Jones character from the movies. Here was a guy who spent his normal workday at the university wearing a tweed suit and a tie. But when it came time to swing into action, he donned his boots, leather jacket, grabbed his bullwhip and commenced to kick ass. When that assignment was done, he went back into tweed and stood behind the lectern. Being a real man, he was as comfortable in one outfit as the other. Phil lit up. He looked sharp that night and was at ease. For a year afterwards, he would walk by me, smile and shake his head and whisper, "Indiana Jones." In truth, he found that one can easily switch from one tribe to another without being disloyal.

Actually, I've concluded that this is the basis for western-cut suits. Men will wear cowboy-style suits and ties, and cowboy boots so that others will know they are rural and rugged in "real life." Those clothes make the wearer a dashing action-figure: worlds apart from bankers, accountants, or other office workers. That's tribalism.

Digging deeper, maybe one's reluctance to switching tribes or affiliations may lie in the basic fear of getting lost and not knowing who you are.

Let's go back to "Ain't Men Awful." A volunteer fire department runs on morale—not money, not policies—morale. Take all the resources away from a volunteer department, and if the department's morale is positive and energetic, the job will get done. But show me a department with all the financial resources it needs, and infiltrate it with a back-biter, sniveler, chronically negative person, and I'll show you a department which can be destroyed within a year. Don't doubt me on this. I'll show you.

Don and Roy were fault-finders by nature. Even though they may have noticed positive things and even appreciated them, they seldom remarked on such actions. On the other hand, both felt that it was their duty to expose flaws that they noticed, often sternly or angrily. Each guy did this independently….until a job change put them together for 40 hours a week. Then they started playing "Ain't they awful" (referring particularly to officers in the fire department). A third department member, who also worked there, at one time said to me, he hated going to work now and listening to that negative yammering hour after hour. It was depressing and suffocating. They kept trying to draw him into their tribe, by each making a negative statement about the department, then waiting for him to add one. He ended up contriving tasks to be away from them.

At our training sessions or business meetings, the effects of this manifested itself several ways. They always sat or stood near one another, exchanging knowing glances. If a firefighter did something incorrectly, or clumsily, or said something stupid, they would smirk, shake their heads, having seen one more validation of their negative view of the department. If they tried to correct the flaw they witnessed, they would do it impatiently at least, angrily sometimes. Soon, people hated to perform in front of them, hated to candidly express themselves. And since they were a team, others were too intimidated to reproach them because the team would flare up.

With the officers, these two would mostly remain silent, not involve themselves in discussions, and never, never affirm support for a novel idea or new project. Reflecting on this later, it occurred to me that supporting the officers that way, would have been disloyal to their exclusive tribe.

One assistant chief (who also had strong tendencies as a fault-finder all his life) said he'd been hearing complaints from Roy. Roy had gone to him with the complaints because whenever he tried talking to me, I wouldn't listen. He was right, of course; I wouldn't listen. There was no way to satisfy him. He wanted a dark and oppressive organization, whose officers were not unlike Puritan clergy of the New England states of the 1600s. With their "Ain't Men Awful" game, morale was beginning to plummet. I didn't know how to deal with the problem, because I didn't know what the problem was. When they criticized someone's performance, they were right. I spoke to them about using positive approach during these incidents. They couldn't (or wouldn't) do it.

Finally, after the department member who was a fellow employee of Don and Roy, told me what their workdays were like, I understood what was going on. I wondered how I could stop it. Negativity was an ingrained part of their

250

personalities. You can't change someone's nature. In a large, paid department, they could be transferred to other stations. But what would I do? Here's what I did, and it worked…

At a full-department business meeting, at the end of the agenda, under Good of the Order, I stood up and told a story. Without indicating that I was talking about Don or Roy, I explained how a "confederacy of negativity" can form alliances of like-minded people. Then, how the game "Ain't Men Awful" is played, and once enough "validation" occurs, members of the confederacy cannot allow themselves to be positive about anything because they would look like "turn-coats." I told it straight forward, and clearly, but not directed at any-one in the room. I ended with my statement that once morale is destroyed, the department is destroyed. It would be better to lose valuable members than to destroy the department. Following that meeting, the change in their demeanor was immediate and dramatic. But that tendency to morale-busting negativity, often is an integral part of a person's personality. So, changes in demeanor may be dramatic, but seldom permanent. Stay on top of it. If you can separate members of the confederacy, it may simplify your problems. Openly spotlight the phenomena of "tribalism." Show how it works. Laugh about it. Satirize it gently. Soft, good-natured mockery works wonders.

An example of that: Many readers will not remember the TV show All In The Family. I not only remember it, I remember the first time it aired. The Archie Bunker character was the satirical personification of the men I worked around all day long. There was only one TV channel in Cordova at that time, so I knew my co-workers had all seen the show the night before. When I walked into the shop the next morning, the place was solemn and quiet. They were embarrassed to have seen themselves in that light.

Another story, same setting. We had a local cop nicknamed "Tiny." He was huge and felt his size made him a celebrity. Every morning when I walked into the shop, he would be leaning against the door jam of the coffee room and telling the same type of tale: "So I warned him, but he wouldn't listen, and he thought he could take me. So, I lifted him completely off the ground and slammed him down into a puddle of water. That cooled him off. Ha….ha." Then he would look around at his admirers, and nod his head, "Yep….yep." One morn-ing I walked in and he was in the middle of another same-theme story, and as I walked passed him I asked, "Who are you beatin' the shit out of today, Tiny?" He joined everyone else in a laugh. But that was also the last morning we had to listen to his stories.

Soft (good-humored) mockery is sufficient to spotlight behavior that needs challenging without making tempers flare. Be fatherly in your demeanor. However, if you mock someone more than once, it becomes ridicule—maybe harassment.

251

NEGATIVITY AND NIT-PICKERS

But, back to negative people. Negative people (individually) have value, just like anchors. You need to have them around, Mister Flamboyant Ideaman. They can keep your ass out of jail by always focusing on why one of your ideas is silly or dangerous. So, give them your ear. But, never, NEVER, give them much authority. Never give them a key role in developing a new project. Negative people never accomplish anything, they are never innovative; they are not experimenters; and you will never read about them in history books. And when things go wrong, just listen to their admonishing "I told you so's" with patient tolerance.

This is leading me into another topic, even though I'm not through talking about tribalism (as it pertains to discrimination and group conflicts). But since I'm not skilled at composing transitional sentences, I'll just jump from one tangent to another.

When I mentioned that negative people have their uses, more specifically, I mean "nit-pickers" have their uses. They also pose a threat. I first became aware of this phenomenon while reading Dr. Francis Schaeffer's (philosophy professor-turned theologian) *Escape From Reason* (InterVarsity Press, IL 1968 In tracing the trends in Western thought and Christian ideology, Schaeffer wrote about the affect 13th century monk, Thomas Aquinas, had on Western thought after writing *La Suma Theologica*. Schaeffer drew a horizontal line across the page, and above the line wrote the word "universals." Below the line he wrote the word "particulars." In his interpretation, "universals" meant things one believes intuitively, or by faith. You know—ideals. Got it? Below the line, the "particulars" are fact-based, tangible, provable specifics about life. So, when Thomas Aquinas (or Saint Thomas Aquinas, as the Catholics refer to him) wrote a thesis in the format of the Greek philosophers, to prove—logically—the existence of God, Schaeffer asserted that Aquinas tried to use "particulars" to prove a "universal." Schaeffer considered this to be a breach of faith. Schaeffer continued to trace milestones in Western philosophy and theology through the next 800 years demonstrating how universal ethics and core beliefs—ideals—were constantly under attack by people who lived below the line, those absorbed in the "particulars" of any topic or endeavor. The "nit-pickers" if you will. And, he added, the Below-the-Liners will always win, will always devour the Above-The-Liners, because the Below-The-Liners have all the tools needed to win and the Above-the-Liners only have their ideals which cannot be substantiated.

* * * * * * * * * *

What's this got to do with running a fire department? Everything. If you have a wild idea about buying an unfunded, expensive piece of equipment and even take part in the brain-storming session on fund-raising activities, you better put a below-the-liner in charge of planning and implementing the fund-raising community barbeque. It's about product versus process. You want the product, and you want it right now. Process-oriented people can get the job done, but it takes them forever to do it. In fact, if you don't stay on top of them, they will never get it done. If given a free rein, these compulsive list makers will write lists and do little more. An illustration: An above-the-liner wanting to drive from New York to California, hops in his car and takes off. He will be beset by one problem after another before he finally makes it there beaten and broke. By contrast, the below-the-liner starts planning and writing his lists. He may research so minutely, even to studying the viscosity of different brands of motor oil, that a year later, he still hasn't left his house and his living room is piled with papers, maps, flip charts, and memberships in auto clubs. He enjoys that aspect of life so much, he would rather do that than actually go to California. He is obsessed with process. These two people make a perfect team. The above-the-liner will drag the below-the-liner to California, and the below the liner (always on the verge of panic) will make sure they both get there with a minimum of problems. Incidentally, there will always be daily strife between these two. Expect that.

These dynamics are true in any organization. But one thing I would insist on: These below-the-liners, obsessed with particulars, should be "support" personnel only and should never have ultimate authority. If placed in charge of an organization, the organization will bog down and never move. If an organization existed only to provide a service or produce a product, then their organization will fail. It will be smooth, tidy, legal, and useless. It will be a polished, smooth-running, well-maintained vehicle that never leaves the driveway. These below-the-liners need to monitor the budget, review all new regulations, and provide means to protect the inventory. These folks are also paranoid about things going wrong. They are fearful of anything dynamic or flamboyant. Disarray renders them frozen and speechless. They fall apart.

Okay, I know that some people can be both, but most people favor one side more than the other. And depending on the topic under consideration, one side will often completely overwhelm and dominate the other. Stay aware of this phenomenon.

PROCESS FIXATION

Here is an example of process fixation. Incidentally, it is also an example of conflict source called "Your Rules vs. My Needs," which I will explain later. When the City of Cordova hired a new finance director, he thought it was his mandate to bring the finance department into the 20th century. To justify all the money he needed to hire more staff, purchase an all-new computer system, and contract with a very expensive programmer to set things up, he used the most recent statements issued by the state auditors. When using those statements of needed changes, he spoke of the auditors as though they were regulators and that if we did not comply with their recommendations, we would somehow be found guilty of financial mismanagement. Hell, every department head (including me) has pointed to the omnipotent "they" as the reason we need to buy this or that. What pissed me off was that it seldom worked when I tried to use it.

Anyway, Jack got the money, purchased the new computer system, hired the criminally expensive programmer (who basically moved in), hired more staff to feed the "required" information into the computer, then began a campaign of memo writing to department heads giving us his new marching orders. He was the new sheriff in town and made no bones about it.

Discretionary spending was nearly zeroed out and staff was cut in Police Department and Public Works to pay for the new positions in the Finance Department. In the name of fiscal responsibility, the direct-services-to-the-public decreased, yet the public did not save any money.

Jack was a number-crunching below-the-liner who had been given authority and free rein. All department purchase orders, even after being approved by the department head, still had to be personally approved by Jack. That slowed the process down.

Also, Finance Department personnel succumbed to sinister changes in personality. Other department heads were treated like subordinates. Even my traditional flattering remarks to the women in that department paid no dividends. Neither did my morning stroll through their work area distributing mochas I'd picked up on the way to work. I was screwed.

Jack decided he would save the city money by buying things "in bulk" and getting bids from vendors. He sent another memo out. All department heads would submit their tool requests to him. The only things I needed to buy were a couple of wrenches, a hack saw, and hack saw blades. I would have just run down to the hardware store and bought them. Instead, he assembled the lists from the departments, compiled them, wrote a lengthy "Request for Bids" (RFB) document and mailed them to numerous vendors.

After a few weeks, he discovered he hadn't written the RFB correctly, cancelled it, re-wrote it and mailed it out again. Several more weeks went by before the bids were in, the winner selected, the proposal was scheduled to the city council. A couple more weeks went by before the shovels, ice chippers, wrenches and my hack saw, etc., showed up. Jack then had them all delivered to a central location, where he went, clipboard in hand, and inventoried them. I finally got my stuff after about three months, and noted the amount of money then deducted from my budget for them. I then went to the hardware store and discovered the price was the same. Naturally, I could hardly wait for the next staff meeting. There, I showed Jack the price he'd paid for my stuff was the same. I had waited three months to get the tools. I asked him to factor in the cost of his time and his staff's time in writing bids, going to city council, inventorying, etc. Then I asked him how that was fiscally responsible by wasting so much of the tax payer's money. He just got mad at me.

The city manager—fully aware of our bickering—sat in silence, wanting to move on to another topic. The finance director had power and the city manager was intimidated. The director's power was a result of all the privileged information he had: all the data in the new computer system, a direct line to the state auditors (the great "they"), his degrees, and his years of background in this business.

* * * * * * * * * *

One last thing, and I almost hate to confess to this, but it's too funny to omit. A few weeks after the new computer system was installed and the programmer was showing the admin staff how to operate it, they were all clustered around as the programmer installed tons of information into one of the computers as a demonstration. I was walking down the hall and saw them. I stealthily stepped into the room where the circuit breakers were for the building and opened the metal door. I saw the one marked "computer," flipped it off then back on again quickly, and walked out. Walking back, I saw everyone spinning around in circles grabbing their heads and the programmer frantically waving his arms having lost all the information they'd put in all morning. I laughed about that all day. Some people never grow up.

* * * * * * * * * *

It took years—literally—to reduce the Finance Department back to a reasonable size and to bring the Public Works and Police Departments back to full staffing again.

IN SUMMARY: those whose lives orbit around "particulars" and details are valuable. They are not visionaries on a broad scale and need to be placed where they can be of use but never be granted too much authority unless you want your outfit to grind to a halt.

At the office gathering, the day Jack reached retirement age after a few years with us, he pointed to me and said the best thing about retiring was that he didn't have to deal with "that sonofabitch" anymore. It was mutual.

* * * * * * * * * *

Oh, I almost forgot. There's one other important factor in diminishing the effects of tribalism or more subtly, your unconscious reaction to someone different. Of course, it requires another story.

My son Jason, the cop, one night made "contact" with a young oriental gang member. Jason was questioning him about some activities, during which the interaction was—well—strained. Then something odd happened. This young man looked a lot like Jason's younger adopted brother, Tim. Jason told me that as soon as he noticed the resemblance, he started warming up to the guy. Almost instantly then, the guy started warming up to Jason. Within a minute or two, they were joking around nudging each other. For several nights afterwards, whenever they spotted one another they would smile and wave. Jason was shocked when he later learned that the guy was a suspect in California of some serious violent gang activities, including murder. Nevertheless, whenever one finds a reason to like someone else, sincerely, it is almost instantly reciprocated.

In any circumstance—any circumstance—when you get comfortable and unintimidated around someone, that person knows it, their demeanor can and will change immediately. It doesn't matter who they are.

So, tribalism can be neutralized by attitude alone. But only if you know it exists and its prevalence. Now you do.

MEAN-WORLD SYNDROME

It's great to think that you understand people so well that you can manipulate them. All leaders do this to some degree. Some do it really well. But, don't get too cocky there, Mister Sigmund Freud. Some folks are too broke to fix.

Set three creatures down in front of you: a dog, a cat, and a bird. The dog views the cat as a defenseless prey, a panicked, cowering victim. The bird sees the cat as horrific predator, the anti-Christ. If there are two diametrically opposing views of the same item, which one is true? The human brain operates under the constraints of antithesis. Antithesis states "If 'A' is true, then 'Non-A' must be false." But we can see by the dog-cat-bird example that both views can be equally true. The cat can be defenseless prey and the cat can be a horrific predator. Incidentally, Zen Buddhists would state that the cat is neither predator nor prey, the cat simply is. But what do they know?…they're foreigners. Normal people operate under the phenomenon of antithesis, which is the same way computers operate, and look at how infuriating they can be.

I remember years ago when some military fatigue-wearing nut walked into a McDonalds restaurant and started shooting everyone there. Later, the cops searched his apartment and found more military clothes, a stash of weapons, and lots of magazines about soldiers of fortune, news clippings about all the rotten things that were happening around the world. He was pretty pissed off about all the crap going on in the world and was not going to be a victim of its orneriness.

Neighbors said he used to yell at noisy kids in the street from his apartment window and once fired shots in the air to shut them up. That was when I first heard the term "Mean World Syndrome." If one focuses on all the shit people dump on one another, it is pretty easy to validate one's view that the world needs a good ass-kicking. And if 'A' is true, 'Non-A' must be false. One simply can not see goodness and beauty and the redeeming honor and character that people have.

Back when we just started providing EMS, we got the individual training we needed but had not yet developed much in the way of squad protocols. Nobody had. That's why it was pure luck that one of our EMTs was not shot dead on this run.

Guy Mullins was a seething man. Tall, and although slim, he was like steel cables and springs. He was a middle-aged, rural man whose jaw was always clenched and eyes were full of anger. He knew the "world" would sometime try to screw over him and he was already mad and ready to show the "world" that it was screwing with the wrong man. When he sat at a bar, he would not sit next to anyone and that was perfectly fine with the other patrons. On some occasions I would sit next to him and say "Hey, Guy, what's up?" He would be civil with me and I felt that he would like to be friendly but didn't know how. Our conversation comprised of me making statements and him grunting back.

He used to drink at the Moose Lodge regularly, but one afternoon they kicked him out and he was so angry he couldn't speak, so he took his Moose membership card out of his wallet, and standing in the middle of the street in a rage, ate it.

I don't know how he got the girlfriends he used to show up with in the spring, but they were always younger and seemed pretty decent…just like the one he had one summer. They rented a small room in an apartment building just across the alley from city hall (fire station, police station). His girlfriend happened to run into a guy from her graduating class who was in town and they went back to the room to play cribbage. That's what they were doing when Guy walked in. Guy grabbed his hunting rifle and shot the young man through the heart.

A neighbor called, and one of our medics, R.J. Kopchak, was nearby and ran to the apartment to render aid to the victim. As R.J. ran up to the apartment, Guy was walking out, still carrying the rifle. Out of impulse of self-preservation, R.J. grabbed the rifle with both hands and he and Guy reefed back and forth on it and slammed each other on the hallway walls. Finally, in desperation, R.J. yanked the rifle from Guy's grip and hammered Guy over the head with it. The cops ran up and subdued him. The young man was dead, still holding his cards.

It's good that Guy is still in prison because it would be impossible to reform him. For him the world will always need to have its ass kicked, and—by God—he's the one to do it.

* * * * * * * * * *

For us, what is an accepted standard for everyone now—rostering EMTs into small teams, scheduling teams to respond only during their shifts (eliminat-

257

ing free-lancing), and waiting for police to stabilize scenes of violence before entering—is a protocol we instituted right after that call.

Another victim of his own world-view was Don Shawback. He differed from Guy Mullins in a couple of important ways: He was not a silent seether, and his anger was directed to specific people. But he was similar in one respect: He woke up angry, was angry all day long, and went to bed angry. Every moment of his life was spent thinking about all the things that made him angry. I'm convinced that if there were a lull in his anger, it would feel like drug withdrawals and he would have to quickly conjure up a scene in his mind to keep the toxins flowing in his body.

Some of his more minor problems resulted from his inability to comply with regulations, and he would purposely "stomp on his own dick"—as Mike Gundlach used to say of him—to prove a point. Don's anger would increase when he learned that his non-conformity had been barely noticeable by regulators yet he was fined, restricted in his endeavors and basically made his own life miserable.... just "to show those assholes that they can't tell me what to do."

Regarding his romantic relationships, he was dangerously obsessive. His divorce was a result of that and so were the restraining orders against him. He harassed his ex-wife relentlessly and spewed out, to anyone cornered by him, all sorts of twisted sexual imagery he attributed to his ex and their daughter. This gut-knotting obsession went on for years. One of the reasons he was in Alaska was because he couldn't be in the same state as his ex and have any semblance of a life.

Unfortunately, his girlfriend was unaware of this side of his personality when she became his girlfriend. Now, I have no idea if his obsessive thinking, his perpetual anger, or his inability to not step on his dick caused her to terminate the relationship, but as soon as she started going out with other guys in town, he started spinning uncontrollably. He could think of nothing else. He stalked her so she got a restraining order. He violated it and went to jail. As soon as he was out, he was back at it again. He would tell everyone he saw how she screwed like a tramp, and he threatened to kill her and her boyfriend.

Her boyfriend, Paul Malek, was an easy-going sort, but when Don forced him off the road, leapt from his car and ran toward him, Paul pulled a .22 revolver and shot. At the same moment, Don spun around to escape, but was struck in his right "love handle." He ran back to his car and drove himself to the hospital.

Paul drove to the police station and told them he just shot Don. Incidentally, because of that shooting, Paul—who was the singer in our band—would never sing the song "Stagger Lee" because it was a song about Stagger Lee shooting a guy over a crap game, and Paul didn't want to appear cavalier about shooting someone.

258

Anyway, after the ER doc plugged the hole in Don, we were to transport him in the ambulance to the airport for a medevac flight to Anchorage. EMT Mike Gundlach and driver Joe Levey lifted Don's rotund carcass into the ambulance and took off. Don was still agitated and talked incessantly. Finally, he lifted himself up into a seated position, pulled up his garment and showed me the wound. It entered the layer of fat above his right hip, harmlessly traveled around his perimeter and came out near his navel. "See that? See that?" he said in excited indignation. "Christ Don," I replied, "if you hadn't been so damned fat, he would have missed you altogether." Joe and Mike cackled unashamedly but I knew Don wouldn't laugh. Even under normal circumstances, he had no humor.

Not long afterwards he was out of the hospital and back in Cordova making threatening remarks about his ex-girlfriend and her new boyfriend. He mentioned to Joe that he planned to kill her, so Joe—feeling guilty about betraying a confidence, or maybe overreacting to nothing more than talk—told the cops. Back in court again. He was in court and cautioned and then in jail on and off for weeks. He couldn't stop himself. He called me at home, in agony, bemoaning how sluts keep screwing over him and how Joe betrayed him and how the police chief was conspiring against him. I tried to explain to him that his entire life has been shit because he either could not or would not control his thoughts. I told him that there must be a million things in the world that he could think about, but he refused to think about anything except his ex-girl friend screwing. I guessed that if he consciously chose to think obsessively about positive aspects of life, he could be filthy rich. I told him to take a test: See if he could think of something nice and positive for one hour during the next week. He never called me back. But he did come into city hall one day to complain to the police chief about everything.

In the hallway, I asked him what—if anything—did he think of for one hour that didn't piss him off. He just blew me off and walked out. His future is already predetermined. Someone is going to die.

I repeated to Don something I heard a motivational speaker say about that. That is, a negative thought cannot be erased, it must be replaced. If thinking of a red house is bad for you, you cannot tell yourself not to think of a red house because that image appears when you say that. Instead, you must consciously think of a blue house. It's conscious thought-control. The trouble with Don was that he could not think of anything nice.....positive. I doubt that a beautiful day had any effect on him whatever. He would probably grumble because he knows it won't last.

Obsessive thinking is more addictive than heroin. Heroin use has restrictions on it, but thinking can be done anywhere, any time. A person can never not think. Sorry about the grammar. Hell, there was even a movie on TV about a guy who became obsessed with agonizing over his girlfriend's past sexual encounters before the two of them met. I never saw the movie but read its synopsis in the TV Guide. I didn't need to watch it. I imagine that she was completely unaware of the dark thoughts that haunted him. She probably would have been confused about why he was obsessing on them.

I had a couple of other real-life examples of mean-world syndrome and compulsive thinking, but I'm tired of writing about it. So, here are some esoteric thoughts to ponder, and then file away as being without any real relevance.

I had often wondered if in these cases it were attitudinal at all. In the novel East of Eden, Steinbeck described the evil, amoral Kate as a mutant. He wrote that she had no environmental influence that made her devoid of conscience or empathy. He supposed that there was some gene deformed in her before her birth just like someone who was born with a physical deformity. One person might be born without a nose, and she was born without feelings.

Perhaps these men were born deficient of some hormone. Their pancreases produced insulin just fine but, after they reached maturity, there was a significant shortage of serotonin in their happy tank. The cure for such would be auto-injector of serotonin, and all they would have to do is push the button when they began to seethe and they would immediately start feeling like Bob Barker—or more likely, Doctor Timothy Leary. After all, clinical depression cannot be eradicated by a simple force of will, it requires medication to change the chemical imbalance. Actually, there were those "reductionists" in California in the sixties that were convinced they could neutralize all the negative characteristics of human nature chemically by dumping LSD into the cities water reservoirs.

Three examples come to mind when I think that these problems may not be attitudinal: A person hears a remark and bristles at being insulted by the stabbing statement meant for him. If a dozen other people were all present and listening, they would never have heard that remark as "cutting." The dozen would be confused at the anger the insulted one shows. The dozen would never convince the angered one the remark was absolutely unremarkable. Yet the angered one heard it that way, the evening is ruined, and the angered one will never forget it. These people live their lives this way, insulted and consequently steeped in malice. In one of the three people I know with this "condition," his father was very much like that, too. Was it conditioning that caused it, home environment, "Nurture?" Or was it "nature?" Was it an inherited gene like one that dictates height or color-blindness? We "normal" people can never honestly believe that dyslexia exists. It is incomprehensible that some can look at a "d", then a "b" and can't see that they are different from one another. Yet it's true. How can someone not see a certain color when it's as plain as day to everyone else. Yet it's true. How can someone hear a phrase like, "No, here's how I believe it happened." And feel they have just been called "a liar." Yet it's perceived that way by some people all their lives. Here's the worst part about that: You cannot convince some people that "b" and "d" are different; that an unseen color actually exists; that an opposing view is not a personal insult. But those insulted people live perpetually in the eye of the storm, and so does everyone close to them. If you were not present to witness this "brush off" or insult, his recounting of it will have you convinced it is accurate, because his recounting will be so detailed and his emotions so legitimate. He is not faking it. It will only occur to you after repeated incidents in the future that he has an affliction. There is nothing you can do to fix it.

Another aspect related to this affliction (this condition) has to do with "personal space." I heard that all of us have what we consider personal space. And this was illustrated by this example: Two guys sitting at a small round table in a bar. Subconsciously, each one imagines an invisible line is drawn down the center of the table. But one guy, by small, unnoticed increments, moves the ash tray across the center line toward the other guy. Next, he does the same things with other items on the table. Pretty soon the encroached starts feeling annoyed, then agitated. He can become angry. I also heard that people have this same space around their cars when driving. Encroachment in that space causes agitation and anger. Apparently some tests were conducted on repeat felons and found that the circumference on their personal space was larger than average, possibly accounting for their propensity to detonate at the drop of a hat leaving the potential victim of their violence astonished at the accusation that "you were in my face!" Was Guy Mullins like this? Probably.

Or maybe Christians are correct. Demons really exist and manage to take hold of people's souls. They take over one's soul by taking over one's mind. And when one ignores the rules of mental discipline and sinks into hatred, drifting away from a connection with goodness, one's soul falls into an abyss.

Maybe "A" is true and so is "Non-A." Or maybe I should just move on to less philosophical ponderings like the Rosenberg experiment.

THE ROSENBERG EXPERIMENT

The adage that a person's success is directly proportional to your expectations of him was proven by an experiment conducted in an elementary school. I don't remember when, where, or in what grade it was done, so I'll just make that stuff up.

Fourth grade students were given IQ tests. A teacher—I think his name was Rosenberg—changed the results of those tests. Rosenberg took the highest scores and gave them to the students who scored the lowest. The lowest scores were given to the students who actually had the highest IQ's. These bogus results were given to the fifth grade teacher at the beginning of the next year. At the end of the fifth grade these students had performed in accordance with the expectations of the fifth grade teacher. Those with the lowest IQs, scored the highest on exams and projects because their teacher expected that of them. Those with the highest IQs scored the lowest because their teacher expected that of them. Apparently, the teacher subconsciously taught to those he or she thought would "get it", i.e.: eye contact, posture, tone of voice, etc. I imagine the eye contact, posture, and tone of voice directed at the students expected to do poorly was different, but almost not discernibly so. I expect the difference could only be detected subliminally. At the end of the year, Rosenberg revealed what he had done and was promptly sued by parents.

What makes it so difficult for us mere mortals to use this principle to get the most from our people is that to work, we must honestly believe our subordinates are capable of great things. It is ineffective to just fake it. Can you look at your one member who is disheveled, unsure of himself, has an absent look in his eye and is, as Dr. Phil describes, a "mouth breather," and actually expect him to excel in academic areas? You have to be an expert at visualization. Here is an example of visualization that I used lecturing parents at Cordova's Community College "Parenting University." Following my explanation of Transactional Analysis that I got from a book (whose title I can't recall), I asked them to try an experiment. "Stand that infuriating 12-year-old in front of you, and try to picture what he will look like when he is 28 years old. It can be done. Then, picture this 28-year-old as being decent, mature, amiable and intelligent. When you have that picture, speak to him, not the 12-year-old in front of you. Speak to the man you want him to be. Speak to the man you are certain he will be. If you can honestly see him that way, that may very well be the person that speaks back to you. The difficulty is in trying to do that when you really feel like pounding the dog shit out of him."

The same principle holds true in dealing with subordinates. But don't get your expectations too high. You may be dealing with people who already have a substandard picture of themselves. One manipulative maneuver like this is not going to miraculously transform this firefighter into a genius action-figure. But, repeated and consistent, it may get you someone you can depend on and who will always hold you in the highest regard because you "treated" him so well. Okay, I want 17th century philosopher John Locke to be right, too. But even though his principle is good, it doesn't work miracles.

Funny, when our daughter entered kindergarten and my wife, Lou, and I went to the orientation evening at the school, the teacher stood at the front of the room and spoke to us parents. Later, I commented to Lou that I hope daughter Cindy could sit on the right side of the room. "Why?" Lou asked. "Because the teacher favors that side of the room." Incidentally, I noticed that about myself, too. Unless I make a conscious effort, I will always teach to the left side of the

262

room, seldom notice those on the right side, make eye contact with them or elicit verbal responses from them.

SETTING THE GROUNDWORK TO MINIMIZE PERSONNEL PROBLEMS

Most (although not all) of the views I recount in this section came from somewhere else. Over the decades, I read about or listened to so many theories about leadership, I reached the point where I can no longer remember whether an idea I am expounding was mine or someone else's. If I remember who had the idea, I will give them credit. If I can't remember, then all I can say is "thank you." However, I only recount here observations and techniques that were successful for me in leading volunteers for 28 years.

Organizations with the highest morale have the highest productivity. Low morale destroys productivity and sometimes the entire organization itself. Personnel problems are the single biggest morale busters of any organization. Personnel problems are a communicable disease. They spread, distract, sap strength and vitality, and eventually destroy. "Success" is a physical manifestation of a frame of mind. Equally, "Failure" is a physical manifestation of a frame of mind.

Organizations that suffer fewer personnel problems are those that loudly tout their organizational philosophies and their organizational standards. That vision of your organization must be set in concrete and displayed in a slogan or motto. Those new people coming in, know they are aligning themselves with those philosophies and standards. Hell, that may be the reason they are coming in.

This shared vision, this vision congruency is not only essential for the foundation of the organization itself but essential in increments of the organization. Many personnel problems are a result of innocent misunderstandings between people due to lack of vision congruency. Those misunderstandings grow to become true conflicts. Here is how that happens.

There is usually a difference between our mental picture of how things should be and how things really are right in front of us. That makes us edgy. On the other hand, when what we see externally closely matches our mental picture of reality, we are at ease, at peace. When the two do not match, our uneasiness and stress triggers our impulse to make them match.

That's the problem with newlyweds coping with the first year (the hardest year) of marriage. The husband, over the years of his life has formed a picture of what a "normal" relationship looks like. It may not even be a pretty picture, he may not even like it, but subconsciously, it embodies his expectations–it's *normal*. The home, the decision-making process of a married couple, the degree of affluence (or lack of it), are implanted in his mind. When the world he sees in front of him right now, does not match that, he will either consciously or unconsciously do or say things to change the external reality to match his internal picture.

The wife has done exactly the same thing, and has formed, over the years, a different picture of what a married life looks like—what is normal. This includes physical surroundings as well.

The difficulties in the early years of a marriage is the struggle of those two people to create the environmental reality to align with their mental pictures. The process during that time is the slow adjustment of those two pictures into vision congruency. If clutter in the home makes the husband feel uneasy, it's because his previous home was spacious and orderly….normal. Even if the wife knows this and logically agrees that clutter complicates life, and decides to remove it, it will creep back, because tidy and uncluttered surroundings make her feel uneasy. It isn't normal. It doesn't match her mental picture of life. Developing vision congruency between these two may take years of teeth-gnashing.

I heard this example of this principle on an Earl Nightengale recording on how this principle applies to organizational goal setting: Let's have someone attempt to accomplish a project for my department. I want Joe to put together a 500-piece puzzle. I will give him 5 or 6 hours to do it and will give him $500 if he gets it done. Then I dump all the puzzle pieces on the table in front of him. So, I have explained the project to him. I gave him the resources to do it (the puzzle and the time). I gave him the motivation to do it…$500. But when I walk away, I take the box with me that has the picture on it. We know he won't be very successful, because he doesn't know what it's supposed to look like when it's done.

More similar to what we deal with in our business, is to have two people work on the project; so, have Bill join Joe and give them the time, the puzzle, the cash incentive. But before walking away, let them know the puzzle is of a dog. Tell Joe the dog is an Irish Setter and tell Bill it's a Dalmatian. Bill and Joe have the desire to accomplish this; they feel cooperative, and they have the resource. But they will not succeed. One will pull toward what he honestly believes the picture is to look like. The other will pull the other way. What they lack is what can only be provided by a leader: A detailed vision. It is not the leader's job to personally put the puzzle together, but to paint the picture. One can even make the visioning a group process through master-planning or brain-storming. In the end, the leader must paint the picture clearly and in detail.

If you don't know where you are going, just wait a little while, and someone else will point the way. If you do not lead, leadership will be taken from you—maybe informally by the group. Then you will have 10, 20, or 30 different visions of what reality should look like and your organization will be crippled by massive goal-diffusion. Of course, you can have multiple goals, but they have to be in alignment. Each goal can be one of the details within that picture, that vision.

You don't have to ask, "Will standards be set?" but "Who will set the standards?" Our ability to retain valuable members increases in direct proportion to our expectations. However low, standards do exist. Having no standards, is a standard. When everyone involved views standards as a covenant, a pledge of quality, then pride in your outfit grows stronger and stronger.

* * * * * * * * * *

It is extremely important to remember that philosophies and standards are based on external needs. Organizations are created and exist and are defined by the product they produce or the service they provide. If, in doing that, the organization also meets some of the personal needs of the members…great. But that is not what the organization is there for. When members of an organization focus more on their needs, at the expense of the external purpose of the organization, it can open the floodgates, and you will drown in a sea of needs. What we hope is that being a member of a revered and admired organization will almost be enough to meet the individual member's internal needs.

When someone proclaims a need for change, determine if the change is external or internal. We all know that we often justify something we want to do by explaining how it is to benefit the public, but we do it because the members think it would be fun or exciting. So? It's okay to bullshit the elected officials or the public sometimes. But it is surprisingly easy to fall for your own line of bullshit after a few days. Don't do that; it can transform you into a pain-in-the-ass, frantic, spittle-spraying zealot. I would advise you to do this: If you are going to budget for a program to elevate your EMS service to a higher level so that your crew can administer cardiac drugs and operate a manual defibrillator, and you know your medics want to do that because they (like all people) have an internal need to advance and grow; and you also know that your run-volume where these skills would be useful is too low to justify it, just remember when you go after this funding, you really are doing it for the crew, not for the public. You never need to say this openly, but you might remind your EMS supervisor from time to time that you are doing this to keep the crew happy. Never fall for you own bullshit, or anyone else's. Meeting internal needs to keep your people excited and growing ultimately does benefit the public. But do not become a victim of chronic internalizing or you will regret it. Maintain a sensible balance between external and internal needs.

Before I leave this topic, here is a trick I pulled once: My EMS captain and crew became preoccupied with the budget, fairness of the dispersion of funds, lack of autonomy and such. This had been going on for several weeks of "training" nights. They were grumbling, dark and morose. Once when I walked into

266

the room, the conversation stopped. So, I opened the file cabinet and pulled out some papers, then stood at the front of the room, lecturer-style.

I said, "You know if a commercial air-liner belly-flops at the airport and we roll out there, the firefighters are responsible first for stabilizing the scene, then—as first-aiders and stretcher bearers—they will report to the EMS Captain here and wait to be assigned to help you medics with patient care. In an event like that, the entire state, maybe the entire nation would be watching you. So," I continued, walking up to the marker board, "Tell me what injuries are considered critical, and what injuries are considered less-than-critical and don't require immediate transport?" It was a struggle for them to get started, but after a while, the triage factors were on the board. Then we began discussing treatments for each of the types of injuries, transmitting patient reports to the hospital from the ambulances, and so on. At the end, I reminded them that no one else in the community is capable of, nor has the responsibility to perform that function. They will succeed or fail, lives will be lost or saved, only by their proficiency or lack of it. "You have extremely important things to think about, talk about, and spend your training nights preparing for. Everybody is counting solely on you." Then I left. I know—I just highlighted that crushing personal accountability phenomenon again, but it was necessary.

Later, I reminded the captain that the readiness of the EMS division was his job. I shouldn't have to conduct a spontaneous lecture to keep his crew focused on their skills. It worked. Remember, refocusing your crew has to be done on a regular basis, or internalizing will creep up on you and morale will drop. Although not always, it is usually the group sniveler that focuses on the internal needs. Those needs may be legitimate and may be helpful to the department and the community. Write them down but do no act upon them immediately. Internalized problems infect groups where there is a lack of conspicuous external goals and objectives. The lifeblood of any organization lies in ideas and creative thinking.

I read in the book *In Search of Excellence* by Tom Peters and Robert Waterman, a statement that it is the leader's responsibility to establish—spoken or unspoken, written or unwritten—what the organization truly values about itself. Each officer in the department has that responsibility. That should be the first criteria in selecting your officers.

SOURCES OF GROUP CONFLICTS

Like any chief, or any leader of a large group of people, I majored in minors. I spent an inordinate amount of my time resolving conflicts within the group. The two major sub-categories were: Conflicts between clearly identifiable groups within the organization or conflicts between individuals. It became easier to get a handle on these group conflicts once the basic sources of these conflicts could be defined. While reading some material from the National Fire Academy, I found they had assembled sources of group conflicts into five general categories. Over the following years, I found their categorizing valid. Here they are:

OLD GUARD VS. YOUNG TURKS:

The Old Guard had traditional values and standards. They had built the organization by hand, agonized over it. They had already made all the mistakes and learned from them. On the other hand, the Young Turks feel they are being robbed of the opportunity to create, to breathe. They want to try new and exciting things, to be innovative, but the "establishment" (the Old Guard) has set the organization in concrete. Conflicts between these two groups seem irresolvable.

MANAGEMENT VS. UNION:

Most group leaders—whether they are official leaders, or unofficial leaders—are either goal oriented or people oriented. Fire chiefs are goal oriented first. It is their responsibility to either set the goals of the fire department or to meet the goals established by the city or the public. He is given that mandate and the resources to do it, and employs his people, leads, and manages them to provide the services. He and his senior officers are Management. Union members are people oriented. Their focus is not on the external goals of the organization, but on its internal workings and work conditions of the people on staff. One can extend these definitions to include the autocratic corporate president whose goal is the financial bottom line and the employees who feel they must protect themselves irrespective of the bottom line. One can carry this off on to a tangent by saying Conservatives are goal oriented and Liberals are people oriented. Hell, we can stretch this even farther by saying Republican conservatives believe "The business of this country is business," and that if a country prospers, it benefits everyone that lives in it. While the Democrat liberals may believe that in order for large corporations to get filthy rich, the common working man will be treated like cannon fodder, then discarded unless people oriented unions protect them. Let's go for a real stretch here by attributing genders to these views. Fathers believe they need to teach their children to cope with unhappy situations and toughen up. Mothers, who are nurturers, try to cushion and protect children.

268

It's a good system to have both masculine and feminine influences in a family. The same is true of a country or a fire department. Of course, this tangent gets harder to hold together the further you stretch it. At any rate, it is easy to spot this dynamic when it erupts into conflict within your organization.

CENTRAL OFFICE VS. FIELD:

Here is where the chief can represent the central office by insisting all operations be conducted according to policies he has written that are so detailed, what he really wants is robots for company officers. Or where the chief represents the "Field" and must drastically alter the fire department's operation to comply with some mandate from city hall. Here's a question for you: Does running an operation by "policy" indicate a lack of trust? A few decades ago, my brother-in-law, who was a captain in a metropolitan fire department, told me that his department had a policy from the chief that a responding engine company must lay-in a supply line whenever they saw smoke in the area. The company officer did not even have the authority to investigate first. I wondered how many one-block stretches of supply lines were laid to dumpster fires over the years. Of course, when I developed policies regarding operations, they were based on years of experience (as the Old Guard), and having made all the mistakes myself, I knew the only proficient way to fight fires. Right. Anyway, Central Office versus Field is the third on the list of the most common sources of group conflicts.

MY TERRITORY VS. YOUR TERRITORY:

Here is where policies need to be clear, and developed after a series of meetings with the involved parties. Do you doubt that? How about this: a vehicle accident with injuries on a highway. EMS is in charge of patient care; Extrication is the responsibility of firefighters; the state trooper shows up because he has authority over disruptions on the highway. Suppose he tells the firefighters and the medics where they must stage their vehicles, and the place he selects is not convenient for the operation. They protest. He becomes badge-heavy. Whose territory is it? Or how about patient care is on-going by your medics when the patient is delivered to the hospital emergency room, the ER nurse starts directing medics to protocols that are contrary to the sequence of care already underway. Whose territory is it?

YOUR RULES VS. MY NEEDS:

When the city finance department requires that you solicit three Requests for Bids or three quotes on a relatively inexpensive piece of equipment, and you want to cut down on paperwork and time? You need to buy a certain brand of equipment to complement what you already have. So the finance departments "rules" conflict drastically with your needs to cut back on useless data collection. Rules have a tendency to breed in dark places. Control freaks are fertilizers.

Again, after watching for several years, I noted that—true enough—most group conflicts fell into one of these 5 general categories. But, to boil it down even further, check this out. There are really only two things to examine when pulling the opponents together for a chat: 1- Where you are going; or 2- How to get there. Those two things—your goal, or the alternatives on how to achieve it—would be your primary topics during the meeting.

In analyzing the conflict, step back to the beginning. Does everyone agree with the ultimate goal? If so, move further down to the last place of agreement and work forward from there. Watch out for hidden agendas. They are everywhere. And oftentimes, the hidden agendas are based on an affiliation with one of the 5 conflict sources: Old Guard vs. New Turks; Union vs. Management; Central Office vs. Field; My Territory vs. Your Territory; Your Rules vs. My needs. See how this chapter is coming together? Cool, huh?

* * * * * * * * * *

So, in your meeting, openly examine the conflict over goals. Seek the origin of the other side's goal. Seek an over-riding goal that both can accept. How about the conflict over budget expenditures between fire suppression and fire prevention?

Having agreed upon goals (where you're going), move to examining methods (how to get there): Develop a list of incremental objectives needed to get there and evaluate each method in terms of effectiveness.

Facts or data will be used when estimating the effectiveness of a proposed method. Share data and sources of data with both sides. Develop a plan for mutual validation of data. How about this? The history of fire runs indicates that 90 percent of all fire runs could be handled by 4 firefighters or less. Or this? Nationally, the number of structure fires has dropped 40 percent in the last 20 years. Does that mean that it is time to re-focus, to re-tool the fire service? Does that mean that the Young Turks are to be given their chance to re-tool this business since the Old Guard was so successful in improving the environment that we all work in?

Gaining consensus on where you're going, then on how to get there, backed by facts or data, unfortunately, will never resolve a problem based on the most infuriating conflict type—the one of value or philosophy. It is the hardest one to deal with in a civil fashion. "A woman's place is in the home, not the fire station." Or, "No one under 21 is responsible enough to…" One statement I heard, once, is kind of Zen-like: "Botanically, there is no difference between a flower and a weed. The only difference between a flower and a weed is a judgment." Someone's ingrained value system or philosophy cannot be changed by debate. Meetings are a waste of time. It's back to writing policies. Yep. If the obstinate,

270

crusty old bastard can't stand the new environment, he'll move on. It's a shame, but I doubt if sensitivity training will change him. You can try it, though.

* * * * * * * * * *

Here is one dimension of conflicts you should be aware of. This dimension I yanked out of *Getting To Yes* by Roger Fisher and William L. Ury (Penguin Books, NY 1983) a book on negotiating. This dimension can be present in conflicts not only between groups (or tribes) but between individuals. Quickly differentiate between animosity over "Cause" or "Purpose." In a nut shell, when two people are arguing, they may identify a cause rather than a purpose. Folks look back in time when arguing over causes, and look forward when arguing over purpose.

"I'm pissed because he did this or that…"

Your question to him, "What is the purpose of your confrontation? Where do you want it to go?" will invariably be met with a confused expression and silence. When people bicker over past acts, they are standing with their heads turned, looking backwards (figuratively). If you were to ask them to state clearly a purpose of this confrontation—looking forward—you will notice they are ill-prepared to do that. And since they are not looking forward, they naturally have no plan to resolve the issue. If you were to ask a group of angry Palestinians why a Palestinian blew up an Israeli bus, they will ramble off a list of causes that go back to the establishment of the of Israel in Palestine. If you asked them the purpose of blowing up the bus—how will that change their situation—the best they will come up with is that it lets people know how mad they are. In your situation at the fire station, you might allow each party to vent, as long as you maintain control, then inform them that if the current situation is intolerable to either side, that side needs to clearly describe what they want the situation to look like. If necessary, have him, or them, write the description down. But in either case, the description must be clear and detailed. They must paint a picture, as it were.

When that is done, the participants are looking forward. Whether the source of the conflict was one of the 5 conflict sources mentioned earlier or if it were a cause vs. purpose conflict, it's time for some consensus-building. Brain-storming, which sounds easy to do, actually has some rules that are essential to its success. I have used it repeatedly over the years and have been very successful at it. Here are the rules and their rationale:

- Participants never face one another across the table. They all sit on the same side of the table(s) and, shoulder-to-shoulder, face the issue which is written on the marker board in front of them. They become partners whose adversary is standing up in front of them (the situa-tion). As an aside, any time you have a group activity that may be-

271

come confrontational, and they can't all face the black board, then tables should be formed in a circle, not rectangle. It helps.

- Explain that all statements regarding a solution will go up on the board, as close to verbatim as possible. Explain that you are soliciting statements that will fix the problem. "I want you to suggest things you want to see, not what you do not want to see." You, or the facilitator, may actually ask the statement-maker to help you compose the statement of the board so that it will be accurate and meet the intent of the person. Having him help you will assure that in the end, it is concise and clearly understood, and later, he cannot say that he was misunderstood. It removes ambiguity.

- Tell everyone to go to the bathroom. There will be no group breaks until the process is complete. It breaks the group "train of thought," and will result in lots of individual discussions and random discussions in the hallway. It screws up the process.

- Be firm when explaining that no one is allowed to make remarks (or cheers or boos) about that statement at that point in the session. This part of the session will be polite, disciplined, met with patience by everyone, and may take a long time. Gotta be that way.

- The facilitator must remain flexible and allow the group to go off on tangents. A silly remark may spark an outrageous suggestion, previously not thought of, that could ultimately be the solution. Joking is okay, in fact, it's helpful for the attitude of the event.

- This segment must be allowed to run its course. You will know when all the ideas have been tapped. It's better to spend a lot of time doing this than rushing it, and having some members feel they have been railroaded. If that happens, you will be wasting more time in the future revisiting this issue.

- When the participants fall silent and just look at you or one another, and all the statements are on the board, ask if anyone else has an idea to add. When they all shake their heads or shrug their shoulders. Let them know you are ready to move on to the next phase.

- By this time, the members may have forgotten which person made which statement. That would be great, but it may not occur or occur completely. The important aspect of this is that when a member disagrees with Joe's statement or suggestion, he is no longer disagreeing with Joe, he is disagreeing with statement #4 on the board. The originator of the statement doesn't feel he is being attacked or disrespected, he can divorce himself from that statement or that position. The one disagreeing is looking at a scribble on the board, not looking at Joe. Joe doesn't get defensive. It's not "My idea versus your idea." So.....

- Each statement in sequence is evaluated by the group as to the likelihood that it could help resolve the problem. Try your best to summarize the groups' feelings about that statement and write that down, also.

- Get a bigger marker board.

- Eventually, the list of viable solutions gets smaller and more manageable as the others are discarded. Routinely, turn to the entire group and ask, "Are we on track here? Is everybody satisfied with the process so far?" Any negative remark or even facial expression needs to be openly delved into, spotlighted. Never allow this unsatisfied member to leave at the end of the session with the ability to complain that his feelings were not considered. If he is one of your chronic troublemakers or faultfinders, now is the time to empty his gun. Publicly remove all the bullets from his gun. If he has hidden agendas for his position and wants to keep them hidden, he will openly agree that this process is something he will support. He may get exasperated and spew out what his hidden agenda or his prejudices are. Good. It's now out in the open. He may revert back to his fixation with cause and you can remind him that the membership has agreed to look for a solution, not re-hash old arguments.

- As the solutions emerge and start to resemble a plan of action or change in policies, or whatever, don't forget to keep taking notes. Eventually, you may wind up with a plan or a change that needs to be implemented. Now, if you want to, you could arrange to have the whole thing recorded and minutes kept, but keep in mind, it is the process itself that will probably diffuse the conflict.

- A word of caution: If, because of your position, your tenure and/or experience, you want to set boundaries on the solutions—for example: nothing will be altered in our emergency responses or diminish our ability to provide a good service to the public, etc. etc.—tell them about the boundaries clearly at the beginning.

- When closing the session—using an upbeat facial expression and body language—ask if everyone is satisfied. Get an affirmative nod from each person. If one person cannot affirm that he is satisfied, openly state that "Everyone has agreed that we have resolved this. Joe ap-

273

parently isn't completely satisfied, but I don't see what else we can do. Do any of you? If not, I hope Joe can live with this." That's it.

PLANNING THE CONFRONTATION

"Confrontation" means meeting a conflict head-on. Generally, we want to avoid confrontations because they are stressful. But confrontations do not have to be traumatic. We are trying to resolve a problem. Careful planning will help you accomplish your goal without all the emotional stress.

Actually, you can plan right now how to respond when the problem is first brought to your attention—how to get at the real root of the problem or complaint. The next step is to analyze the situation for solutions. Then you can plan and conduct the confrontation.

When listening to a complaint, show total attention. Make eye contact, square your body toward the person, eliminate distractive behavior and have an open, anticipatory posture. If the person is hesitant to expound, use what I call the "reporter's stare." We have all seen that hundreds of times on TV. When a person concludes their sentence or statement, stare at them as though you are waiting for them to complete an incomplete statement. On TV, those being interviewed invariable add something else. Mostly, it's just a recap of what they have already said, but sometimes it really is additional information. Incidentally, here is a trick for good eye contact: Good eye contact can be facilitated by just noting the color of the speaker's eyes. But don't let that distract you from hearing what he is saying.

Try to forget who is speaking and listen to what he says. At times, that's tough, but try it anyway. You may have the inclination to say to yourself, "Oh, it's him again. He always hates (this or that)." With a chronic complainer or a spokesman for a specific "tribe," you may have a tendency to hear only yammering, rather than what he is actually saying.

I combined two techniques that work great. The first one I picked up while on an ambulance run trying to communicate with a Japanese guy. His accent was so thick, I couldn't understand him. Suddenly, I pictured that he was on a TV screen, and everything he was saying was being typed out on the screen like sub-titles. In less than 5 seconds, I could understand him perfectly, because I was reading his statements. I have used that numerous times since then. I even practice that from time to time when watching TV to make sure my mind doesn't wander. When you do that, you can then repeat the statement back to them, affirming that you understand them completely.

Second, when his statements become long and involved, I use those worn-out old memorizing techniques that have been around to centuries. I mentally take

274

a walk through my house. When I confronted one female EMT with examples of her disturbing and disruptive attitude toward other medics, and asked her to explain it, she rambled off a long list of examples showing that she was being treated less than respectfully and professionally.

She complained that whenever she screwed up on a drill, it would evoke either moans or snickering from the others. On a run, she was not allowed to make a patient report on the radio, and so forth. She named more than half a dozen examples of disrespectful treatment. Her first statement about screwing up on a drill, I pictured walking into my house and seeing her working on a patient in my foyer, fumbling around and surrounded by laughter. I mentally stepped into my kitchen, but I had to step over a radio that had been dropped in the middle of the floor. She had so many examples, I had made it through my house and out to my back porch before she was done. Hell, I could have walked through my house and continued all the way through town if needed. But when I repeated her examples back to her, it was almost like she had been exonerated. Well, her behavior hadn't been exonerated, but I did stop the game of "blood spot on a chicken" that was occurring.

Incidentally, "Blood Spot on a Chicken" is a game that starts out to be funny and often ends viciously. When chickens get agitated and start pecking at one another, and one of them gets pecked hard enough to draw a little blood, the other, uninvolved chickens see the blood spot and they all jump on and peck it to death. Within a group, it starts out with someone making a rude and demeaning comment directly to another. It evokes a little laughter as the victim tries to ignore it. Seeing no ramifications for such behavior, another member takes a jab at them. The victims of this are usually pretty lousy at verbal retaliation. A violent response is usually the only thing that stops the game, but then you have a more serious situation to remedy. So, Mister Sophisticated, we humans aren't that much more advanced than a coop full of chickens.

Anyway, our victim chicken really was an underachiever on the squad. But the game-playing hampered any hope for her improvement. What was needed was mentoring, not ridicule. I told her that her behavior needed to be changed, but that I would tend to the way the squad was treating her. I told the entire squad that I also don't want to be embarrassed by substandard patient reports over the radio from the ambulance to the hospital, and I certainly don't want substandard care for our patients. But when someone is not up to par, the squad is to help that person get better. If they just can't cut it, inform them that they are not quite ready to perform that task just yet, but do it in a nice way. I said, "There will be no 'Ain't Men Awful' behind her back, and no 'Blood Spot on a Chicken' to her face. You will not ridicule people who cannot perform a task." Yep, I actually do explain these "Games People Play" and refer to these interactions by their names. But, back to the point: Suspend your frame of reference when listening to someone, and use these tricks to literally hear what they are saying.

When responding to their statements, do it in two parts: The "affect" the incident had on the complainant; and the "Meaning" or why it had that affect. Ketchikan's Chief Gene Fisher talked about this technique; he said he actually had a list of emotional response words he kept in the lap draw of his desk. That way he could not only affirm that he heard the comment the firefighter made, but understood it's affect on him. "So, you feel resentful (Affect) that your company officer doesn't seem to have faith in your skills (Meaning)." Or, "You appeared amused (Affect) when he stumbled and fell (Meaning)." Gene had words divided up into categories that listed degrees of intensity. Under the "Anger" category, were words like irritated…bitter…enraged…livid…frustrated." He had "hurt…disappointed…despondent." Other categories included Fear, Happy, Disillusioned, Jealous, Discourage, Concerned, Remorseful, and Regretful. I was never that verbose. But, I used the technique anyway.

Okay, if your analysis is accurate ("You feel frustrated because you feel you can never perform up to par in the eyes of your peers"), it can open floodgates of insight. Check this out: "Tom, you seem hurt because you weren't put in charge of that team." Answer, "You're goddamn right I'm hurt! But, not because I wasn't given that team, but because nobody even asked me if I was interested!" Tom had been reduced to someone of no significance. Tom didn't want the team, he wanted the respect of being recognized.

Here's one for you: All the members had gone home that night, and you were putting a few things away. Jeff, who had been quiet all night, is still there by the pool table. He won't look up, preferring to stare at the floor, and says haltingly, "You know, you bust your ass all your life to do things right, and you screw up just once, and it's all been for nothing." He then walks out of the room. You don't know what his problem is, but he is nearly despondent over it. He is a victim. The "world" took a big dump on him, and he is at the end of his rope. By tomorrow, he may be over it. Nevertheless, he would not have stayed behind specifically to say that to you if he didn't want some response from you. Better give him a call.

* * * * * * * * * *

Okay, in summary, when analyzing the cause of people's actions:

- Demonstrate attentive listening.

- When called for, use the "reporter's stare" to glean further information.

276

- To assure you're hearing accurately, read sub-titles just under their chins or use memorizing techniques for accuracy in your verbal feedback.

- Suspend your frame of reference by paying more attention to what is said than who is saying it.

- Respond back with statements formed around your perceived Affect and Meaning

I imagine there are tons of techniques published for doing this sort of thing. They are probably all good. But, these are the ones I used—and they work.

By the way, if you are suddenly presented with a problem or a complaint and you are not prepared to deal with it spontaneously, do not be manipulated or forced to be reactionary. Say that you need time to think about it and schedule a meeting for later. Do your homework. Investigate the conflict thoroughly, then plan the confrontation.

Remember, we confront issues, not people.

Also remember, this confrontation has a purpose, rather than a cause.

Oh, and if I forget to mention it later, its okay in some situations to just tell people to shut-the-fuck up. I'm not kidding.

CONDUCTING THE CONFRONTATION

So, you have talked to the people involved in this controversy, and you have said to all of them, "I need some time to think about this." Even though you have bought some time, it is not appropriate to drag your feet any longer than necessary to establish in your mind the elements of the confrontation. One or two troublesome people can control all our thoughts, occupy your mind way out of proportion, while the other 90 percent of our people get ignored. That's not fair to the department, and it's not fair to you. Confront your problems and get it over with. You set the time; you set the place; you set the mood; you call the meeting; you maintain control.

Now, nobody wants to feel manipulated (although we all are); what we do is "influence change."

Take out a tablet and write down the following seven questions I got from Chief Fisher:

1- What is your goal?

2- Whose problem is it?

3- What are the consequences?

4- What will be the intensity of the confrontation?

5- What will I expect from this?

6- What are the standards?

7- When, where, and who will attend this meeting?

EXPLANATION:

1- What is the goal, what do you want, what product, behavior or service? Sound simple? It isn't.   If you cannot describe specifically what kind of behavior you want to see, how can you possibly explain it to someone who is on the defensive? Can you write it down? Do so. Remember, you are not confronting a person, you are expecting a change in behavior.

2- Whose problem is it? Is he despondent? Are you angered? Does he feel fine but is pissing off everyone else? Maybe this guy just doesn't fit in. Do not cop-out with statements referring to a "bad attitude." Be specific: "Your behavior tends to devalue other people. You are overly critical of other members. That frustrates them. It makes them feel they must be accountable to you…dance to your tune." In this case, it's the problem of several other members.

3- What are the consequences (of this continued behavior)? The consequences to you, the senior officer, is that you have to listen to a lot of complaints and worry about morale and attrition. The consequence to the department is that morale has dropped, and attendance has dropped because no one wants to be around him. The consequence to him is that people avoid dealing with him. And, he has angered you.

4- What is the intensity of the upcoming confrontation and the intensity of the problem? Is it low-keyed frustration or very angry?

5- What will you expect from this confrontation? What behavioral change do you expect? Describe it. Make a note of it.

6- What are the standards? How will you know when he accomplishes it—has he made the changes? No more new complaints about him in the next 6 months?

7- When, where, and who will attend this meeting? If you expect it will be a low-intensity meeting, go to him, or meet for lunch in a diner, or go for a walk. It's a funny thing about talking to males. I saw a study of adults trying to engage adolescent boys in conversations. The only way to get boys to start talking candidly is to be doing something at the same time like drawing or shooting a basketball. Walking or eating works well for adult males. However, if it is going to be a high-intensity meeting, have it in your office—your territory. Territory is important. If you are confronting a firefighter, then his company officer (or immediate supervisor) should be present. Do not go around or reach a "contract" without the company officer knowing about it…unless it's extremely personal.

Now you have a written plan and that's about all you can to do to prepare for this meeting.

There are different types of confrontations. The type you choose may be determined by the problem. Set the tone and the pace of the interview. Maintain control. Don't have him come in, then you start out with chit-chat about the Super Bowl. That's small talk and everybody knows it. Stay on the topic. Incidentally, sometimes it is effective enough just to tell someone that you want them in your office at 9 a.m. the next day. They will stew about it all night. Then when they walk in, just tell them that you are disappointed in what they did. Then excuse them. But for a more involved confrontation, work on the following ways of forming "Facilitating" or "Discrepancy" sentences:

A Facilitating sentence: "I'm really disappointed by your lack of attendance at meetings." Period. Reporter's stare. Explore the problem. Low intensity. Use your attending skills. Speak in I messages and you messages.

A Discrepancy sentence: "The reason I asked you here is because I'm confused. On one hand, you say you want to be the best in this business, but on the other hand, what I see is (blah, blah, blah). And I'm concerned about that." Silence. Wait. Low intensity.

A little higher intensity might be something like this: "I'm really pissed that you didn't chock the wheels on the truck and it rolled down the hill and struck a car." You have a right to your feelings about his behavior (his behavior—do not devalue the person). "I will have to deal with my own anger, but now we're going to deal with your behavior."

Or, "I'm really angry at the way you conducted yourself at the meeting last night." Affect.

"What did I do?"

"While I was up there teaching, you talked through the whole thing." Meaning.

"What I need you to understand is……" (insert Standards or Consequences). Send I messages.

High intensity can begin when he first walks in to your office. You say, "Sit down" not "Have a seat" because "Have a seat" is an invitation not a directive. "Listen, I have a problem with your behavior, and I'm going to tell you what that is. I'm going to listen to what you have to say, then I'm going to tell you what we're going to do about it." Tell him your problem, then really listen to what he has to say. Use all your attending skills. Then tell him what the consequences of his behavior will be on you or the department, if continued, and inform him that it won't be tolerated. You keep control. If he tries to interrupt, you say "Now, just keep quiet for a minute, you'll get your chance to talk."

Don't forget body language. You don't need to take a course, just remember a couple of things about your hands. When using your authority to make a point, palms face downward. Remember Donald Trump's "You're fired" mannerism? Well, don't point at his face. Too challenging. Just palm down, minute movement, barely tapping the desk with your second or third finger. You could use all your fingers like lightly playing a piano. Don't use your index finger to tap the desk to emphasize words in your sentence, unless you want to raise the intensity. If you do that and make thumping sounds with it, this guy may bristle up instead of being intimidated. When you make I sentences, you can use five-fingers to tap your own chest lightly. But, again, if you use just your index finger to stab yourself in the chest while raising your voice, the intensity of the confrontation raises. And if you poke your index finger at him while ranting, your office could get messy.

Anyway, when it has all been aired, tell him exactly what you want from him…just the way you wrote it down the day before. Start with phrases like: "I would like you to consider…," or "I suggest you consider…," or "I need you to understand…," "I demand that you…," and then add the tangible consequences for non-compliance.

If he is obstinate, he may want to go back in the conversation to some previous statements, or deflect your attention from the ultimatum you just gave him. Don't allow him to take control of this play by writing a new script. Be assertive. Even though you understand where he is coming from, don't accept overbearing behavior from him. Be polite but firm. Keep your voice soft. You have clearly defined your objectives so there would be no doubt, no misinterpretation, no misunderstanding. In your own mind, and in advance, you have decided what trade-offs you would be willing to offer—if any—to get your objectives, knowing that oftentimes you don't get something without losing or selling off something.

Be prepared to play "broken record." If necessary, re-state your position as often as necessary until it is finally accepted by him. Not "why." Don't argue pros and cons indefinitely. Just repeat (even in exasperation) until he finally hears—and accepts it. If he wants to negotiate, don't be afraid to say "no" politely but firmly.

You need to get a commitment from him, and if he waffles and won't be held accountable, crank up the intensity. Pin him down with a bottom line. "I feel really frustrated because I can't get you to understand…" If he is unwilling to comply with the standards you have set, then, "I need to know if there is anything we can do to solve this?" That should be your final statement, followed by a pause, no rambling, no expounding on it. In effect, it is the final ultimatum. If the meeting gets to this point, don't leave the meeting with the issue unresolved. In your planning for this meeting, you have already decided your actions if he refuses.

Maybe I should have mentioned this at the beginning of this section, but it's not too late now. Your bearing during these events doesn't have to be like you have seen on TV. You don't need to scowl, snarl, raise your voice, or act like a prick. A neutral facial expression works fine. Just be business-like, unless your button gets pushed. On the other hand, don't be overly friendly. Oftentimes, those uncomfortable with confrontations act whimsical because their greatest fear is of being disliked. If that's you, and you can't get over that, then you shouldn't be wearing those trumpets. You should be able to get over that by practicing (roll-playing) with a friend.

There's a big difference between planning this confrontation and fantasizing about a blistering admonition like you are playing the part in a movie. You are not going to be arguing with this guy, you are going to calmly tell him how things are going to be different.

Don't let yourself be manipulated by letting him shift focus to another problem. That's why you made notes when planning this meeting.

Here is another employee manipulation he may try: "I will if he will." Or "I will if…." (if something else changes). He is negotiating. You don't have to negotiate or bargain. If he starts ranting, and it's relevant, let him empty his gun. When he is out of ammo, reply with Affect and Meaning sentences, having memorized what he said, give him feedback to confirm that you understood him. If it added nothing new or relevant, then go on with describing what behavior changes you expect to see.

I'm not advising you to be a nit-picker, but you should address personnel problems before they get too big. You can do it in front of a group—if it is a group problem—or individually. Ultimately, you are showing that you are aware of what is going on in the department. Just showing that is enough to resolve many conflicts. Here is an example of that, in an experiment that was later tagged "The Hawthorn Affect."

I may get this a little wrong because I never read about it, someone just told me about it. The details aside, the point is a good one.

In the 1920s, the Hawthorn factory wanted to improve productivity, so the managers came up with a hair-brained idea. They thought their assembly room was too bright and might have been uncomfortable to work in. So they walked around the room and unscrewed every fourth light bulb to dim things a bit. Productivity went up, but after a couple of weeks, it slumped back to normal. So they went around the room unscrewing more light bulbs. Productivity went up again, but soon, dropped back down. After a while, it got so damned dark in there, people could barely see.

So this time, they went around screwing light bulbs back in and productivity went up. At the end of this whole episode, they finally determined that the light level had nothing to do with productivity. What caused the increase in productivity was that the employees knew that the managers were interested in them and their work....period.

It's really the same principle I became aware of during the rehearsals of our jazz band. We would work out new songs, make mistakes, work some more, etc. But whenever anyone walked by, stopped and watched us, we would play songs we knew well and performed much better. It's a natural reaction. So, watch what is going on. Watch your people perform. Now, I'm not talking about spying. That's underhanded and will never be forgotten. You are a nurturing father, not a detective working with Internal Affairs.

* * * * * * * * * *

Before I summarize this section of the book, I want to just throw in a couple of other things I have used successfully in defusing a situation or dropping the intensity of a face-to-face confrontation when I was caught by surprise. I got these ideas from one of the many seminars I have attended.

Let's say a member of the public bursts into your office, mad as hell about something the department (or department member) did. His face is red and he breathing like he's been running. He stands in front of your desk, and even though you offer him a seat, he wants to stand (because he intends to pace and flail his arms).

When he starts his rant, interrupt him and ask, again, his name. He'll spit that out and start his rant again hoping to not lose momentum. Pick up your pencil and ask him to spell his name. Be sympathetic and kind when you do this and apologize for interrupting. By this time his complexion and breathing is almost back to normal. Offer him a seat again and something to drink. The whole idea here is to remove the emotion (affect) from the discussion.

Since you have your pencil in your hand anyway, while Mister John Q. Public is explaining his problem, without looking down at your tablet much, start jotting brief notes as though he were dictating a memo to you. He will watch you write and may even crane his neck a bit to read it. Of course, he can't. The event switches from being an admonition to a report.

Have you ever been trapped in your office by someone who wouldn't leave? Of course there is the old "stand up, extend your hand for a handshake, and thank them for coming by" routine when the visit was somewhat formal. But here is one that Ernest Hemingway developed for the "friend stopping by" incident. While living a Paris, his hotel room was his office—where he went to work each day. But friends never saw it that way. A friend would come knocking at his door, pour a cup of coffee, relax on a comfortable chair and commence to chat away the day. Hemingway was at work, but the friend saw him as being at home. After Hemingway had visited long enough to be polite, he would suggest they go across the street to a sidewalk café for a drink. There, they would chat a while longer over drinks, Hemingway would then stand and excuse himself by saying he had to get back to work, and leave.

LEADERSHIP SUMMARY

Of all of the seminars I have attended and all the books I have read on leadership, many of them were very useful and many of them consisted of not much more than wishful thinking by the author. Most of them had some merit. I eventually settled on theories that I found worked. I became especially fond of Transactional Analysis but didn't try to explain it here because I would never have done it justice. What I have described worked well for me over the years.

The book Games People Play allows one to articulate common sources of personnel problems. "Tribalism" is not a phenomenon you will find described anywhere else, but its existence, as an integral part of human nature, allows the leader to address root causes of many group conflicts. The Glass-Is-Half-Empty negativity; Product vs. Process; the Visionary vs. the Nit-Picker, if recognized for what they are, can be put in proper perspective and save the leader a lot of anguish.

I inserted a couple of stories illustrating that the "Mean World Syndrome" (the pinnacle of negativity) cannot be remedied by the organizations leader, or—quite possibly—by anybody else.

I included the general categories of group conflicts, thanks to the National Fire Academy, as a pinpoint of organizational Tribalism and how "brainstorming" can help eradicate some of these problems.

And finally, I described an effective system for planning and conducting confrontations.

I hope they serve you well enough to minimize morale-busting and time-consuming personnel problems so you can spend your time fine tuning your service to the public.

Section V

Training

NOT ENOUGH HOURS AVAILABLE

There are outside agencies and organizations that think they know a hell of a lot more about what your fire department needs than you do. They begin by setting "standards" for training and equipment, then soon toy with the idea of making them mandatory. Most of the people in these agencies are well educated except for math. Math is a problem for these people. Let me give you an example.

Some departments require their firefighters to meet NFPA Firefighter-1 standards which require a course that can take up to 160 hours. Smaller class sizes take much less time than 160 hours. Cordova has provided the course annually since 1979. EMT-1 class—which in the beginning took about 80 hours (and that included perfect CPR performed on a "Recording Anne") now takes 120 hours, which does not include the CPR portion. After those courses are provided, your department may then provide a 3-hour training session each week, and a volunteer attending 50 of the 52 weeks per year, can acquire 150 hours of continuing training.

Now, enter the "I-know-what's-best-for-you" agencies that, with staunch self-righteousness, declare that "You shall provide training in Hazardous Materials."

Okay.

"Thou shalt also provide training in Infection Control (bloodborne pathogens)."

Okay.

"Thou shalt also provide training in responding to terrorist acts."

Okay.

"By the way, organize a Local Emergency Planning Committee (LEPC) and conduct regular meetings."

Okay.

"Thou shalt provide and practice the 'Accountability System'."

Gotcha.

"And don't forget the Rapid Intervention Team drill that goes with the new '2-in/2-out' requirement."

Un-huh.

"Oh, and P.S..."

I'm listening.

286

"Schedule a couple of nights for SCBA Fit Testing."

You betcha.

You ask, "How much training time should I dedicate to HAZMAT?"

"You can pick the: 8-hour, 24-hour, 40-hour, or the never-ending nuclear physicist level."

And you ask, "How much training time should I dedicate to Infection Control?"

You get the picture. If, out of the four meetings per month, one is a business meeting and one is an equipment and apparatus maintenance night, then actually, refresher training is down to 75 hours a year. With the ever-increasing demands placed on our time from outside agencies, where do we squeeze in Engineer training, Company Officer training; training for specialty teams like High-Angle Rescue, Dive Rescue, remote Search and Rescue operations, Wildland Fires, Shipboard Firefighting, and other topics?

Another complicating factor is the peaks and valleys of seasonal attendance to weekly training. In Skagway, Alaska, everybody's gone in the winter. In Interior Alaska, summer construction or logging season (or in coastal Alaska, commercial fishing season) depletes attendance at weekly training. Some things you just can't change.

And if it's a bad idea to fudge on your bloodborne pathogens training or your HAZMAT training, for fear someone will snitch you out to OSHA, then what other training do you fudge on? Well, that's your call, but I will tell you this: there are ways of conducting refresher training that squeezes value into every minute at the station. Whether you have 150 hours available to you or only 75 hours available, you cannot afford to waste one minute of it. Over the years, I used several techniques to maximizing training time. I will describe some of them.

MAXIMIZING TRAINING TIME

Regarding suppression training, think about it this way: When a high school basketball coach assembles a bunch of first-year students, he teaches them how to dribble, pass, and move the ball up and down the court. They learn individual skills. After they are adept at ball handling, then they are absorbed into a group, or team, where they can now practice team skills. That's where they learn the "plays." Like learning how to square dance, they practice the choreography of the "dance."

A basic course in firemanship, whether it is a purchased course to meet NFPA's Firefighter-1 requirements or a homegrown course designed specifically for the needs of your town, these students are learning ball-handling. They are learning the individual skills they must have before they can be turned over to a team (and engine company or truck company) to rehearse the "plays." There is a choreography to a hose evolution preliminary to an initial attack. This new player must know what position he is to play and what tasks are expected of someone in that position. The department training officer did his job by training the new man in those individual skills, and now the "Team Captain" must do his job by showing the new guy the "Plays," reinforcing the message that for the play to work, the team must work as a unit. Smooth. One of the best ways to assure uniformity in this portion of training is to establish Company Performance Standards.

Under normal circumstances, how long should it take for an engine company to make a forward lay, then be flowing 400 gpm's on a fire?

How long should it take for the truck company to have an equipped crew on a roof for ventilation?

After numerous drills, the company officer can write these things down and use them as a bench-mark for on-going training. Everything is a competition, even against oneself. The purpose is not to try to continually move faster, not to become reckless and unsafe, but to establish a "feel" for how an operation should go when conducted by motivated and skilled personnel.

Now, when an Incident Commander has a sense of the time required for three or four companies to set up for an attack, he can then estimate what stage the fire will be at when the troops are poised for an attack. That's when the training moves up to the highest notch, the full department or multi-company attack. It is then that the chief officers can begin assigning companies by objectives. "Engine Company 4, take the floor of the fire from Side A. Engine Company 2, take the floor above the fire from Side C. Truck 5 ventilate." That's the test of your training philosophy.

Here are a few examples of maximizing the effectiveness of training time:

I just looked at an old photo showing our evening's instructor standing next to the pump panel on an engine, pointing to the proportioning valve of the engines foam system. He was explaining its use to the firefighters standing in a group in front of him. Would you care to guess how many of the students heard and fully understood everything he said? Probably the closest three guys. Hell, it was just a challenge for the speaker to be clearly heard above the sound of the overhead blower, let alone the constant yammering of the guys in the back of the group talking about things completely unrelated. Or the guy in the middle of the group coughing every time the instructor had an important point to make. In addition, there is really no way to satisfactorily assess how well the students absorbed the information.

SOLUTION

The solution to this common problem is simple. Actually, it's better than simple, is saves the instructor work and time. The maximum size group for this

288

technique is three. If three firefighters comprised a team that was to get re-fresher training on—for example, the cleaning and maintenance of the engines pressure relief valve—and a brief note on the process were scotch-taped to the pump panel, those three could take ample time to disassemble it, clean it up, and reassemble it. Even done leisurely, it only takes a few minutes. Better yet, they would all remember how to do it. If you had a total of ten tasks to perform, located at different places around the apparatus room, and assigned a 3-person team to each one, 30 firefighters can be trained effectively once each team rotates through all ten locations. And the department instructor(s), having previously set up the 10 locations, can relax, stroll around to assist and answer questions. It is important to remember: This is not for initial training, because it doesn't allow for background information. This is for refresher training, and generally consists of manual skills and equipment use.

TRAINING IN STATIONS

Training in Stations is similar to EMT final practicals except there are no proctors.

The example below is for Cordova's EMS division and is accomplished in the training room alone. Here is one evening's refresher training on nine skills. Arranged in two-person teams, eighteen medics can easily be accommodated during this "Pediatric Night." EMS Captain Joanie Behrends supplied me with these nine sheets of paper. Go to the front of the room to "Station #1" and sit down.

STATION   #1

THE LAW AND YOU

TREATING MINORS

- Please review this 8-minute video tape. When finished, rewind it for the next group!

- Please EAT goodies so I don't take them home and eat them!

(Joanie always provides snacks for the EMS training night)

The table over by the window (for example) is "Station "2"

STATION   #2

PEDIATRIC ASSESSMENT

- First, please the review info at this station

- Then, please do a **SAMPLE** history on the patient, get the pt. weight, get heart rate, take blood pressure, assess cap refill (remember it works in kids), take respirations, and get temperature.

STATION   #3

WHAT IS ALL THIS STUFF?

- Please refresh your memory about the pediatric equipment here that you don't get to use very often
  - Pulse oximetry on peds
  - Broslow tape and entire new kit!
  - New Teddy bear oxygen masks!

STATION   #4

PEDIATRIC INTUBATION

- Please INTUBATE!

(obviously, we have a pediatric intubation manikin)

STATION   #5

PEDIATRIC BAGS

- Please help put together a new pediatric bag and familiarize yourselves with it. Make a training bag, too (both bags exactly alike if possible).

STATION   #6

SIDS and CPR

- Please read the review information about SIDS. Then pretend that you have been called to the home of a SIDS patient (someone be sure to play the part of a parent)…and perform CPR on the infant

STATION   #7

PEDIATRIC IV's

- Please read the tips for starting IV's on pediatrics!
- Please start an IV

290

STATION   #8

SPINAL IMMOBILIZATION OF PEDS

- Please immobilize your patient using the new pediatric unit

STATION   #9

PEDIATRIC TRACTION SPLINT

- Please ACCESS AND TREAT your patient

The medic can focus on each skill without distraction, and if it's complicated, his/her partner can help figure it out. There is no way for the medic to "fake it" or avoid getting it completed. It's extremely effective. It's also very efficient. Eighteen medics mastering nine important skills in less than three hours, is the epitome of efficiency. And the icing of the cake is this: Instead of the instructor busting his/her ass diagramming things on the marker board, pacing and pontificating, praying for clarity, while the students sit on their glutes pretending to be interested, this night the students do the work, read, discuss with their partners and keep busy every minute while the instructor relaxes for three hours.

Doing this two or three times a year is a nice break in routine.

Firefighter stations could include knot tying, hydraulic extrication equipment, donning SCBAs while blindfolded, laddering, setting up Positive Pressure Ventilation (we tape strands of toilet paper from the ceiling of the room we will "ventilate," so people can see when it is being done effectively), and any number of things that 2-person or 3-person teams can do.

TACTICS SCENARIOS

Next, is an evening of Tactics scenarios. We do the 3-person team concept again but the teams do not rotate from one station to another. Following are seven scenarios that I used one evening. If you have more than 21 people show up for the evening, form some 4-person teams rather than 3-person teams. Here are the instructions I typed out for the 7 scenarios:

Your team is to go to the following site of the emergency. Look over the area layout, topography, or special problems, and develop a plan of attack to deal with the problem. Write down—step-by-step—exactly how you would:

- Place apparatus
- Assign all the resources you might use
- Plan to get additional resources you might need

291

Upon returning to the training room, draw a diagram of your plan on flip chart paper and present it to the other members. There are no wrong answers. Remember, your scenario may be drawn out of the hat, and you will act as Incident Commander for a real "wheels-rolling" drill after the classroom demonstrations.

HERE ARE THE SCENARIOS:

TEAM #1.

Time and Place: It is 1 a.m. at the Ocean Air Apartments on Browning Street at 3rd.

Situation: A pre-flashover fire. Smoke is seen at all openings in the building, probably coming from a fire somewhere in the lower portion of the building. On side A you can see one occupant on the top floor standing at the right-hand window. There are several cars in the parking lot.

TEAM #2

Time and Place: Noon at the intersection of First Street and Adams, by Hoover's service station.

Situation: A passenger car "T-Boned" a fuel truck with sufficient force to create a leak in the fuel truck. The placard contains the number "1993." Fuel is running down the gutter. The occupant of the car cannot open the door because of damage to the car. A small fire has started under the hood of the car. Because of damage and positioning, you cannot open the hood of the car.

TEAM #3

Time and Place: 2:30 p.m. at the fueling station at the Orca Oil tank farm

Situation: You are responding to a tone-out of a fuel spill, and you arrive to find that the driver of tank truck had been filling the tank truck with gasoline. He was unable to release the trigger on the nozzle, so the gas overfilled and overflowed and is running copiously down the side of the truck and onto the ground. So he just ran away down the street. You met him half a block from Orca Oil. He is soaked with gasoline.

292

TEAM #4

Time and Place: 11 p.m. at Laura's Liquor Store, First Street

Situation: Upon your arrival, you note smoke in the area, primarily coming from the entrance of the liquor store. You are met by the clerk who tells you there is a fire in the store room in the back. During your size-up, you notice that open flame has penetrated through the store room ceiling. Now, you have been informed that the apartment above the store looks "a little smoky" inside.

TEAM #5

This is strictly a mental exercise and doesn't require leaving the station.

Time and Place: Noon, Saturday

Situation: You have received a page that the northbound Alaska Airlines 737 has contacted Cordova's Flight Service at the Mile 13 airport and reported that they are experiencing difficulty in controlling the aircraft and that they see smoke inside the aircraft. The airport fire department has called for mutual aid. The ETA for the 737 is about 25 minutes.

TEAM #6

This is strictly a mental exercise and doesn't require leaving the station.

Time and Place: 3:40 p.m., a Tuesday in January. Copper River Highway, below Wolverine Bowl

Situation: An avalanche occurred from the area on the Heney Ridge known as Wolverine Bowl. The snow, mixed with broken trees, swept across the highway. Your page stated that persons were covered by the avalanche. Upon your arrival, you are informed by a motorist, that he was driving behind the school bus (school was just let out) and he saw the snow hit and cover the bus. You notice about 6 inches of a tire jutting out of the top of the snow pile.

TEAM #7

This is strictly a mental exercise and doesn't require leaving the station.

Time and Place: 2 p.m., Saturday in April

Situation: In a tone-out for the ambulance, the dispatcher stated that the vessel Lucky Dollar called on radio channel 16, and reported that a crewman aboard was suffering chest pains. The Lucky Dollar was anchored in Orca Inlet about a quarter of a mile off the ocean dock. The call was made by the only other occupant of the vessel, who is now in the process of pulling up the anchor to proceed to the ocean dock. Upon your arrival at the ocean dock, the dispatcher informs you that the crewman of the vessel called again. The man with chest pains went "unresponsive" and stopped breathing. The other occupant is now doing CPR. You notice that the Lucky Dollar is slowly drifting south with the tide.

* * * * * * * * * *

There are no bad answers, no criticism, just applause and "nice job" at the end of each team's presentation. And, yes, brand new members who have never responded to a fire also get assigned to a team.

After about half an hour of peace and quiet in the training room, the teams return and prepare their presentations. When they are done, I draw a number out of the hat. If it were #1, the Ocean Air Apartments, for example, I give that team a portable radio and have them go back to the apartment building and have them tone out the department to that location. When we roll out of the station, they start radioing assignments to the apparatus, which the companies comply with. Supply lines are laid, attack lines are stretched, ladders are hoisted, the whole thing.

This is a night of thinking; of playing "what if?" It doesn't matter that the students are not company officers. Even the newest member can see the total picture of a fireground choreography and the part that each person plays in it. They become familiar with terrain and topography, hydrant locations, conceivably—if two hydrants are in close proximity to one another, but are on different water mains—learn a bit about the inter-relationships of hydrant flow capabilities, a building's fire flow requirement, and an engine's pumping capacity

SINGLE LARGE SCENARIO

The next example involves all the firefighters in a single scenario, but they are divided into their three separate companies to deal with different aspects of the scenario. The task was to calculate the amount of foam needed to fight a fire at our larger tank farm. One company was to calculate the surface area of each tank in the tank farm (they were not all the same diameter), then calculate the foam application rate and total concentrate needed for each of the tanks. One company was to calculate the foam application rate and total concentrate needed for the diked area if a tank were to fail. The third company was to calculate some important information about the use of the airport's foam crash truck. I gave the company officers the formula for calculating it, then a written example.

Find the area (square feet) of the top of a tank. I use diameter squared times .8. A tank 36 feet across is 36 x 36 x .8= 1,036 ft².

- Gallons per minute application rate if the fuel is burning: .16 gpm per square foot. 1,036 X .16 = 166 gallons per minute of "foamy water" applied either sub-surface injection or over-the-top.

- One needs to plan for that application to go on for 50 minutes (although it would probably be extinguished much faster). So 166 gallons per minute for 50 minutes = 8,288 total finished foam ("foamie water") to knock the fire out. Our department was using 3% foam concentrate, which means that 3% of those 8,288 gallons of that froth we shot on to the fire was foam concentrate, found in those very expensive 5-gallon buckets. 249 gallons of concentrate.

- So, Group 1, using the formula…. "Area X .16(gpm) X 50(min) X .03 (3% concentrate) = concentrate needed to be on hand before suppression begins…. for each different size tank in the dike.

- Group 2 used the same formula to calculate the amount of concentrate needed for inside the dike if a tank failed.

Those figures went into our pre-fire plans and initiated an agreement with Alyeska Pipeline company in Valdez for flying foam supplies to us if we should need it.

Of course, the airport fire department has lots of foam, but having a contingency plan is good. Speaking of the airport DOT fire department, Group 3 was tasked with figuring out another operational aspect.

I typed the following information for them:

- The airport's crash truck is a one-man operation and can move very close to apply its product from its turret. It carries 200 gallons of 3% concentrate in its foam tank and 1600 gallons of water in its water tank. The side of the dike that this engine can access has no hydrant there. The engine can discharge its finished foam in about 2 minutes, then has to go get more water for its tank.

- How many tanksfull of water will it go through before it runs out of concentrate?"

The answer was "four," but that's not the point. This is the point: A certified Firefighter #1 knows about educting foam and delivering it on to a flammable liquid. Refresher training like this took one night to develop a complete pre-fire plan, reinforce a mutual-aid agreement, generate questions and answers about how to educt 166 gallons per minute of foam solution or greater. However, since we need to reinforce the manual skills regularly, here is the sheet of paper I handed to our company officers the next week:

With the information you developed last week, deal with the following scenario:

- A lightning strike ignited Tank #12 at the Orca Oil tank farm. It is full and the top surface of the tank is burning.

- There are two immediate exposures: Tank #11 and Side B of the warehouse.

- Consider foam application "over-the-top" for Tank #12 and set up to do it.

- Consider master stream exposure protection for the warehouse and for Tank #11 and set up for it.

- Consider that by cooling Tank #11, the dike is filling up with water. Locate the drain.

- If extra manpower is necessary for engine companies, use truck company people.

- This is a wheels-rolling exercise. Tie into the water main system and attack the problem. Remember, don't really fill up the dike with water, direct streams into a harmless area.

Sometime before applying foam (this is pretend foam, Mister Big Spender), tally the amount of concentrate you will need to have on site, and the numbers and sizes of hose(s) you'll need to deliver the proper flow rate and the capacity of all the eductors you have available.

The Deputy Chief is out of town, and the Chief has diarrhea. You're on your own.

* * * * * * * * * *

Exercises like this not only sharpen manual skills, it allows you to quietly sit back and observe which individuals appear to have valuable leadership potential and reasoning skills.

There are tons of ideas out there to maximize the benefits of limited training hours. How about the "Relay Night"? Send a company to one location where the proctor explains the pretend situation in front of them. The company attacks and resolves the problem, then drives to the next location and is met with another one. So on and so on.

Following one such night, a firefighter came up the idea of using a fire hose as a flotation device for a water rescue. He had an inch and a half male adapter fitted (brazed) with an air chuck. He screwed that onto the hose and capped the other end. Using an air tank, he inflated the hose to be pushed or thrown out into the water for someone in need to grasp. Never used it for that, but sometime later, we got called out because a small dog had crawled into a small culvert and got stuck half way down and didn't have enough sense to crawl backwards to get out. After lots of unsuccessful ideas, the backhoe arrived to dig up the road so we could cut the culvert. Then I remembered the inflatable hose idea. We stretched the hose out, inflated it and slowly inserted it into the culvert toward the face of the dog. As we gently moved it in, the dog had no choice but to start backing out. All the way out. Cool, huh?

Bottom line, you will never have enough time to become skillful in all the tasks required of you. Carefully select the ones of greatest consequence for your department and your residents. Get creative so you don't waste any of that non-renewable resource...time. There's no getting around it, Chief; you're out on a limb.

BASIC STRUCTURAL FIRE TRAINING

Photo: CVFD

AIRCRAFT CRASH AND FLAMMABLE LIQUIDS TRAINING

Photo: Cordova Times

297

WINTER RESCUE TRAINING, ASCENDING AN ICE FALL

Photo: Cordova Times

HIGH-ANGLE RESCUE TRAINING

Photo: CVFD

PART VI

EXPANDING POLITICAL POWER

FIGHTING THE POLITICAL BATTLES AT STATE LEVEL

FIRE SERVICE ASSOCIATIONS

One doesn't have to be a paleontologist to study the evolution of organizations if one can watch with the clarity of semi-detachment. Now, you can't be indifferent when you are totally immersed in meeting the objectives of an organization, but monitoring the changing objectives (over the years) is monitoring its evolution. As a member of the Alaska Fire Chiefs Association for nearly thirty years, much of its history is my history. Much of its evolution is mine also—even if it is, at times, unfortunate and superficial.

What will a reader, gain by slogging through such a large chapter that is devoid of excitement or adventure found in stories of emergency responses? Like any organization, we have training officers that don't train, we have managers that don't manage, and we have leaders that don't lead. But most depressing, we have creatures that don't create. A *craftsman* uses the same raw materials available to everyone, but creates something unique. "Creating" something is as exciting as kicking in doors or climbing mountains. It defines the creator, the artist, or the composer. It is possible to transform an organization from one that supports mostly its own members, to one that also becomes a major power broker across the state.

Frenzied and tedious as it was, it was probably the most fun I'd ever had. In addition, the principles that I used to manage and expand influence are applicable to any organization. When Machiavelli wrote *The Prince* in 1532 he wrote an excellent book on management and leadership. Generally, the critical remarks that labeled him a self-centered manipulator, were made by people who never read *The Prince*. This transformation of the Alaska Fire Chiefs Association was an "Extreme Make-over" complete with organizational plastic surgery and acting lessons. Later it involved political wars, intrigue, and manipulation. You know....fun stuff.

In addition, I learned something about myself. I came to realize that something I had always looked at with disdain—the game of political manipulation—turned out to be something I was really good at, and eventually came to crave. At the beginning of my two-year Association presidency, I was a starry-eyed idealist. I was a knight who would defend the victims of fires and disasters, whether it would be on the streets or in the halls of the state legislature. By the end of my tenure, I realized that the banner of my noble cause was a façade. I was just looking for a fight.

Nevertheless, every story I relate here is an example of a principle in leadership or management. Some of these principles can make your favorite organization one that must be reckoned with.

300

One must remember that two major facets of an organization must be under your care at all times: the internal and external facets. All organizations are created specifically to accomplish a goal, such as produce a product or provide a service. The act of creating an organization is existential in its purest form. The organization is defined by the purpose it serves. The creators of an entity organize it solely with effectiveness and efficiency in mind. The internal processes of running the business of the organization are established to maintain order, operate with relative ease and frugality—efficiency. But one can only gage the effectiveness of the organization when the results of its activities are compared to clearly defined, measurable goals.

By that definition, the Alaska Fire Chiefs Association was not really an organization when I first joined it. We were an association whose purpose was for members to network with one another, advise one another, discuss common interests and problems. For example: How many hours are be required to train new department members in basic firemanship? Or more specifically, how might a department schedule such a program for volunteers to be able to attend in their spare time; or what about the overtime costs for the paid firefighters? Naturally, legal issues were always discussed. Administrative dilemmas always hit the table. And, of course, personnel problems were always hot topics. The common thread that ran through our discussions was "What can I do in my department about this problem?" The first issue I recall that approached us from *outside* Alaska's fire service that could have impacted us significantly, grabbed my attention so thoroughly that I wanted to be one of the guys to tackle it. It was the problem of shipboard fires at sea, brought to our attention by the disastrous fire and sinking of the cruise ship Prinsendam in the Gulf of Alaska. Association President Bud Sands, chief officer from Fairbanks UAF, asked me to look into it

I was asked to research the question of jurisdictional authority for ship-board fires at sea. When the Prinsendam caught fire, 120 nautical miles south of Yakutat on October 4, 1980, the Coast Guard air station in Sitka dispatched two H-3 helicopters. Fire Chief Marty Fredrickson of Sitka took a couple of his firefighters, radios, airpacks, and other equipment and boarded one of the choppers. When the chopper was over the area, they radioed the ship and offered firefighting assistance from Marty and his guys. The ship's captain conferred via his radio with the ship's owners, the Dutch-owned Holland-America cruise line, who declined the offer, stating that the engine room fire, having been isolated and flooded with carbon dioxide would burn itself out.

However, the fire that started about 1 a.m. was completely out of control at 6 a.m. when the captain gave the order to abandon ship and evacuate their 320 passengers and 190 crew members. The ship was taken in tow and the still-smoking vessel headed south across the gulf for Seattle. The cruise line expected the fire would be out long before arriving at the shipyards in Seattle. In the meantime, a bunch of us chiefs had gathered at a code-revision meeting in Anchorage when Chief Marty Fredrickson was called out of the room for a phone call from the mayor of Sitka. Upon returning, Marty said the mayor had received a call from the Prinsendam owners requesting permission to enter Sitka's harbor, so that someone could attack the stubborn fire. It was still extending. Discharging CO^2 into a burning compartment and sealing it off is an acceptable and universally used suppression technique on ships. However, it is no guarantee that the heat in the compartment won't heat the surround bulkheads (walls) and overhead (ceiling) so hot that combustible materials adjacent to them will ignite, spreading the fire outward and upward from the compartment of origin.

The fire in the Prinsendam was spreading up from the engine room. "What did you tell the mayor?" Ketchikan's Chief Divelbiss asked. "I told the mayor that the opportunity to fight the fire with a small force was lost. This fire would take more resources than we can get. Plus, when this thing is destroyed by the fire, it might sink in our shipping channel and we'll never get rid of it. We would be screwed." We went back to our meeting, and the Prinsendam continued its trip south.

Later that afternoon, Chief Divelbiss was called out of the room for a phone call. It was from Ketchikan's mayor. The Prinsendam owners had called asking if it could be towed into Ketchikan. The fire was completely out of control. "What did you tell the mayor?" we asked. "I told him to tell them 'Hell no'."

"What happens if the Coast Guard's district office in Juneau tells the Prinsendam to head for your port?" someone asked. "They're not bringing that burning sonofabitch into Ketchikan," Divelbiss reiterated. No one contradicted him, yet we all wondered if obstinacy was legally defensible. There was a brief discussion regarding our lack of knowledge about the law and about ship fires before we got back to reviewing proposed code revisions.

After the meetings we left Anchorage, but for six days the $50 million vessel wallowed in tow. The bridge was gutted, the hull scorched, and smoke continued to trail the horizon. It took on more water through its blown-out portholes, and just after daybreak on October 11, it rolled over and sank in nearly 9,000 feet of water.

It was at our Association conference, a few days after that, when President Bud Sands asked me to do the research into marine law and jurisdictional authority. I found it very easy to do. One of my volunteer firefighters, Steve White, was an attorney, and—as it turned out—he majored in marine law. When he completed his research and report, I submitted it to the Association and it became the basis for city fire departments responding to shipboard fires. Of course, since then textbooks have been written and courses provided, bringing in Coast

302

Guard officers, ships masters, and others involved in these events. But that was an example of critical networking needs the fire service needed.

* * * * * * * * * *

As an aside, I'm not convinced that jurisdictional lines are all that clear even to this day. When I had a burning fishing boat towed out of the harbor to another spot where we could fight it, the local commander of the Coast Guard ship told me not to do that. He didn't want it sinking in the shipping lanes. Understandable. But I didn't want it sinking in the harbor becoming a submerged hazard to navigation. I also didn't want another sunken boat releasing all that fuel in an enclosed harbor, creating a floating bomb. He got mad at me for towing it away after he disallowed it. Afterwards, we both ran to our respective libraries of legal documents to see who was right. We never did resolve it.

* * * * * * * * * *

Emergency operations have clear objectives. Support functions tend to grow fuzzy. Here is an example of "goal diffusion." Maybe "mutation of motivation" is just as accurate a description.

The following year I brought a training issue to the Association. I told the then Association President, Anchorage Deputy Chief Jim "Babe" Evans, that we have placed so much emphasis on technical training for firefighting, we can only hope that the guy running the show has some idea of what chief officers are supposed to do. I gave the example "If you are on the fifth floor of a non-sprinkler wood-frame hotel and are awakened by a fire below, you look out the window and see all the firefighters running around in the street and notice the one wearing a white helmet, you have no idea if he knows what the hell he is doing. You will assume he does because he is the chief. But really, in many of our departments, the only criteria for being the chief is that you stayed in the department long enough and it's now your turn."

I went on to say, "If you want to be a dentist, you can go to school and become one. Aspiring welders can go to technical schools to learn that trade. To be a fire chief, you just have to wait your turn. And if you wanted to go to school for this, you're out of luck because there are none. We need to write a course, the same way we wrote one for firefighters."

"Okay, go ahead." Chief Evans said.

Two years later it was done. I was always a great procrastinator. Based on the National Fire Protection Association's standard 1021, I wrote the course to be delivered in four 40-hour increments. It covered all fire officer standards from Fire Officer-1 through Fire Officer-6. Since then the standards have been compressed from six levels to four levels. But since there was no money in State Fire Service Training to provide such a course, I applied for a legislative grant, and after testifying in Juneau—citing the same examples I used on Evans, we got $80,000 for conducting the first year's course. We got chiefs from all around the state, from every imaginable type and size of department to take the course. I got another grant for the next year and we graduated a bunch more chiefs.

Conceptually, it was an easy project. But many years later, when I recommended we revise the course to keep it current with the times, then provide it again, it was impossible to do so. The original objective of developing and conducting such a course was simple: Teach a chief or potential chief the things he or she will need to know to do the job. When I was in the process of revising the course, numerous persons questioned me about how this course would mesh with other programs now in existence. Fire officers involved in the University of Alaska (both in Fairbanks and in Anchorage) wanted to know if their students could get credit for Fire Science courses they had taken toward their Associate degrees? Others who had taken a 40-hour IFSTA course for company officers (Fire Officer-1) wanted to know if they would have to take the entire 160 hours or only 120 hours. Other people wanted to know if the course would be "accredited" by the National Fire Protection Association, since the course was based on those standards. Still others wanted to know if the National Fire Academy would endorse it as one of their outreach courses so that students could receive a certificate held in such high regard. When the Anchorage Fire Department Training Officer Craig Goodrich called me, suggesting I fly to Anchorage and address a meeting he was hosting for instructors from the National Fire Academy, I went. In response to my request that the National Fire Academy take my revised course—once I'd revised it—and endorse it as their own, they were vague and evasive. It was obvious that I would not get their support. Hell, the National Fire Academy didn't even exist when we had delivered our first two courses.

In the end, between the Fire Science classes at the universities, the IFSTA company officers courses, The National Fire Academy, and NFPA, there were too many individual objectives to be met. For me, this resulted in "goal diffusion." It was like someone dropped a handful of marbles on a tile floor. I spent so much time trying to scoop them into a single, manageable pile, there was no time to play marbles. Without a single, common goal—an agreed upon vision—you will have goal diffusion. Student objectives had changed dramatically. It went from learning the skills of the job, to chasing certificates as steps in a career ladder. Students in any endeavor often study and cram, not to learn, but to pass a test. It was a mutation of motivation. The focus went from protecting your residents to personal advancement and certifications. After about six months of discussions, meetings, phone calls, and research into everything *except* course content, I informed the Association executive board that I would not be able to do this.

Charles Darwin convinced the world that creatures went through species-changing evolution. Organizations also go through evolution. This is only one example. By the way, in case you haven't noticed, they never go back to what they once were.

After I had spent about six years on the Chiefs' Association (AFCA) board of directors, I held the position of vice president. The president, Chief Al Judson from Juneau, had been doing a terrific job on the political side of things, working with our lobbyist at the state legislature as his tenure was up. When I was nominated for the president's seat, I got cold feet and asked Al if he could run for another year, and he agreed. But by the '85 conference in Sitka, I had a game plan to reorganize the association and had the confidence that comes with having a good plan. I was elected at our last meeting in Sitka and I explained my plan. Previously, I didn't look like the personification of a chief, and was not articulate unless I prepared what I would say. When I was younger, and got excited, I would flail my arms and rattle off like Beretta on crack. I was a chief of a little volunteer fire department. So—I explained—I would become the Wizard of Oz and create an illusion of power.

We have neither money nor a large voting block. Why would anyone listen to us? The only way we can become politically influential is by creating "an illusion of power." Individuals and organizations have become powerful and influential based on "technical" information that they alone were privy to. As an organization and as individuals, we can become the *Resident Experts* in this field. It must be clear that our motives are noble and magnanimous. Gandhi, Martin Luther King, and others found international recognition based on this. The National Organization of Women (NOW) has very few actual members, but since they have lauded themselves as the voice of women in America, they have created the illusion that they speak on behalf of millions of women. As a result, it is only because the media asks them their opinions, and prints them, that they have power. You only have power when people give it to you. If your words influence others, you have power. Regardless of why people give you power, it's important to never abuse it. Use it carefully and always demonstrate your worthiness. Never lose credibility. Be painfully accurate, then concise. Once caught in being deceitful or self-serving, your power—the gift they gave you—will be taken back. Credibility and accuracy is all our organization has to offer. The only man who deserves power is the one who—every day—shows his worthiness.

But initially, to gain attention and respect, we must create an 'illusion of power'. The Wizard of Oz had tremendous influence and power because of the façade created by the booming voice and 'face' of Oz. But when Dorothy's dog pulled back the curtain, there was some very ordinary-looking man back there pulling the strings creating the illusion. That will be me.

Well, we need to be everywhere. Nothing should happen in this state affecting the fire service that we are not aware of from the onset. Then we need to be prepared to exert influence to correct anything that will jeopardize us or the people we protect. We need to be able to react quickly, but we also need to be more proactive. Rather than just attacking things we don't like, let's create things we do like.

* * * * * * * * * *

I had already prepared to make 12 staff assignments that would spread our influence across the state. Ten were new and were to become the workings of the wizard. Each person was selected, based—in part—on appearance (must look like a chief, in a 3-piece suit, a little gray hair, be articulate and have impeccable mannerisms that would instill confidence in his words and opinions. No ranting zealots. I was convinced by Machiavelli's *The Prince* that representatives of your group (ambassadors) must be the personification of the image you want projected. They must have full authority to speak on behalf of your organization, which means you must select carefully those people whose opinions you would trust fully.

FIRST:

We needed quick access to the governor and members of the legislature in Juneau. Al Judson, chief in Juneau, was personally acquainted with Governor Sheffield, so Al agreed to be liaison with him and the legislature. Al was the Paul Bunyan of the fire service. He had an easy laugh, and was an imposing figure in a navy-blue blazer and red tie. Al Haag, our lobbyist, would spend the time walking the halls in the capitol building doing recon, and, together with Judson, they would handle that activity.

SECOND:

Second, was Chief Charlie Lundfeld from the North Star Borough, which covered the rural areas outside of Fairbanks. He could be smooth and persuasive, but push his button and he became Yosemite Sam. In fact, years later, a new city manager in Cordova who hailed from the Fairbanks area recounted the first time he'd ever seen Charlie. Charlie and a home owner were rolling around on the ground pounding the snot out of one another while Charlie's crew was extinguishing a fire in the man's house. Anyway, Charlie spent lots of time wearing a yellow shirt and carrying a canteen on wildland fires, so he would liaison with State Forestry. Particularly, he would work on the Urban/Wildland Interface project. This focused on the ever-increasing populating of wooded areas with homes, making the firefighting of wildland fires more complicated and critical. Within a year, he would be instrumental in the delivery of a Wildlands training "boot camp," incorporating personnel from DNR's State Forestry and the Interior Fire Chiefs Association (where most of the states wildland fires occur). The boot camp ran eight tactical and logistical stations, and the whole thing was covered by the media. It was a great success.

306

THIRD:

Chief Bill Schecter from the University of Alaska Fire Department in Fairbanks, was to be our representative to the Alaska Municipal League (AML). I told the group that we had gained nothing by dropping out of the AML years ago after a disagreement with them. During our absence, AML had made some bad policy decisions because they lacked our input. Rather than being outside looking in, we needed to be in the group where we might be able to assert some influence. We needed to be a valuable information resource to them. For example, Schecter later attended the AML's Finance Officer's committee and gave them accurate estimates on the impact the federal Fair Labor Standards Act would have on cities with paid fire department staffs. In return, we got from them a position statement in support of a proposed amendment to the Public Employees Retirement System (PERS) which turned out to be very beneficial to fire service personnel. AML also supported an EMS resolution which improved Emergency Medical Services across the state. And that was just on his first meeting. He did impeccable homework on all the issues. In fact, he met with State Public Safety Commissioner Sunberg prior to the AML meeting to discuss fire service issues. He went totally prepared as AML's "resident expert" in fire and rescue. An interesting note on how an illusion of power spreads: If you are seen speaking to a person of influence in an informal setting, the witness wants to be seen speaking with you, too. People are suddenly nicer to you.

FOURTH:

Alaska Department of Health and Social Services' Emergency Medical Services division (HSS/EMS) had established an advisory committee of physicians, instructors, and bureaucrats to develop policies about local EMS services. However, none of the members were actually in-the-trenches responders. Their policies affected those of us who deliver the services, provide the training, and establish budgets for them. Assistant Chief Robert Purcell from Homer was a collegiate, profoundly articulate person who had a softly convincing manner when he spoke. He was as studious as Schecter. He took a seat on the HSS/EMS committee to make sure that they kept their feet on the ground and regulations were attainable. The EMS standards remained high, but the process of meeting the requirements was as doable for volunteers as for full-time career medics. Remember, whether you are a fire officer or a fire service organization, if you do not lead, you will be led. In the past couple of decades there has been an insurgence of governmental bureaucracies who write procedural regulations without personal knowledge of the detrimental impacts they have on our operations. If the intent is good, support them. But if the process they dictate is silly, they must be challenged. For that, you need political power.

FIFTH:

To create an illusion of power, an organization must be widely accepted as having their shit together. Even if you have your shit together, it means nothing if people don't know about it. Public image is the responsibility of your Public Information Officer (PIO), and my fifth assignment. Here, not only is style and abilities important, but image is vastly important. Anchorage Fire Department Deputy Chief Jim "Babe" Evans was perfect. In his fifties, he had the successful executive image: He was tall enough, and filled out a business suit amply. With a mustache and his black hair graying at the temples, he was the CEO and father figure we needed. I asked him to become a familiar figure around media offices. Newspaper and TV station staff should know him well. In fact, having an occasional friendly lunch with them (at Association expense) would be a great idea. He was to be alert for opportunities to make statements on behalf of the association. I just wanted to see him on TV from time to time with a caption on the screen saying "Alaska Fire Chiefs' Association." For example: The morning I heard about the devastating earthquake in Mexico City, I called "Babe" suggesting he explain to the media the earthquake preparedness in Alaska.

Anyway, a heroic act by some civilian should result in a certificate or commendation to this person from us in the presence of TV or newspaper crews. He was to assure that our name was synonymous with the insightful father-figure that was needed to develop our power base. Machiavelli would have been proud of this selection. Evans developed a promotional flier for new membership for the AFCA. Later, in December of that year, he was contacted by the Governor's press agent who was interested in a statement from us about the proposed State Fire Commission. The press agent discussed it with the governor, told Evans what the governor said, and Evans then had the newspapers print a statement about the governor's support for the establishment of the Commission. Once I wrote a speech I wanted delivered at the annual EMS symposium held in Anchorage. I sent it to Babe. In fact, I recorded myself reading it because I was so particular about voice inflections and pauses. It's a fetish of mine. In the presence of hundreds of people, including Governor Sheffield, he delivered the speech which got a standing ovation.

SIXTH:

The sixth new position I created was assigned to Chief Tom McAlister of Valdez. You know, the AFCA was not self-serving, but that could not always be said about architects, engineers, building contractors, or building owners. They could be very powerful people, and were tough adversaries of the State Fire Marshal. Whenever people like these would launch an attack against the State Fire Marshal, who might be in the process of approving new fire, life-safety, or building codes, and the Marshal needed some support from the fire service, he would tell Tom. Tom would brief me, then he would inform the attackers what our stand would be on that issue. Tom had the authority to speak on behalf of the organization. He would be the guy in charge. Remember, no one wants to spend time talking to a "messenger." That's an important principle. Each time I got a report from Tom about code issues, I would send copies to the board of directors to discuss during teleconferences. He agreed to remain in close contact with the State Fire Marshal regarding any issues on fire codes, life-safety codes, and building codes.

SEVENTH:

Eielson Air Force Base had its own fire department primarily for military aircraft. But they were also structural firefighters for all the buildings on the base near Fairbanks. Chief Bud Rotroff, during his vacations, would be on "Overhead Teams" for wildland fires. As a member of these management teams, he was a real expert in the Incident Command System (ICS) used—at that time—primarily by forest firefighters. Most city fire departments at that time only dabbled in it. Very few of us were familiar enough with it to allow us to work well with forestry. However, all fire departments saw its worth in working with other agencies when responding to disasters. Alaska Division of Emergency Services (ADES)—our state civil defense organization—although familiar with it, was not using it. Law enforcement, particularly the state troopers, saw no need for it. Law enforcement's use of coordination-based, multi-agency tactical operations were pretty much limited to operations like a barricaded gunman, a hostage situation or civil riot. Those were not routine. Rotroff was to encourage the use of ICS by all fire departments and by state agencies like ADES and the state troopers. He was to brainstorm ways these different agencies might train together.

Since there was a similarity between Rotroff's ICS implementation task and Lundfeld's Urban/Wildlands Interface project, those two worked together on several projects. They both attended an ICS Simulator school and reported that the instructors—from Idaho—would be willing to put on that school for the Alaska State Troopers (AST) and ADES. They would talk to the State Commissioner of Public Safety about arranging a school for AST, ADES, fire officers and even city managers substituting scenarios like hazardous materials incidents, search and rescues, city-wide disasters and such. They would try to get ADES to fund it.

EIGHTH:

Ketchikan Fire Chief Gene Fisher, who participated fervently in any executive-level training, would continue to recommend and arrange for training for us particularly in leadership and management. Technical skills and administrative knowledge were also important, and he agreed to try to supply us with the latest programs. He would also try to use—as much as possible—technical experts from other state agencies because: 1) they were free; and 2) the other state agencies would know we were out there and learning our stuff. Plus, it's a good way to meet people that you may wish to phone sometime regarding some issue. Just keep broadening your power base and expanding your Rolodex.

NINTH:

The Alaska Vocational Training Center in Seward was being required to write a Marine Fire Protection curriculum for mariners who were striving for required licenses. Seward's Chief John Gage agreed to monitor that and assist them in making those programs truly valuable in saving lives at sea and reducing fire losses at our docks and harbors.

The Home Sprinkler Coalition, spearheaded by Ralph Crane, Anchorage Fire Department, was trying to convince NFPA to review their standards for home sprinklers, believing that 250 gallons of stored water was more than needed to extinguish an incipient fire in a home. If a series of tests could prove that to be true, and NFPA could modify their standards, rural dwellers might be more receptive to installing smaller units. Ralph was seeking grant money to fund a series of tests to prove that less water would suffice. I told him that we would help him any way we could, and that he could use the Association as a reference in his grant application.

TENTH:

Valdez's Chief Tom McAlister accepted the tenth assignment and agreed to help Crane on behalf of the Association. A couple of years later they acquired a $50,000 grant to perform the tests.

Related to this topic, on his own, Chief Tom Opie—chief of the Barrow Fire Department and chief of the North Slope Borough (comprised of Barrow and seven other villages at the top of Alaska in an area more than half the size of California)—bought one self-contained, electric-powered sprinkler system. The unit was about the size of a large, stand-up freezer. To demonstrate how it worked in his fire station, he ran a video camera and narrated while his deputy chief set some paper on fire in a waste basket. The waste basket was positioned right under a sprinkler head they had rigged up. The head released and dowsed the fire in just a few seconds. It was a real "home-made" type of video that Tom took to his Borough Council's meeting. They gave him $1.5 million to start in-stalling the systems in homes in the borough. The first village to be completed was Atqasuk. The following October, at the AFCA conference, Tom showed another video taken in Atqasuk. The surface-mounted system released in the middle of the night, wetting down the sleeping occupant who did not know that a fire had started in the room.

310

One of the neat things about Tom was, he didn't care about NFPA standards for residential sprinkler system installation. Anything that would work and get the job done was fine with him. He designed water tanks that would fit under beds or on-site designs for using domestic water supplies, and the water tank was kept full by a pump operated by a toilet tank fill valve. "If NFPA doesn't like it, they can kiss my ass. I've got people to protect here." Tom—6'5"—shoulder-length white hair, a Buffalo Bill as tough as the arctic itself, would not be intimidated by urban "technologies." Hell, besides that, once Tom (I know I'm getting off the topic here, but it's my book) refused to allow the State of Alaska EMS Department to decline certifying his EMTs just because they couldn't pass the state's written exam. They could pass all the skills, had the knowledge, and were excellent EMTs, but they didn't have good enough command of the written language to decipher the questions as written in English. So Tom flew to Juneau, stormed into the EMS office and told them to write the questions in Inupiat so his Eskimos could read them. Otherwise, they should be allowed to have someone ask them the questions and be prepared to explain the questions. Juneau sent a certifying officer up there to give the exam verbally. They all passed.

Back to my organizing of the AFCA.

ELEVENTH:

The State Emergency Response Commission was established to assure that departments (communities) were properly prepared to respond to catastrophic hazardous materials releases. I was unable to assign someone from the Association at that time, but shortly afterwards appointed Nikiski's Chief Al Willis to this position.

TWELFTH:

Because of funding problems, State Fire Service Training (FST) had stopped sending people to other towns that had requested an expert in a certain field to come and help them. That was unfortunate because that Technical Assistance Team program was neat. Before there were BSA Explorer Posts (or at least before I knew of them) I wanted to establish a "cadet" program for teenagers to learn firefighting and such in Cordova. I called FST and was told that Sitka's Chief Marty Fredrickson had developed one. FST sent Fredrickson to Cordova and helped me set up ours. When we were done, our kids got high school voc-ed credits for joining the department's cadet program. To do that, I submitted a training curriculum, complete with written exams. The help I got from Fredrickson didn't cost me anything. Anyway, since the FST technical assistance program was no longer funded by the state, Tom Monk agreed to try to manage the Association's Technical Assistance Team program. He had to figure out how to do it for free. With this service, local communities became familiar with the AFCA "resident experts" and our magnanimous dedication.

Example: Skagway is a small town that has only one source of income—tourism. Cruise ships flock to that community every summer to walk around the turn-of-the-century buildings that were built during the great gold-rush era. These old, meticulously maintained wood-frame buildings could have caught fire any time, spread uncontained throughout the downtown district and within a year, the town would have disappeared off the map. Dick Groff traveled to Skagway and explained the state's Sprinkler Incentive Act which the AFCA was instrumental in having enacted. To make a long story short, Skagway took the idea and modified it a bit, and had all of those historic buildings sprinklered. At a town meeting, the city offered the building owners a neat deal. The city said to the building owners, "These buildings are your property and if you sprinkler them, your insurance premiums will drop so much, you will recoup the cost in just a few years. But, since the city's only income is from the tourists that come to see these buildings, we will pay half the cost of sprinklering them." The entire community agreed to the idea and it was done. All Skagway needed was an idea from the "resident experts" of the state.

* * * * * * * * * *

Anyway, those twelve assignments touched just about everyone who was anyone of influence in the state: governor, legislators, state departments and divisions, associations comprised of local mayors and city managers, the state-wide media, private businesses such as architects and engineers, and about everyone involved in developing and enforcing codes and standards.

Like a lot of people, I have read books on management principles. I read The One Minute Manager, too. What a pile of dung. The book opens up with a description of this One-Minute Manager, standing in his office looking out the window. He has nothing to do. He managed his staff so well they get everything done. Wow! The principles of his management style are Zen-like in their simplicity and as profound as a religious revelation. Total hype. Even with the most competent staff in the world, managing an organization requires work and unrelenting focus. It begins with this: Paint a verbal picture of what the end product will look like. Get an agreement on that. Meet individually with each staff member and after a discussion, jointly write down a job description. So now, you think, you go on auto-pilot and don't involve yourself in it much anymore. Wrong. You call every week, about the same time, and talk to your staff member. Get written reports if the project warrants it and send copies to everyone else in the Association that should see it, and press, press, press.

But back to Sitka: My first attempt at creating this illusion of power was the last day of that conference. Governor Sheffield was scheduled to come to Sitka and address the fire service. Since I had to leave town early, I asked Al if he would continue the role of AFCA president until the governor left. I spoke to the ASFA board and the Sitka chief. I asked if we could move out of the huge auditorium and into a smaller room in the building, and get all the conference attendees and their spouses, the local firefighters and their spouses (or as many as possible), and everyone we could find to pack the room. I wanted a jolly, convention atmosphere, and the governor to have to squeeze through a throng of people to get to the front of the room. He showed up and the effect was terrific. But if you put 200 people in a room that hold 1000, it's not very impressive.

As an aside, I knew Governor Bill Sheffield from the time we almost killed him in Cordova. You wanna hear about it? Of course you do.

Governor Bill Sheffield, a repeat visitor of our annual Iceworm festival in Cordova, was feeling particularly spry—as we all tend to feel when the winter's day is high and clear and crisp—and was almost giddy as he made city officials aware that he would look best in the parade riding on the tailboard of our retired but operational '42 Ford convertible fire engine.

For several days department members were also feeling spry, and yes—giddy—for our own reasons. We tend to get that way whenever we feel something closely akin to a practical joke is in the mix. At that time, there were several large empty lots next to the fire station in which we dug a large, shallow pit and filled with water. We had the routine planned flawlessly to protect against delays and dead time, which could present opportunities for the squeamish to torpedo our plans to provide the governor with a memorable visit. "Pit fires" are real confidence builders for firefighters. They look horrendously dangerous—silver suited and hooded humans marching into a roaring ball of orange fire one to two stories in height, 30 feet in diameter, disappearing from view into that hellish sphere—but it isn't. But since they look so hellish, we hoped that neither Governor Sheffield or any member of his entourage had ever seen one before. Otherwise, it would be very difficult to get him into one.

JoAnne Havens would drive the old engine for the governor. Yes, we had forgiven her for her for her previous driving problems. And she had become expert at "popping the clutch" on a still-coasting vehicle whose engine had stalled. A valuable skill with the old engine. Timing, however, was a problem. She thought that since she had stopped the engine in front of the grandstand and the governor had dismounted the tailboard, that his intention was to go to the grandstand and watch the rest of the parade. That was not his intention. We were all surprised to see him turn back and grab the handrails and step back up on the tailboard. We wish that JoAnne had also seen it as she "popped the clutch" to restart the stalled engine on the slightly coasting vehicle. The onlookers were stunned, the governor's entourage was horrified, and the firefighters were hysterical with laughter as the governor was dragged a few feet behind the engine before the spry old bastard showed what he was really made of by refusing to let loose, and instead gave a mighty yank, bounced on one foot and bounded back onto the tailboard. This was met with unanimous applause.

As the governor rode on down the street, I stopped and spoke to his aid. I informed her that he will be driven directly to the fire station after the parade…for a surprise. "What kind of surprise?" She asked. "A big surprise," I answered. She looked toward the state trooper (his body guard), who only looked back and shrugged. I liked that guy.

We'd already had the pit filled with water, the apparatus positioned and the hoses laid out and charged. As I helped the governor into his silver proximity suit and hood (leaving the gold-colored visor up) the crew was emptying 55-gallon drums of gasoline into the pit. "Is this safe?" the aid asked with apprehension. "Of course," I replied. The TV news cameraman was almost salivating and was pacing, his eyes were glazing over…it was almost sexual. The trooper stood stoically at attention. However, his right eyebrow did raise a bit. I knew we'd better get this started before his expression changed more dramatically. One glitch. In the proximity suits, we all looked alike. I dashed into the station and immerged with a spray can of black paint, spun the governor around and painted a large "G" on the back of his coat….for the cameraman.

"I don't know….." the aid was about to say as the lit flare soared overhead toward the pit. The gas fumes exploded into a fire ball whose concussion nudged me forward. "Jesus!" I muttered, hoping immediately that no one heard. The aid did not hear that, she was plastered against the wall of the fire station. The governor was completely mellow and calm as he slammed his visor down and took his place on the right side hose. The seven us marched into the flames, sweeping them away from us as we advanced. Of course we had the entire department on back-up lines, walking the perimeter of the pit.

The fire was snuffed and we backed out. The cameraman was practically cart-wheeling, the aid was digging for her nitro tablets, and the trooper was shaking his head almost imperceptibly. Back in front of the camera, visors up, we spoke to the lens and made stupid newsworthy remarks about fires—or something. Camera off, hoods removed, the governor combing his hair, he looked oddly fulfilled yet oddly pale. Well, it was hot in that fire. His pallor never registered with me until I learned shortly after that his ashen color was accompanied soon after by chest pains which prompted a direct flight to Seattle and immediate by-pass surgery.

Governor Bill Sheffield is a cool guy. You know, a guy can have an M.I. watching TV or conducting a meeting: Often dying from them. Had he keeled over in the pit and expired, I suppose I'd still be hearing about how I killed this guy. The truth is, he had a ball, and so did we. There are worse ways of getting chest pains.

314

Also, before the conference was over, I explained that we needed to invest some of our money—limited as it was—into image building. We need stationery, like real executive officers. The membership went along with this idea. Of course, at the time I didn't know how much this would cost us. It was several weeks later when Gene Fisher and I were both in Anchorage a short time and went to a stationery shop. It was a nice place with lounging chairs and elegant decorations. I knew we were in the right place. When the woman came out to wait on us and asked us what we needed, we were completely honest with her. Gene told her that we wanted stationery that was dignified and subtly powerful. "And envelopes to match," I added. She paced and talked, mostly to herself, and asked us some questions. She concluded that heavier than normal paper was needed. Gray was a nice executive color, and for power we needed royal blue and gold. The AFCA emblem was embossed with our name in royal blue, and centered were the five crossed trumpets in gold. My mailing address was below that so that when a new president was elected, the address was all that needed changing.

Next, I explained to the members, that whenever the President of the United States spoke, his podium sported his symbol. In fact, many viable organizations had podium emblems for the dais. So I had one made in the same colors as the stationery. It was really bland. Then I had a long banner made, to be hung on the wall of wherever we were meeting. It looked like shit, but nobody said anything. Battleship gray is not very eye-catching. Deputy State Fire Marshal Gordon Brunton from Juneau wanted to have lapel pins made also. They looked better with only blue and gold with no gray. We sold those to members, so that paid for itself. I had business cards printed up for several of the staff positions. What little money we had left, we put into a Merrill Lynch checking account.

With the objective in mind and the organization arranged to meet it, the game was on. To this day, I am surprised at how quickly the AFCA became a power to be reckoned with. Nothing that affected Alaska's fire service could occur within the state that we weren't immediately aware of and prepared to deal with.

MANAGEMENT PRINCIPLES

Following are some other important management principles to keep in mind:

Never believe your own baloney. We human beings are so talented at crafting illusions, the craftsman himself begins to believe his own fiction.

At one point, much later, I heard of an influential man in Juneau who was going to oppose a project of ours and I flared up, paced circles around a fire engine, planning an attack on him. I'm gonna do this, and I'm gonna do that! I stopped dead in my tracks and remembered, I can't do anything. Our power is an illusion. I had fallen prey to my own illusion. It's easy to do, and that's the point. If you grab someone by the front of his shirt, you better have the meat to back it up. I would have done the Alaska fire service a terrible disservice by launching an attack against this person. I reiterate: Never fall for your own bullshit. Politics shows no quarter.

For example, take the saga of the Anchorage unions vs. the mayor. Publicly opposing a candidate who later wins an election validates to the candidate, and everyone else, that we do not have the size voting block to make a difference to them. Years later, the city of Anchorage witnessed this gross misjudgment perpetrated by unions.

After several long years under Mayor Fink, the Anchorage Fire Department—which had experienced a hiring freeze, and no contract, and bitter fighting, especially between the union steward, Joe Albreich and Mayor Fink—the firefighter's union had had about enough of the conservative administration. The police department union, although less confrontational, agreed. Of course, the International Association of Electrical Workers (IBEW) disliked the small-framed, hard, forceful Fink immensely. Toward the end of Fink's last term, conservative Rick Mystrom would run against the more liberal assembly member Mark Begich. It was obvious to Mystrom, as it was to everyone in the state, where the lines were drawn, but he met with representatives of each of the three unions and asked them to not actively oppose him. Of course people could vote for whomever they wanted, but please, he asked, just spotlight the candidate they supported without attacking the one they opposed.

The International Association of Fire Fighters (IAFF) and the police union backed off a hair but let the entire 300,000 residents of Anchorage know in every way they could, that they supported Begich. The IBEW openly opposed Mystrom and contributed heavily to Begich. These unions, as well as others indirectly affected by a mayoral election were certain that they had not only a large enough voting block, but enough influence on the residents that Mystrom would go down in flames. Begich got stomped. So what do you suppose happened next?

An old associate of Mystrom's, previous Anchorage City Manager Ron Garzini, who was in Cordova, told me that after the election, Mystrom visited each union rep again with the old "well, it's over and I'm the boss now, I hope we can work as a team" speech and got comments back from the police rep that was not very team-spirited. "Watch," the old friend told me, "Rick will have to do something unpleasant to remind the cops that he's the boss." Before the week was over, I saw on the TV news Police Chief O'Leary in front of a microphone about to make a statement. Mystrom stood behind him, hands folded in front, never looking at the camera or making any statement. Mystrom made his statement by standing where he did. The chief stating that, unfortunately, in an attempt trim back the budget, the police officers who had been working four 10-hour days a week—which they loved—would be going back to five 8-hour days to achieve these savings. This change dealt the police officers a tremendous blow. That work schedule remained in effect for a couple of years I think. Before long, the cops became more "team" oriented. As a result of that, recruitment efforts increased, academy classes were run twice a year, and new equipment was falling out of the sky. The IAFF still hadn't learned. The hiring freeze continued, which is why my son (who had tested and made the hiring list twice) could not get on the department. So he joined the police department.

Organizations like ours feel compelled to involve themselves in local and statewide politics. It's our responsibility to fight for the safety of our citizens by whichever means are available. But my advice is to pick your fights carefully. The Anchorage unions had their bluffs "called." Actually, they didn't realize they were bluffing. They honestly believed they had the voting and influential power to swing the vote. Yet they publicly displayed their lack of muscle. They were no longer a force to be reckoned with. In other words, politicians could now ignore their demands without trepidation. They have no power and everyone knows it. Remember to bluff your adversaries without deceiving yourself, for you may be called to "put up or shut up." Don't be revealed as a blow-hard with no real substance, or it's all over.

Hell, while I'm telling this story, listen to how convoluted this next problem became. There were several fire chiefs during these times of trouble between the city administration and the union. One of the chiefs sadly became so ineffectively stuck in the middle, he could do nothing for either side. He had run out of ideas. It got so bad that the Anchorage city manager would hold meetings with the union president and discuss fire department business. The only reason the chief knew what was discussed was because the union president would go to his office and report on the meetings. First, you're out of ideas; next, you're out of the loop.

Taking advantage of our new prominence in the state political arena, it was time to try out our influence. We were to do this at our conference in Kodiak. Kodiak was already a high point for me because I shared the banquet dais with the candidate for the U.S. Senate seat held by Frank Murkowski. Glen Olds had been in politics for years, and in mannerisms and intellect, reminded me of Adlai Stevenson. He was gracious and soft spoken, yet so intellectual, people had a hard time following what he was saying in his speeches. Under President Kennedy, he had been Ambassador to the U.N.'s "World University" project, a university president, and an adjunct professor of philosophy. Since he was the keynote speaker, I spoke to the attendees first. In my speech, I compared what we in the fire service do—politically—with what author and paleontologist Loren Eiseley called "the spectral wars": I was quoting Loren Eiseley after explaining that he was "one of my favorite dead people."

When Olds got up to speak, he said how stunned he was that anyone would quote his old friend Loren Eiseley. He revealed that years ago, he—Glen—and his father used to go with Loren to his cabin in the mountains on long fishing trips.

Glen and the rest of us at the dais were really pleased when each of us paid $10 to buy the last $100 ticket for the $10,000 raffle and won $1,000 each. As an aside, I wrote ten checks on our Merrill Lynch account that Saturday night and discovered the following Tuesday that all the checks bounced, because Merrill Lynch did not transfer our money from one account to another in a timely fashion.

Anyway, the point of this story is not about candidates for U.S. Senate, but about the gubernatorial race. The Republican gubernatorial candidate, Arliss Sturgulewski, accepted our invitation to attend the conference and explain why the fire service should support her. The Democratic candidate, Steve Cowper (pronounced Cooper) did not attend.

So a bunch of us met for breakfast Sunday morning before we had to catch our flights out of there and discussed this matter. State Fire Service Training's Tom Take said he had a friend in the Cowper "camp"—an Anchorage firefighter named Jim Sellers. I suggested that Tom call Sellers and tell him how disappointed the fire service was that Cowper didn't care to discuss our concerns with us. I instructed, "Imply that we are sitting on 5,000 votes" (that's how many firefighters there are in Alaska). Gaylen Brevick, president of the Alaska State Firefighters Association (ASFA) was there, and I asked if it was okay for me to speak on behalf of them as well. He said okay, then we discussed what I would say if Cowper contacted me.

It was 7:30 p.m. the Tuesday night before the elections when Cowper called my house. I discussed several of our concerns with him, but my emphasis was his appointment of the Commissioner of Public Safety. The State Department of Public Safety is the agency for state troopers, fish and wildlife protection, Department of Motor Vehicles, which comprise the biggest divisions in it and they are all law enforcement functions. Also in the Department of Public Safety is the Division of Fire and Life Safety and, within that, State Fire Service Training. That division employed 18 people. The head of that division—the State Fire Marshal (appointed by the Commissioner of Public Safety)—supervises his deputy fire marshals who conduct building plan reviews, do building inspections around the state, and investigate suspicious fires. The State Fire Marshal also appoints the supervisor of Fire Service Training. There are only three people employed to run that activity.

Since the Division of Fire and Life Safety is so small, its critical job is often underrated by the commissioner who is traditionally an ex-state or local cop.

318

Consequently, Fire Protection always becomes the red-headed step child of Public Safety. I explained to Cowper that much of what is accomplished by this division was done only because members of local fire departments volunteered to help. Things like writing training programs, petitioning for legislative grants, teaching classes for free or nearly free. We wanted the division to get more respect from the commissioner and, in fact, could see no reason why a commissioner should not be from a fire service background instead of law enforcement background. Why not? I reminded him that Anchorage—Alaska's largest city—had a Director of Public Safety who supervised the police department, the fire department, and the local civil defense. Anchorage Mayor Tony Knowles appointed John Franklin to that position. Franklin had been Anchorage's fire chief for many years before his new appointment. It was common knowledge that Franklin was doing an excellent job.

Cowper explained to me that he had no one in mind yet for this appointment and could not guarantee that his appointment would have a fire service background, but would guarantee that whomever it was, he/she would give that division more support than previously. In fact, before making an appointment, he would check with us for our input. He also said our organization could have a representative on his transition team immediately after the election. As soon as we hung up, I called to Brevick and the Chiefs board of directors to start spreading the word throughout the fire service.

After he won the election, I appointed Charlie Lundfeld to the governor's transition team and he had a ball intimidating the bureaucrats who ran the obscenely huge departments in the state. What was most fun was that all these other agencies and politicos in Juneau wondered why the hell the AFCA had someone in—what turned out to be—such an influential position. The illusion was growing daily.

The appointment of the commissioner went as promised. First, I called Pat Wellington, in charge of security for the Alyeska Pipeline Company. Pat had been police chief in the City of Juneau back when Bill Bagron worked there. They both worked as guards for Governor Hickel, and later, Pat was appointed as Commissioner of Public Safety. All that time, Pat was a member of the Douglas (neighboring town to Juneau) Volunteer Fire Department and was active in politics for the fire service. When Pat answered the phone, I told him that Governor Cowper was going to let us help him in selecting a new commissioner. "Great," he said. "Would you like to be commissioner again?" I asked. He just laughed and asked me if I knew how much money he was making now. He had a full retirement from the state and was working for a very sophisticated corporate organization, in charge of security for the hundreds of miles of the Transalaska pipeline. But, he added, he would help locate someone.

A few days later, he called me to tell me some of the people being considered by the governor and the transition team, then added, "I know someone who would be interested that I've known for years. Art English owns a private security company, would be excellent, and would work closely with you." I replied, "If you think he's the right guy, that's good enough for me." Wellington called the governor's office while I called Charlie, and Art was appointed in January. He and I spoke numerous times on the phone by the time several of the board members and I had a chance to meet him in April at a board meeting. He flew from Juneau to Anchorage to have lunch with us.

But this was not just a one-way street. Since he did everything in his power to support Fire Prevention and Fire Training, we threw our influence behind him in rallying support for his entire department—particularly the funding for Search and Rescue, which the state troopers are statutorily responsible for. When funding for Search and Rescue was threatened, we sent hundreds of letters and made scores of phone calls to the legislators to get the money reinstated. When the same under-funding threatened the Trooper Academy in Sitka, we did the same. This alliance between the fire service and state public safety worked for our mutual benefit and resulted in a tremendous rapport between the two services. What was really funny was how baffled state legislators were when the local fire services came down so hard on them whenever they threatened the funding of the state troopers.

Not as a political ploy, but just what a professional should do, is to realize that we are all in the same business. When I started looking outside of my own department, I started noticing how neat, hard-working, self-sacrificing and courageous some of these troopers were. And it was about this time that when I saw a highly skilled trooper we worked with on a Search and Rescue operation bust his ass and take some real risks. I wrote a glowing letter to the commissioner saying so. While firefighters mostly work in spectacular operations in front of the public and the media, state troopers usually work alone and no one sees when they do something truly remarkable. Now, I don't just make things up, but I sure as hell will let a commissioner know when I see something honestly commendable. It wouldn't kill you to look outside your own service from time to time. You might be pleasantly surprised to learn there are some pretty neat people who are not in the fire service. Then say so. Caution: If you do it too often, the letters lose their value, and you look like someone desperate for attention.

* * * * * * * * * *

Governor Cowper and I weren't buddies, or anything. We just had occasions where we talked about things.

December 7, 1988, a devastating earthquake struck the Soviet Republic of Armenia, burying thousands of people in the rubble. I watched the news cover-

320

age that Wednesday evening. Thursday morning, in the shower before work, it occurred to me that Alaska could send some search dogs and handlers over there to help locate trapped victims.

When I got to work at 7:00 a.m., I called the governor's office. He wasn't working yet, so I told the woman who answered the phone that he might consider offering the services of the Juneau-based SEADOGS (Southeast Dogs Organized for Ground Search). I had used their services here in Cordova a few times and knew their veteran handlers, Bruce Bowler and Jeff Newkirk (a Juneau firefighter). I explained to the governor's aid that it might be a good way to pay the Soviets back for a couple of favors they did for us earlier that year: They had successfully searched for seven missing walrus hunters from Gamble who were adrift on an ice flow, and also had sent an ice breaker to free two trapped gray whales in the ice pack near Barrow.

A few hours later, Cowper called me to say he had first contacted the U.S. State Department who told him that there were plenty of search dogs already in Armenia and that the SEADOGS would not be needed. "After I hung up, I said 'screw it', and called the Soviet Foreign Minister, Edward Shevardnadza, who said they could use the dogs. I called Bruce Bowler who said they would be glad to go, and that 'We owe 'em.' Then I wrote to Soviet Ambassador, Yuri Dubinin, to make the offer official." He continued to explain that the U.S. State Department would give him a bunch of shit over this, but he didn't feel he needed any permission from the federal government. He was just going to do it. Cool.

The same day, Bowler wrote me a quick letter saying, "Thanks for the plug with the governor. We've been training for this type of disaster for 12 years, and we appreciate your vote of confidence more than we can say. See you in Leninaken."

That was on a Thursday. They couldn't leave right away because the state department declined to give the searchers its official sanction and the searchers had to obtain expedited passports to make the trip. Cowper explained to the press that he didn't care about the state department, and "We're so accustomed to dealing with the Soviets, that that's what we did."

On Monday, Governor Cowper saw the searchers off at the Juneau airport where they boarded an Alaska Airlines flight to Seattle. There, Bruce, Jeff, and Reilly Richey (an elementary school teacher) and their dogs, boarded a chartered Flying Tigers 747 cargo plane with doctors and supplies from the Pacific Northwest. The group went through New York, then Frankfurt, West Germany, and arrived in Yerevan, Armenia, on Wednesday. It was the first 747 to ever land in Armenia.

There were already more than 200 search dogs at the scene, but they were worn out and fresh dogs were needed. In the week since the earthquake ripped through the region, toppling hundreds of buildings, more than 18,000 people had been pulled from the rubble. By the time SEADOGS arrived, there was little hope of finding survivors, but recovering the dead was also important. It was estimated at that time that about 60,000 people had been killed in what was one of the 10 worst quakes of the century in terms of lives lost.

Living in a tent in sub-zero temperatures, they would search one building in the mornings and another building in the afternoon (usually getting to bed around 2 to 4 in the morning). They located about 100 bodies in the time they were there.

To leave Armenia, an ex-arms dealer put them on his personal Boeing 707 after he had several silk couches removed and replaced them with airliner seats. They landed in Ireland for fuel and were bussed to a pub and supplied with drinks, even though they declined to get up and sing as requested by the patrons. They refueled again in Canada for the last leg to Miami. Before leaving the 707, the 18-year-old son of the plane's owner invited Jeff to his wedding in Hollywood. Jeff regrets having declined. Anyway, Miami security personnel walked them aboard a commercial airliner for the flight to Seattle; no tickets, no check-in, nothing. In Seattle, Alaska Airlines donated the flight for them to Juneau.

They arrived Christmas Eve morning capping a 12-day, 29,000 mile trip. Governor Cowper met them at the airport when they returned.

* * * * * * * * * *

Every year the fire service needed to launch a statewide campaign rallying support for the State Fire Marshal's office and Fire Service Training budgets against funding cuts proposed by the legislature. Gaylen Brevick, President of the Alaska State Firefighter's Association, believed the solution would be a heretofore untapped revenue source.

Because of a proposed cut in funding of FST from $436,000 to $188,000, January '87, Gaylen presented the results of his research into the franchise tax that the State of Alaska charged insurance companies to operate in Alaska. The state used to charge 3 percent, but to encourage companies to operate in Alaska, the state reduced to tax to 2.7 percent. At that time, this 2.7 percent generated $23 million. Gaylen discovered that most other states still charged 3 percent, and should Alaska increase its tax back to 3 percent, an additional $2.5 million would be gained. After Galen met with the governor, the governor was impressed that the fire service thought of another funding source, but the legislature would have to buy into the idea. Legislators Duncan and Hudson suggested Gaylen talk to Ron Larson, chairman of the Public Safety Finance committee, because of Larson's political power and finesse, and he was in the majority. Gaylen should convince him to submit legislation.

322

Larson considered doing that but thought he should get an opinion from Judiciary regarding the legality of "dedicating tax revenues to a particular department." Of course, logic has nothing to do with anything, but we believed money generated from insurance companies should be considered "user fees" since all codes enforced by the fire marshals and professional training provided by FST reduces the fire losses the insurance companies must pay off. We considered it reimbursement for services rendered just like plan reviews, inspections, fire cause and arson investigations, and training to suppress fires that keep their profits up.

By February, everyone was getting into the act: The state's FST training committee members supported the idea and they would inform their legislators. Senate President Jan Faiks complained about the complications in raising the tax. The deputy commissioner of public safety and the state fire marshal were testifying at a joint house and senate subcommittee. Individual legislators were contacted: Peter Goll liked the idea and was ready to go for it. After the governor's special assistant on public safety reviewed and liked the idea, the governor sent the idea to his bill writers. Pat Wellington, past Commissioner of Public Safety and friend of the fire service told us to assure that the bill states the legislative intent is to fund fire prevention and training. Gaylen thought that after the bill was introduced, it could be amended to include the wording of intent. Peter Goll told Gaylen to talk to Department of Commerce's Commissioner Smith, since Commerce collects that tax money. Smith agreed to look into it. John George, Juneau volunteer firefighter and insurance executive would look into appropriate language like "user fee" or "program receipts." Smith should be prepared to forward the extra .3 percent over to State Fire Marshal's office. I would contact Public Safety Commissioner Art English who would contact Commissioner Smith about the procedures for doing that.

Calls were coming in about support for the House Bill 230/Senate Bill 224. The third week of April, the bills were to be on the floor of both houses, and the attorney general said that the wording of intent was legal. Everything was rocking. Faxes were rolling in of copies of letters to Commissioner Samson of Department of Labor looking forward to some consolidating of inspection activities between DOL and DPS with the new funding source. Then in mid-May a message from Gaylen saying HB 230 got hung up in Tim Kelly's Labor and Commerce Committee, because Kelly said it was 'very complicated" . Even though the bill was not dead, it would have to wait until the next January. Years later, Tim Kelly's allegiance to the insurance industry in the state became clear and explained much. I called Art English with this latest development.

By the time next January rolled around, it was clear that the bill would never get past Tim Kelly and several others opposed to irritating the insurance industry. Yep, the politically influential insurance industry had been long aware of our proposal to increase the state tax on their premiums. And even though .3 percent doesn't sound like much, it's still $2.5 million out of their pockets. Yet all they really needed to do was to influence a few legislators who were in key positions. All the legislators needed were a few (couple) of defensible clichés or statements to stand behind like "No additional taxes," or "Clever wording aside, it is against the state constitution to 'earmark' money" or "Without tax incentives, many insurance companies would have no impetus for doing business in a state with such a small population…there's no money in it for them."

By this time, the AFCA had so many people in Juneau sitting in key spots in and around the legislature, I knew every time someone sneezed. As imperceptible as they tried to be, they were casting detectable shadows. The issue aside, it was so cool pretending I was Vito Corleone or a spy or something. I loved getting phone calls telling me who was talking to whom and who that person may be representing. I had no idea this would be so much fun. But ultimately, Gaylen and I had to talk about whether to recommend to our Boards that we drop the idea. This whole thing was Gaylen's baby and if the ASFA wanted to pursue it, and desperately needed the AFCA to go along, I would have had a hard time pulling out, but I explained to him that the Chief's Association had created only an illusion of power. We still had no money, and very few votes. The only reason we had political influence was because people believed we had influence. I said that, of course we could try to exert our influence, but if we lose in such a visible way, it's as though our bluff had been called and it would become obvious that that's all we are…"bluff."

I told Gaylen that the Chiefs Association must choose its battles carefully. We cannot afford to lose. We still looked at our options. Perhaps we could drag the media into this. They might take the side of the poor, beleaguered fire service against the big, indifferent, greedy, cigar-chomping insurance industry. Nah. One interview with a legislator explaining that with no tax incentive, no insurance companies would operate in Alaska, and the typical resident would say, "Fuck the beleaguered fire service." We even considered letting the insurance industry buy us off. For a nice annual tax deductible donation to the fire service (which would make its way to State Fire Training programs), we would drop our push to increase the franchise tax. But we figured that they knew they had already won. Why donate anything? We agonized over the question for several days afterwards before we agreed not to try to save the bill and informed all of our friends in the legislature not to agonize over it and they could use their capitulation on this as chit for later issues.

Besides the obvious principle of choosing your political battles carefully, and knowing when to pull back, in retrospect, I think we could have gotten something out of this. The news media in this state is slightly to the left of center. They love to side with David against Goliath. We really should have gone one step

324

further and sought grants or contributions from the insurance companies after informing them that the media would champion the fire service....especially the volunteer fire service. We could have informed the companies that the media would be asking pro-insurance company legislators to explain their positions, which legislators don't especially like to do. After all, if insurance companies contribute generously to political campaigns of "friends," they could contribute to us in exchange for us shutting the hell up. All we really needed was $250,000 to off set the proposed cuts to the State Fire Service Training. That's a lot cheaper than the $2.5 million threatened tax increase. Yep, we should have stayed the course just a little longer.

* * * * * * * * * *

At the same time the franchise tax issue was boiling, so was this other mess:

In mid-January 1987, Fire Marshal Sam Neal called me and said that the State Department of Labor was planning on taking over the Fire Marshal's office. Consolidating or shifting functions from one state department to another was not uncommon in the dynamics of experimenting with cost savings or simplifying state functions. But this was odd. The last move of fire service personnel was to move Fire Service Training out of the Department of Education, where it was born as an experiment, to the Department of Public Safety's Division of Fire Prevention (the Fire Marshal's office). But Sam said this was different; this did not originate within the state agencies or governor's office. The state Department of Labor was reacting to recommendation of the International Conference of Building Officials' (ICBO) Anchorage chapter and Juneau chapter.

The state's Department of Labor is a huge agency with over 500 persons in it. State Public Safety's Division of Fire Prevention had 18 people in it. Those 18 people not only included the fire marshals (investigators, inspectors, and plan reviewers), it also included the three Fire Service Training people and all the clerical staff. Fire Service Training would have been left behind in Public Safety (probably in some little corner somewhere...dying a slow death). Of the paltry FST budget, a substantial portion of it came from plan review fees charged by the marshals. That would be lost. After further analysis, we believed ICBO members, made up of a substantial number of architects, engineers, and contractors would appreciate reducing the statutory authority of fire marshals to give the builders more liberal ears to their complaints of code enforcement being too stringent. So, we worked from the premise that it was the building trades trying to make their lives easier and businesses more profitable by emasculating the fire marshal's office. And since the fire service had such a tiny state contingency—which we had always tried to protect—we were not about to let it be destroyed so that building designers and contractors could fill their pockets, while the lives of our citizens become a little less safe. In a nutshell, if DOL absorbs the inspectors/plan reviewers, arson investigation could go to the troopers, fire training could be considered a local responsibility (or be absorbed by the trooper training academy in Sitka) and the state fire protection focus disappears.

What was ironic was this "consolidation" of plan review functions of the fire marshal's office and the Department of Labor which must also approve building plans, originated with us. To simplify and speed up the plan approval process for architects and engineers—so that contractors could start building sooner, in this land of short building seasons—we had thought that a one-stop shopping system of plan reviews would be a great idea. But to give the DOL plan reviewers/inspectors more authority, we'd always thought that they should be incorporated into the fire marshal's office.

Two days later we heard that the Juneau Chapter ICBO had a bill drafted for this. Within a week we had planned to ask the AML for supporting consolidation under the Fire Marshal's office and approached the Commissioner of Public Safety to discuss the full scope of DPS responsibilities to include public safety for occupants in buildings.

I filled out membership applications for the AFCA to join all three chapters of ICBO in the state. Judson agreed to represent us in the Juneau Chapter, Tom McAlister, Valdez, said he would represent us in the Anchorage chapter, and Charlie Lundfelt in the Fairbanks chapter. The objective was that we needed to be informed of all conversations and actions taking place in these chapters so we would not be surprised. Also, the three chiefs would try to convince the ICBO that consolidation should be within the fire marshal's office. We no more than got into the chapters when the president of the Anchorage chapter, Ron Watts, sent a copy of the original bill to Senate leader Jan Faiks. Some chiefs surmised that since the state never had a "building official," Ron Watts—currently the building official for the City of Anchorage—wanted the state to have one, and wanted to be it.

The politicking, cajoling, and maneuvering went on daily. Tom McAlister wrote numerous papers regarding consolidation under the fire marshal and I sent copies of those to Senator Jan Faiks, and I called numerous times. But she had a legislative council write the rough draft (SB 300) keeping consolidation under DOL. In addition to that, the writer recommended the State Fire Council be placed in DOL also. The bill had 31 pages of enclosures. I wasn't sure if ICBO's Ron Watts was behind all of this, if Jan Faiks misinterpreted what we were saying, if—as some people whispered—she personally disliked Sam Neal because she was being influenced by a friend who worked in Sam's office and disliked him. Maybe she honestly thought it would be less disruptive moving the smaller division into a larger department, or maybe—it occurred to me—it was a late maneuver by the previous Public Safety Commissioner Nix to get even with me

326

for writing a blistering letter to him which I sent copies to the governor and the entire legislature because he had pissed me off so much.

By now it was May and we knew the bill couldn't make it through this session which gave us time to plan. One plan was that we form a "non-partisan" committee to review Faiks' bill: Two people from AFCA, two people from ASFA, one person from DOL, one person from fire marshal's office, and one person from ICBO. Bill Hagevig (the guy who initially created Fire Service Training in Alaska) had an idea that would not require a bill. He recommended that a central location for building service functions be established for the convenience of the public. This office would contain inspectors/plan reviewers for each agency and they could review incoming building plans like an assembly line, approving them in record time. The office manager position could rotate between the senior reps of each agency. This approach, Bill said, could be done by the governor with no muss or fuss.

We were still dinking around with these two approaches by mid-May when the troops were getting really edgy. I was getting calls from chiefs from around the state. Judson said he would take the problem to the governor and inform him that we were organizing a review committee for SB 300. I thought we should tell Faiks to prevent bad feelings, and Judson agreed to tell her what we were up to. In the meantime, we would draft a formal recommendation based on Hagevig's idea so we could submit it immediately to the governor if the committee thing didn't work. A week later, the committee sent me a copy of SB 300. It was no misinterpretation by Faiks; there was a real conspiracy to dissolve the fire marshal's office. I called Art English and updated him. I told him one of the major problems with SB 300 was that if the franchise tax passed, that extra $2.5 million would go to DOL.

It was the first week of June when I read an article in Chief Fire Executive Magazine how code enforcement was consolidated under the fire department in Mt. Prospect, Illinois (a suburb of Chicago). It was accomplished under Fire Chief Larry Paritz (he was the incident commander when a jumbo jet crashed at Chicago's O'Hare airport) and was a speaker at a conference in Juneau about six years before. We had chatted a bit in Juneau and I gave him a volume of my poetry. Anyway, I called around and finally contacted him in Hallondale, Florida. I told him about SB 300 and the committee to defeat SB 300. He said he would help by providing information about the advantages of consolidating under the fire service. I then called Al and told him to expect a call from Paritz about the procedures for consolidating.

When I got a copy of Juneau's ICBO newsletter supporting the franchise tax, but supporting consolidation under DOL, I sent a copy of that to Lundfelt who would try to get the Fairbanks ICBO to break ranks and support consolidation under the fire marshal. By mid July, shit was falling apart. The Wizard of

Oz was running out of levers to pull. However, a couple of days later, I got a phone call from the City of Juneau building official informing me that things were abuzz down there because the Anchorage Chapter ICBO was rewriting SB 300 to put consolidation under the fire marshal. I immediately called Valdez Chief McAlister praising him for whatever he had done in that chapter. But the word about this change was out and things were happening quickly. Tom assured me he would sit on Ron Watts' committee and monitor the rewriting.

Finally, in desperation, I did something I would never do again. I got a supporter from the House of Representatives to submit a bill on consolidation. I gave him what amounted to SB 300 with only one difference: We transposed DOL with DPS. It was an identical bill except it put DOL inspectors and plan reviewers into the Fire Marshal's office. When this bill head-butted with Faik's SB 300, the legislators got really pissed off at everyone involved and threw them both out. It worked, but it left a legacy of bitterness, and our "friend" in the House was angry about being used and made a fool of. That was a tough thing to overcome, over time, but it was a desperate situation. We did save the State Fire Marshal's office, but I feel like a real shit for having done it that way.

AUTHOR ADDRESSING THE ALASKA STATE FIREFIGHTERS
AND FIRE CHIEFS ASSOCIATIONS

Photo: Leigh Gallagher

Who We Are

The world is constantly changing. Hemingway died and was replaced by dweebs. Steinbeck is patronized by men of no substance who run their lives by bluster and bluff. Profound thoughts are replaced by trite quips. The old jazz geniuses have been replaced by techno back-ups and their solos are technically perfect but have no soul. The vision is that all problems can be solved by a system or a process. That's an illusion of the collective mind.

Here are things that never change: The people in your city have placed a trust in you. You defend them by whatever means you have; whether it's on the streets, the back country, the chambers of your city council, or the legislative halls of your capital.

This country's volunteer fire/rescue service has been doing its thing for 275 years. Our toys have improved, but nothing can improve upon who we are.

You know, you can follow mankind's bloody footprints from the caves to the settling dust of the Twin Towers, and not find much that speaks well of human nature. Yet you can visit a fire station and note the character of those who will not stoically watch the suffering of others, will not turn a deaf ear to their needs. And even though the media may electronically inject into our brains the sights of Ground Zero, one can look inside any fire station and see 275 years of a steady, increasing testimony of what's good in man.

If people cannot depend fully on their fire service, then nothing is sacred. Remember, if a tragic event puts your neighbor up against the wall—go get him. That's who you are.

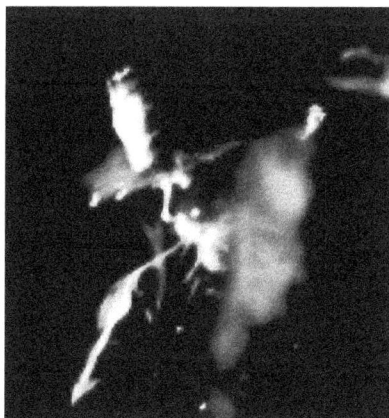

Calling it a day.

Photo of author by Robert Varnam

329

ORDER FORM

   I would like to order my own or another copy of the book *Fire and Ice* by Chief Dewey G. Whetsell. Please send me:

books x $19.95 per copy =   _____

+ Postage (first class) & Handling @ $5.50 book: _____

TOTAL ENCLOSED $ _____

We accept cash, check, or money order made out to Northbooks, or VISA, Mastercard. Prices subject to change without notice.

(You may phone your VISA/MC order to Northbooks at 907-696-8973)

VISA/MC card # ☐☐☐☐ ☐☐☐☐ ☐☐☐☐ ☐☐☐☐

Exp. Date:___/____ Amount Charged: $ _____

Signature: _____

Phone Number: _____

Please send my book (s) to:

Name: _____

Address: _____

City: _____ State: _____Zip: _____

Fill out this order form and send to:

Northbooks
17050 N. Eagle River Loop Rd, #3
Eagle River, AK 99577-7804
(907) 696-8973
www.northbooks.com

www.ingramcontent.com/pod-product-compliance
Lightning Source LLC
Chambersburg PA
CBHW031233090426
42742CB00007B/183